BUILDING A \

Heritage, Culture and Identity

Series Editor: Brian Graham,
School of Environmental Sciences, University of Ulster, UK

Other titles in this series

Bosnia and the Destruction of Cultural Heritage
Helen Walasek
with contributions by Richard Carlton,
Amra Hadžimuhamedović, Valery Perry and Tina Wik
ISBN 978 1 4094 3704 8

World Heritage in Iran: Perspectives on Pasargadae
Edited by Ali Mozaffari
ISBN 978 1 4094 4844 0

Edible Identities: Food as Cultural Heritage
Edited by Ronda L. Brulotte and Michael A. Di Giovine
ISBN 978 1 4094 42639

Who Needs Experts?
Counter-mapping Cultural Heritage
Edited by John Schofield
ISBN 978 1 4094 3934 9

The Making of a Cultural Landscape
The English Lake District as Tourist Destination, 1750–2010
Edited by John K. Walton and Jason Wood
ISBN 978 1 4094 2368 3

Cultural Heritage of the Great War in Britain
Ross J. Wilson
ISBN 978 1 4094 4573 9

Many Voices, One Vision: The Early Years of the World Heritage Convention
Christina Cameron and Mechtild Rössler
ISBN 978 1 4094 3765 9

Partitioned Lives: The Irish Borderlands
Catherine Nash, Bryonie Reid and Brian Graham
ISBN 978 1 4094 6672 7

Building a World Heritage City
Sanaa, Yemen

MICHELE LAMPRAKOS

Routledge
Taylor & Francis Group
LONDON AND NEW YORK

First published 2015 by Ashgate Publishing

Published 2016 by Routledge
2 Park Square, Milton Park, Abingdon, Oxon OX14 4RN
711 Third Avenue, New York, NY 10017, USA

First issued in paperback 2017

Routledge is an imprint of the Taylor & Francis Group, an informa business

British Library Cataloguing in Publication Data
A catalogue record for this book is available from the British Library.

The Library of Congress has cataloged the printed edition as follows:
Lamprakos, Michele.
Building a world heritage city : Sanaa, Yemen / by Michele Lamprakos.
pages cm. -- (Heritage, culture and identity)
Includes bibliographical references and index.
ISBN 978-1-4724-3540-8 (hardback) 1. San'a' (Yemen)--Buildings, structures, etc. 2. Historic buildings--Conservation and restoration--Yemen (Republic)--San'a'. 3. World Heritage areas--Yem-en (Republic) 4. San'a' (Yemen)--Civilization. I. Title.
DS247.7.S36L34 2015
363.6'9095332--dc23

2014043000

ISBN 13: 978-1-138-30845-9 (pbk)
ISBN 13: 978-1-4724-3540-8 (hbk)

Contents

For the ustas*—master builders and craftsmen—of Yemen, past, present and future;*

and for Steven

List of Figures

Note on Transliteration

I have transliterated Arabic using the *International Journal of Middle Eastern Studies* system, with a few adjustments. For certain words used frequently in the text—like *usta* (builder), *maqshama* (garden), *samsara* (caravansaray)—I add an "s" to the Arabic singular to indicate the plural, for ease of reading; at the first appearance of the word, I indicate the Arabic plural form in parentheses. For words commonly used in English (Sanaa, wadi), I have used the common spellings. When the *lam* of the definite article assimilates with the following consonant, I ignore the assimilation. Finally, if individuals use certain English spellings for their Arabic names, I have endeavored to use those in the text.

Acknowledgments

This book presents a very recent history, indeed, one that is still in the making. I am grateful to all those in Yemen and elsewhere who shared their personal and professional histories with me. The list of those who contributed to this book in ways large and small is very long; I hope that those who do not see their names here will forgive me.

Above all I would like to thank Ronald Lewcock, Saba al-Suleihi, Yassin Ghalib, Tarek al-Hammadi, Amat al-Razzaq Jahhaf, and Julia Thielebein, who shared their great knowledge of Sanaa with me and became good friends. Ronald Lewcock also provided me with documentation and images, read and commented on the manuscript, and took me on unforgettable tours of the old city. ʻAbdullah al-Hadrami and ʻAli ʻOshaish were very generous with their time and provided helpful documentation. This work is also indebted to Selma al-Radi, who has sadly passed away. Numerous builders and craftsmen discussed their views of architecture and conservation with me, on site and at *qat* chews. I would like to thank especially Ahmad al-Rawdhi, ʻAbdullah al-Rawdhi, and their family for hospitality and insights. Thanks also to Yahya al-Dhahbani, ʻAtiq Ahmad Saʻd, ʻAbdu ʻAli al-Tuwaythi, and Muhammad al-Anisi; and in Shibam-Hadhramawt, Salih Bahadaila, Khamis Baraman, Saʻid Basowaytin, and ʻAli Murbash. I am grateful to the many families in Sanaa, Shibam, and other towns who welcomed me into their homes.

I was privileged to speak with a number of government officials in Yemen who were instrumental in launching the UNESCO Campaign in the 1980s. Among these are two former prime ministers, ʻAbd al-Karim al-Iryani and the late ʻAbd al-ʻAziz ʻAbd al-Ghani; Yusuf ʻAbdullah; Hussein al-ʻAmri; Qadi ʻAli Abu Rijal; the late Qadi Ismaʻil al-Akwaʻ; and Muhammad al-Tayyib. The General Organization for the Preservation of Historic Cities of Yemen (GOPHCY) and its research arm, the Center for Architectural and Technical Studies (CATS), provided me with assistance and access to documents. I thank ʻAbdullah Zayd ʻIsa, Nabil al-Monasser, Muhammad al-Dailami and their staff, especially Samir al-Shwaffi, ʻAli Hussein Mahmoud, and the inspectors who allowed me to join them on their rounds of the old city. I am also grateful to Khalid Jubari, Salih ʻAbbas, Harbiyya al-Hamyari, Jamal Kushih, Muhammad al-Namer, ʻAbd al-Hakim al-Sayaghi, Amal al-Shaiba, and Anis Qa'id. ʻAida Rishan assisted me with research in GOPHCY's archives. My visit to Shibam-Hadramaut was facilitated by Jamal Bamakhrama of GOPHCY and by ʻOmar ʻAbd al-ʻAziz Hallaj of GIZ (Deutsche Gesellschaft für Internationale Zusammenarbeit), who provided insights, contacts, and documentation. Officials at other organizations and agencies in Sanaa were generous with their time. In particular, I want to thank ʻAbdullah al-Dailami, Nabil al-Maqalih, and their staff at the Social Fund for Development (SFD) for speaking with me on numerous occasions, and for allowing me access to project files. At the Ministry of Endowments, I was assisted by ʻAbd al-Rahman al-Qallam and Muhammad Hubeish.

I am grateful to a number of local scholars who took an interest in my work: at the University of Sanaa, ʻAbdo ʻAli ʻOthman, sociologist; Muhammad Saʻd, urban geographer; Sultan Sallam, ʻAbd al-Raqib Tahir, and Nadia al-Kawkabani, architects; ʻAbd al-ʻAziz al-Zariqa, political economist; and at the French Centre in Sanaa for Archaeology and

Social Sciences (CEFAS), Jean Lambert, Muhammad Jazim, Sabrina Kadri, and Vincent Martignon. ‘Abd al-Karim al-Akwa‘ tutored me in Sanaani dialect, and helped me transcribe tapes from dialect into standard Arabic—with helpful interpretation and insight.

In Paris and in Sanaa, a number of current and former officials from UNESCO and the World Heritage Center were generous with their time: Mounir Bouchenaki, Gérard Bolla, Giovanni Boccardi, Véronique Dauge, and Sa‘id Zulficar. Anna Paolini arranged access to files at UNESCO's Office for Arab States. Thanks to UNESCO archivists Jen Boel and Mahmoud Ghander, and Christine Carbonnel Saillard at the Bureau of Public Information and Audiovisual Resources. Daniele Pini and Franca Miglioli kindly allowed me to participate in a training seminar held at CATS in July 2005. I benefited from conversations with them, and with team leaders Saveria Teston and Luca Lanzoni. Fernando Varanda and Maria Luisa Neri helped reconstruct the events of the 1970s and '80s. I consulted Dr. Neri's files from her years with Studio Quaroni at the Olivetti Historical Archive in Ivrea, where I was assisted by Enrico Bandiera.

I would also like to thank the following individuals: Marcos Abbs, Hassan al-‘Aini, ‘Abdullah ‘Alawi, Yahya al-Anisi, Sa‘id ‘Alawi, Marc Armansin, ‘Abd al-Mailk ‘Azzan, Rémy Audoin, Marilène Barret, Gianni Brizzi, Bob Burrowes, Paolo Costa, Jamila and Wafa' al-Daili (and family), Ursula Dreibholz, Dick Der Steege, Jacques Feiner, ‘Alaa al-Habashi, Engseng Ho, Deborah Howard, Yahya al-Ibbi, ‘Abd al-Ghani al-Iryani, ‘Ali al-Kamali, Hasan Uddin Khan, François Leblanc, ‘Abd al-Karim al-Maqalih, Marco Livadiotti, Jamal Ma‘jam, ‘Abd al-Karim Mohsen, ‘Abdurahman al-Mu‘assib, Jamal Mubarak, Muhammad Murayt, Trevor Marchand, Muhammad Musleh, Tawfiq al-Othmani, Sarah Philips, Hassan Radoine, ‘Ali Sabra, Hadi Saliba, Amal al-Shaiba, Jamil Shamsan, Jamil Sari‘, ‘Abdullah Soueid, Mujahid Tamish, Muhammad al-‘Udayni, Tawfiq al-‘Usabi, and Khalid Wahhas.

Various institutions have contributed financial and logistical support for research, writing, and production. These are the Fulbright Commission; the Social Science Research Council; the American Association of University Women; the Graham Foundation; the Barakat Foundation; the Aga Khan Program in Islamic Architecture and the Martin Family Society of Fellows for Sustainability, both at the Massachusetts Institute of Technology; and the School of Architecture, Planning, and Preservation at the University of Maryland-College Park. The American Institute for Yemeni Studies provided funding for a short reconnaissance trip to Sanaa in the fall of 2002, and facilitated my research there. I would like to thank, in particular, Maria Ellis, Christopher Edens, and ‘Amar al-‘Awdi. The Yemen Center for Studies and Research in Sanaa granted me an affiliation, which enabled me to conduct fieldwork.

Nasser Rabbat, Heghnar Watenpaugh, Michael Herzfeld, Frank Matero, and Nancy Um read and commented on versions of this manuscript. Steve Caton and Najwa Adra offered critical insights at key moments during my fieldwork. Horst Kopp and Paul Bonnenfant kindly provided documentation and images. I thank Adam Chamy for his drawings (figures 2.1, 2.9, 7.1); Anne Kopf for help with research and graphics; and Hagar Attia for her assistance with proof-reading and transliteration. Valerie Rose and Kayleigh Huelin at Ashgate were wonderful and efficient editors. Last but not least, I would like to express appreciation to family and friends who supported me during research and writing. In particular I want to thank my parents, Blanche and Peter Lamprakos; my aunt, Angela Sansotera; and my husband Steven Kramer, who read and edited the manuscript, offered many insights, and supported me throughout the process.

The fieldwork on which this book is based was conducted over a period of 15 months in 2004–2005. If not specified, photos are by the author, taken during these years. Likewise all translations from Arabic, French, and Italian are mine unless otherwise noted. Following common ethnographic practice I have omitted most names in the text, with the exception of public officials, authors of published and built works, and those who asked to be named. I have also included the names of a number of individuals—architects, builders, and others—in order to develop a sustained narrative of their work and ideas. Excerpts from this book were published in *Traditional Dwellings and Settlements Review*, volume XVI, no. 11, spring 2005, under the title "Rethinking Cultural Heritage: Lessons from Sanaa," pp. 17–37.

List of Abbreviations

CATS	Center for Architectural and Technical Studies (the research and documentation center of GOPHCY)
CEFAS	Centre Français d'Archéologie et de Sciences Sociales a Sanaa
EOPOCS	Executive Office for the Preservation of the Old City of Sanaa (now General Organization for the Preservation of Historic Cities of Yemen, or GOPHCY)
GIZ	Deutsche Gesellschaft für Internationale Zusammenarbeit (German Society for International Cooperation; formerly Gesellschaft für Technische Zusammenarbeit, GTZ)
GOAMM	General Organization for Antiquities, Museums, and Manuscripts
GOPHCY	General Organization for the Preservation of Historic Cities of Yemen
ICCROM	International Centre for the Study of the Preservation and Restoration of Cultural Property
ICOMOS	International Council on Monuments and Sites
IHS	Institute of Housing Studies (Rotterdam)
MOC	Ministry of Culture
PDRY	People's Democratic Republic of Yemen (South Yemen; joined with North Yemen in 1990 to form the Republic of Yemen)
SOAS	School of Oriental and African Studies (London)
SFD	Social Fund for Development
TOPOCS	Technical Office for the Preservation of the Old City of Sanaa (predecessor of EOPOCS)
UCHP	Urban Cultural Heritage Project (sponsored by the Netherlands; now dissolved)
UNDP	United Nations Development Program
UNESCO	United Nations Educational. Scientific, and Cultural Organization
UNOTC	United Nations Office of Technical Cooperation
YAR	Yemen Arab Republic (North Yemen; joined with South Yemen in 1990 to form the Republic of Yemen)

Chapter 1
Prologue

For some Arabs the past is an ephemeral existence on the fringes of desert or sea; for others it begins with the Pharaohs or the Mediterranean of Phoenician and Classical times. Yemen is different: it is one of those rare places where the past is not another country. I have eavesdropped on tribesmen visiting the National Museum and heard them expressing surprise, not at the strangeness of the things they see, but at their familiarity. Asking the way to al-Qalis, the site of Abrahah's ecclesia *in Sana, I have been quoted the Qur'anic account of his defeat on the Day of the Elephant as if it had happened yesterday. There is a feeling in Yemen that the past is ever-present.*

Tim Mackintosh-Smith, *Yemen: the Unknown Arabia*, 2000

Yemenis don't get it: they don't understand that we're not on the same time line.

World Heritage Center official, May 2004

Now everyone is aware of heritage. Before it was something they just did.

Sanaa native and wife of a builder, June 2005

The city of Sanaa has inspired travelers, historians, and architects for centuries—from medieval geographers and travelers, to present-day devotees of "vernacular" architecture and heritage sites. As in other fortified towns in Yemen, the basic building element of Sanaa is the tower house—here made of stone and brick, its openings decorated with exuberant plasterwork (figures 1.1, 1.2). Rising majestically from an agricultural plain, flanked by palace precincts and several low walled suburbs, Sanaa presented a striking image to the traveler, as captured by the off-cited saying:

> One must go to Sanaa no matter how long the journey
> Though the hardy camel droop, leg-worn along the way.[1]

The image of the city remained more or less intact through the middle decades of the twentieth century. Today, visitors to Sanaa experience a very different place. The old quarters are embedded in a sprawling metropolis, which apart from certain traditional decorative features, looks much like cities in other developing countries. Within the space of a few decades, the Sanaa of history and legend—the everyday city that middle-aged Yemenis remember from their childhoods—has become the "old city." The walled core has been assigned to what preservationists call the "historical timeline": a place representative of a past which, at least in theory, has been superseded.

1 Adapted from R.B. Serjeant and R. Lewcock, eds, *San'a', an Arabian-Islamic City*, London: World of Islam Festival Trust, 1983 (overleaf). The saying, cited by the geographer al-Hamdani and others, is often mentioned in conservation literature to refer to the stunning image of the city. But it was originally a directive to study with renowned scholars in Sanaa; see G. Vom Bruck, *Islam, Memory, and Morality*, New York: Palgrave Macmillan, 2005, p. 3.

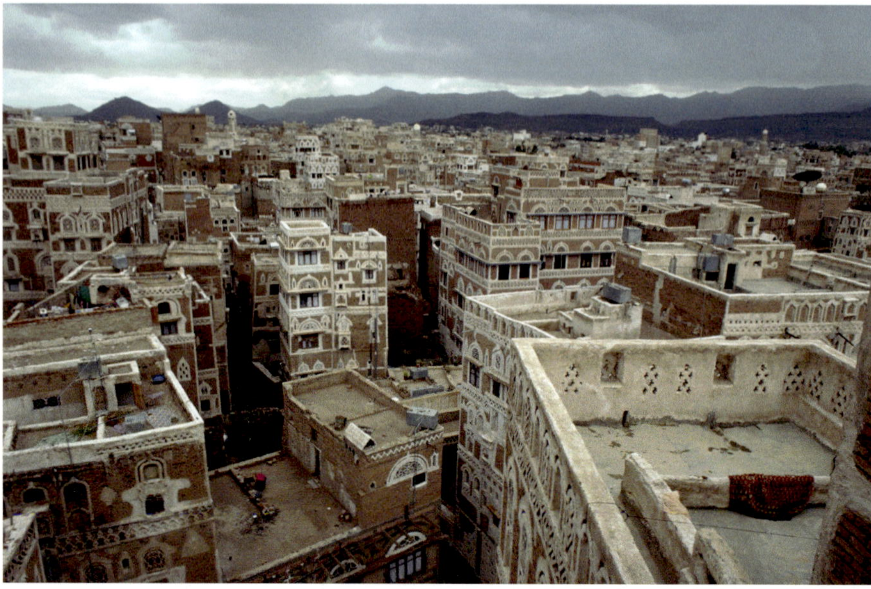

Figure 1.1 View of the old city of Sanaa, 1995

Source: © Aga Khan Award for Architecture/Courtesy of the Architect.

Conservation (or historic preservation, as it is called in the U.S.) can be understood as a complex of ideas, practices, and institutions that are unique to the modern era. The aim of conservation, as described by practitioners, is to save and preserve artifacts of the past so that they may be transmitted to future generations. While all cultures have selectively maintained and preserved elements of the past, the modern era is distinguished by an *ideology* of conservation constructed in opposition to, and in many ways mirroring, the ideology of progress. The contest between the two has been played out in the physical and social fabric of cities: while grand schemes and urban renewal have destroyed historic districts, conservation policy has tended to reify them; both strategies have isolated and circumscribed the traditional within the modern.

In many ways, the formation of a heritage discourse in Yemen has followed a similar trajectory: the old city of Sanaa and other heritage sites are portrayed as symbols of the past, as the country strides confidently into the future.[2] But what is unusual about Sanaa, and other towns in Yemen, is that they have come to be seen as "historic" within the space of a single generation.

Modernization is relatively recent in much of Yemen, at least in terms of its material and economic impact; this is especially true in the northern highlands, where the city of Sanaa is located. It did not begin in earnest until the end of civil conflict in the late 1960s

2 For example, a 1998 government advertising supplement in *Newsweek*, entitled "Poised for Growth into the Next Millenium," depicts the old city of Sanaa alongside construction workers in hardhats and the modern port of Aden. See also P. Dresch, *A History of Modern Yemen*, Cambridge: Cambridge University Press, 2000, p. 211.

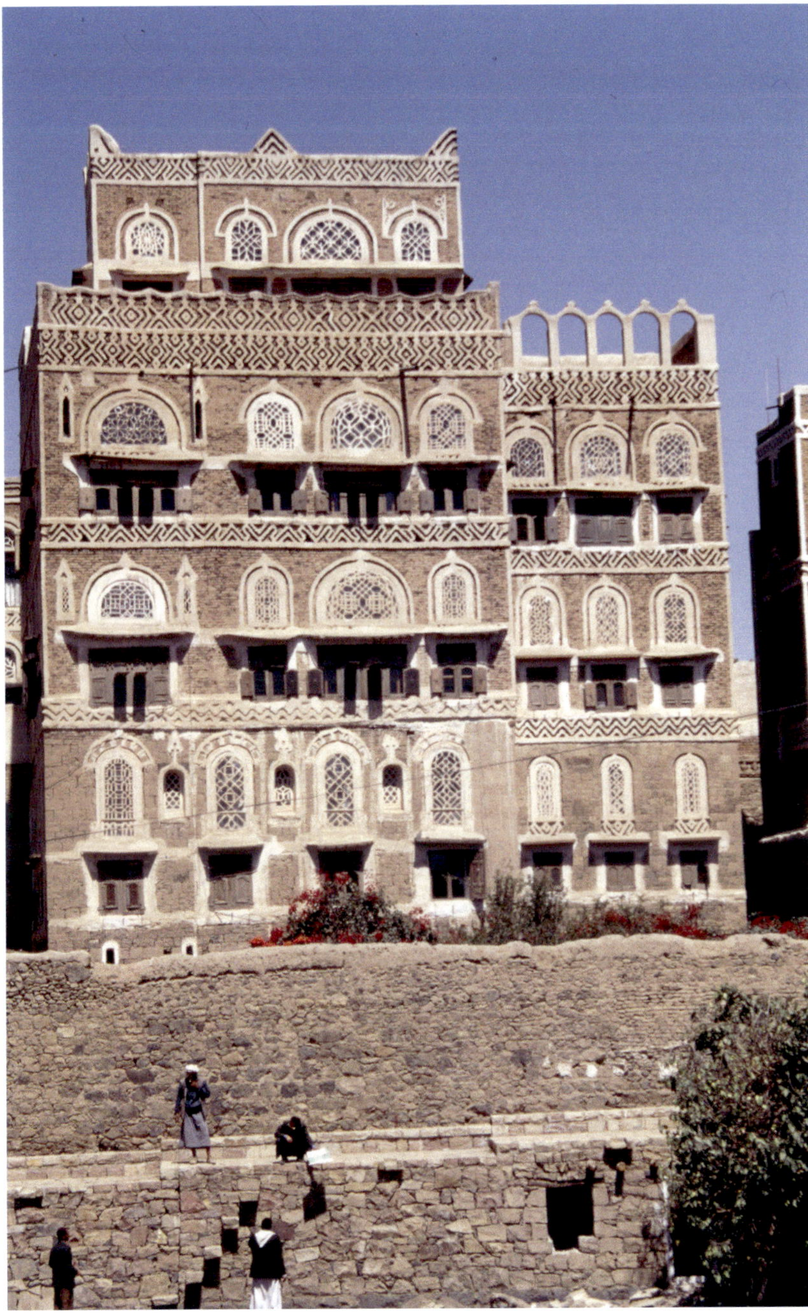

Figure 1.2 Al-Qasimi House, overlooking one of the old city's many urban gardens

(following the 1962 revolution); it rapidly gathered momentum with the mass migration of Yemeni workers to the Gulf during the oil boom of the 1970s, and the influx of cash remittances. Until this time the city of Sanaa consisted of the walled core, and adjacent suburbs and palace precincts. Now the city was engulfed by development, and stretches of wall were demolished to facilitate it. Conditions in the old core deteriorated as residents left, trash accumulated, and the installation of a modern water system swelled the groundwater, destabilizing the massive tower houses. Foreigners working in the city, along with several local professionals, warned that the "old city"—a kind of living relic—was in danger and that something had to be done to save it. The World Heritage Convention had just been adopted by UNESCO in 1972. Planning began for an International Safeguarding Campaign for the old city of Sanaa, the capital of what was then North Yemen, and also for Shibam-Hadhramaut, in what was then South Yemen.[3] European and Arab advocates saw in these places outstanding examples of traditional vernacular settlements, embodying qualities that modern cities had lost.

The UNESCO Campaign for Sanaa has been internationally recognized, most notably by an Aga Khan Award for Architecture in 1995. At least initially, it emphasized urban upgrading and revitalization rather than monuments conservation. The authors of the Campaign and Yemeni policymakers insisted that the old city was both living and historic: it should not be turned into a museum. As such, conditions had to be improved to keep people living there. The Campaign set the tone for conservation activities throughout the country: conservation and modernization were seen not as conflicting operations, but rather as complementary.[4] It is important to note that when the Campaign was launched, there were no planning mechanisms to guide new development; decades later, they exist but are deficient. This makes the effort to safeguard the old core even more remarkable: today it is one of the best-serviced sector of the city, although it makes up a fraction of its area.[5]

Conservation is part of a wider, and remarkably vibrant, public discourse about architecture, building practice, and public space.[6] It is seen as a new, imported idea and yet also as strangely familiar—in part, because the protected buildings and places are not fully part of the "past." In contrast to other parts of the world where conservation was introduced at a later date to remediate the effects of modernization, in Yemen the two processes have been more or less concurrent. This has resulted in a paradox: unlike many other countries in the Middle East that abandoned traditional construction practices in the nineteenth century, in Yemen these building practices have not died out: builders now in their forties and fifties, trained as apprentices to old *usta*s (master builders), or having acquired "traditional" skills after working in modern construction, continue to practice. At the same time, the social

3 As discussed below, these "twin campaigns" mirrored geopolitics: North Yemen, a Republic allied with the West and Saudi Arabia, and socialist South Yemen, allied with the Soviet Bloc, each had its own iconic, historic city. Shibam was added to the World Heritage list in 1982 and Sanaa, in 1986.

4 M. Yoshida, "A Concept on 'Conservation-Based Development' for Historic Cities—Problems Related to the Preservation of the Old City of Sanaa in Yemen," report for the United Nations Centre for Regional Development, 1994.

5 Indeed, some authors have charged that attention to the old city has been inordinate, given the rate of urban expansion and the severe problems facing the city (see below, Chapter 2). On the current state of planning in the city, see the World Bank/Cities Alliance report, *Sanaa, a City Development Strategy*, 2009, pp. 19–20, 35–6, http://citiesalliance.org/node/787 (accessed 12/29/13; hereafter cited as "CDS Report").

6 On this point see T. Marchand, *Minaret Building and Apprenticeship in Yemen*, Richmond, Surrey: Curzon Press, 2001, p. 21.

forms that helped shape, and were shaped by, the built fabric persist in both the old city and the new districts. Middle-aged and older individuals see the houses they grew up in exalted as "heritage." Yemenis, in effect, are not separated from their heritage by an historical divide; rather, living traditions are to be conserved. Yet this everyday environment is increasingly seen as "historic"—and it is this passage that is the subject of the present work.

The rapidity of development, and the great contrast between the old and new districts, reinforced the image of old Sanaa as a medieval city frozen in time. The trope is common in conservation literature and in discourse about the city in general:

> The Old City of Sanaa is unique in Yemen and also in the World. Its value lies in the unforgettable impression made by the whole; an entire city of splendid buildings combined to create an urban effect of extraordinary fascination and beauty. The Old City is untouched by time; it is a functioning city of the Islamic ages.[7]

The idea of the old city as "a functioning city of the Islamic ages" still resonates with many in Sanaa. Like much of the highland plateau, the city was insulated from development pressures until the mid-twentieth century, largely because of the isolationist policies of the ruling imams. When the country opened to the global market, it had the force of a sudden shock. For historian Muhammad Jazim, who worked with the UNESCO Campaign, the changed conditions of materials life led to a new conception of the past:

> Before modernity, everything related to the past, not the present—the ancestors (*al-ajdad*), and so forth ... As for *turath*, heritage, it's a modern usage, Egyptian or Lebanese in origin. None of the old [Yemeni] documents refer to crafts, buildings, or customs as *turath* ... Material life was part of religious and social life. Once the material effects of modernity happened, people came to see these things as part of the past.[8]

These material changes correspond to shifting social relations and cultural practices. It is a complex and nuanced process—as suggested by the opening quotations. Individuals and groups are confronted with representations of the historic past—but they often see themselves as representing and embodying this past, or traditional practices and ways of life. They confront these representations with ambivalence. The heritage of Yemen is now celebrated on the world stage: Yemenis from all walks of life, working within the heritage field and outside it, are proud of this. Yet for some, heritage recalls a legacy of oppression and backwardness that continues to color the present. When the builder's wife quoted above bumped her head on a low ceiling in her house in the old city, she gestured toward the ceiling and laughed: "This is *turath* (heritage)!"—meaning that what outsiders see as heritage can only be backward and inconvenient.

But in many cases, Yemenis of various backgrounds are protective of "tradition." One example is the taste for neo-traditional architecture—what some builders call "new heritage" (*turath jadid*). It can be seen throughout the new districts, and in the old city itself

7 "Preservation of the Old City of Sana'a: Project Findings and Recommendations," UNESCO mission on behalf of the UNDP, Restricted Technical Report UNDP/YEM/88/006, 1996, p. 1. This trope persists: the 2009 CDS Report states: "Until 1962, Sanaa could be considered a medieval city stalled in time" (p. 10).

8 Muhammad Jazim is a researcher at the Centre Français d'Archéologie et de Sciences Sociales de Sanaa (CEFAS). Jazim was part of the team that conducted a survey of traditional crafts for the National Handicrafts Center, to be discussed in Chapter 6.

(Figure 1.3).[9] This new *turath* prompted the remark we heard earlier by the World Heritage Center official:

> Yemenis don't realize that the toy is broken. In the past, techniques developed over very long periods of refinement and adjustment; building was firmly grounded in hundreds of years of tradition. Rulers of the past could not go wrong: people knew what was good and what was bad. But now we don't have the same confidence: the magic touch is gone. We no longer have the capacity to transform our historic cities. It's a different era: we are not on the same time line.[10]

The implication is, of course, that traditional forms should be abandoned in favor of modern ones (an experiment which, incidentally, was tried under the Egyptian occupation of the 1960s, and largely rejected). The official added that Yemenis are attached to their heritage, even though they conceive of it differently. "The strange thing about Yemen," he mused, "is that it didn't go through a conceptual process." The past, in effect, is too close to be fully "other."

The heritage enterprise is part of the larger project of history-making—a project which has, over the last several decades, come to hold increasing interest for scholars. Because of the country's relatively recent modernization, we have a unique opportunity to observe how modern understandings of the past build on older forms.[11] Sanaa is a fascinating case study of this process—revealing a texture and richness that is usually missing in current coverage of the country. But the relevance of the case study is wider: it allows us to gain distance from our own assumptions and practices, and to denaturalize them.

Our "modern" view of the past developed under specific cultural-historical conditions in early modern Europe. It is the product of a developmental view of history, based on the notion of perpetual progress that leaves the past behind—irrelevant to the present and the future. This radical "otherness" of the past has had, and continues to have, profound effects on the collective psyche and culture. It has been inscribed in our built environment, in the ideologies of both high modernism and conservation, which in many ways mirror each other.[12] Reinforced in our laws and building codes and in design education, these twin ideologies have made it difficult to marry the concerns of development, stewardship, and sustainability. Like other aspects of Western epistemology and practice, this view of the past has been presented as universal—and has been disseminated as such, through international networks, protocols, and projects. We have, in effect, exported our own institutionalized approach to the past, under the ostensibly benign process of "knowledge transfer."

---

9 See M. Lamprakos, "Old Heritage, New Heritage: Building in Sana'a, Yemen," special issue of *Viewpoints*, Middle East Institute, November 2008, pp. 34–7 http://www.mei.edu/content/architecture-and-urbanism-middle-east (accessed 3/5/15)

10 Interview, Sanaa, July 2004.

11 Benedict Anderson's call to study the relationship between nationalism and the "cultural systems that preceded it, out of which—as well as against which—it came into being" (*Imagined Communities*, London and New York: Verso, 1991, p. 19) relates to the construction of a national history in Yemen. The quote is cited by Messick as an inspiration for *The Calligraphic State: Textual Domination and History in a Muslim Society*, Berkeley: University of California Press, 1993, p. 2.

12 On this point, see A. Colquhoun, "Newness and Age Value in Alois Riegl," in *Modernity and the Classical Tradition: Architectural Essays, 1980–1957*, Cambridge, MA and London: M.I.T. Press, 1989, pp. 213–22. On the high modernist project see J. Scott, *Seeing Like a State*, New Haven: Yale University Press, 1998, Chapter 4.

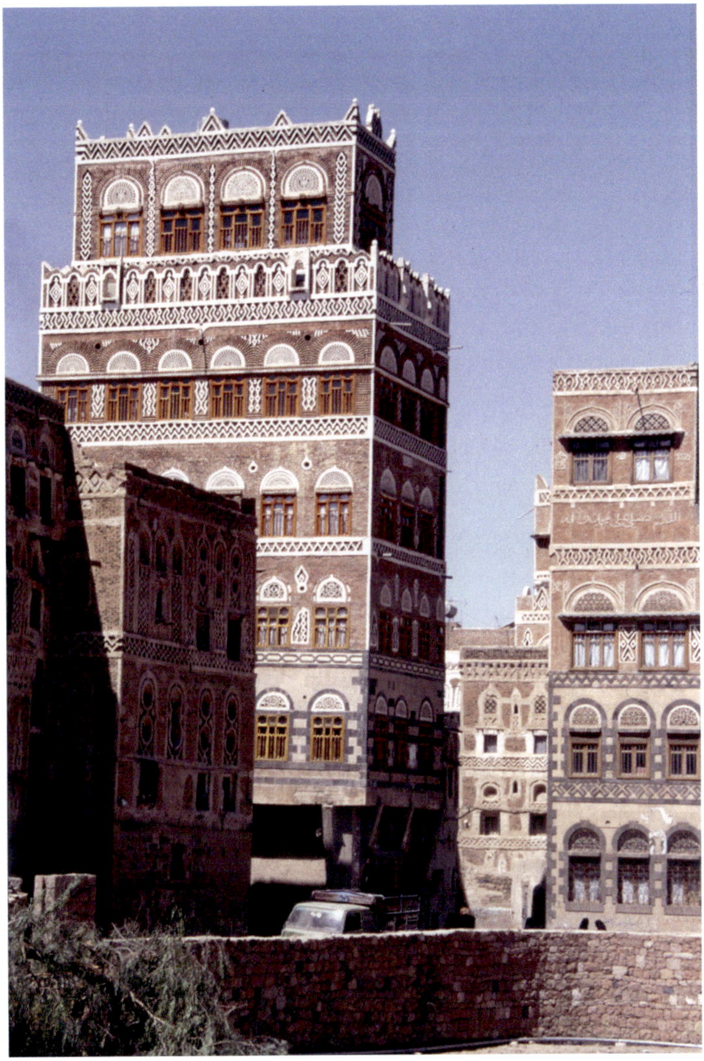

Figure 1.3 **The upper floors of this house in the old city were built by Ahmad al-Rawdhi, an *usta* who specializes in traditional construction. He calls this type of work "new heritage" or "modernized heritage" (*turath mutahaddath*)**

Throughout this work, I explore how these universalizing assumptions about the past have been received, appropriated, and transformed in Sanaa. As such, I refer to the theory and norms of what I call "international practice"—that is, conservation as represented by official bodies like ICOMOS, the International Council on Monuments and Sites (the advisory body to UNESCO on matters of cultural heritage). An international practice of conservation—sometimes referred to as "classic" monuments conservation—emerged in the late nineteenth and early twentieth centuries, in Europe and its colonies. As such, I consider this the wider context of the project "site."[13] As will be clear in the chapters that follow, international practice is far from monolithic, but it is often represented as such in official discourse. The following discussion sets the stage for this study, presenting certain key concepts—in particular, the notions of authenticity and the "traditional city." This section provides background on the origins of the UNESCO international safeguarding campaigns for cities in the Middle East/North Africa region. It concludes with a discussion of two key international documents, the Venice Charter and the World Heritage Convention, which will be referenced later in the text. Readers who are less interested in this general discussion may want to proceed to the next chapter.

The Past as Modernity's "Other"

Conservation discourse assumes that modernity has fundamentally changed our relationship with the past, as expressed by the respected theorist Paul Philippot:

> …. The word preservation—in the broadest sense, being equivalent in some cultures to conservation or restoration—can be considered … as expressing the modern way of maintaining living contact with the cultural works of the past. This way of maintaining contact evolved after the outburst of the Industrial Revolution and the development of an historical conscience brought an end to the traditional link with the past, which may be said to have lasted, in various forms, from the origin of civilization to the end of the eighteenth century.
>
> Indeed, since the rupture, the past has been considered by Western man as a completed development, which he now looks on from a distance, much as one looks at a panorama. On the one hand, this new historical distance has produced the conditions necessary for a more objective, scientific approach to the past in the form of historical knowledge …

Like other realms of modern life, the past must be managed by a class of experts according to "scientific" principles and standards. Citing modern developments in aesthetics, criticism, and philology, Philipot argues that

> [e]ach work of art, each piece of decoration, each historic document is unique and cannot be repeated without faking. It is like a dead language … The survival of traditional crafts should not mislead one here. What survives of the craftsman's tradition in the new industrial world is its practical skill, and while this skill can certainly be of great use in conservation, *it is no longer a genuine expression either of the past or of the present.* To ignore this would mean to close one's eyes to the fact that modern historical consciousness has irreversibly broken the historical continuity and therefore, would lead to a faked expression …

13 See Gupta and Ferguson, "Discipline and practice: 'the field' as site, method, and location in anthropology," in Gupta and Ferguson, eds, *Anthropological Locations: Boundaries and Grounds of a Field Science*, Berkeley: University of California Press, 1997.

The "confusion" of past and present is motivated by nostalgia, Philippot writes. This has led to unfortunate consequences, like the reconstruction of historic buildings and "fake" revival styles in new architecture.[14]

At a distance of more than three decades, these statements are jarring, in particular, because they are assumed to be universal. Today, the language would be softer: conservation discourse is now imbued with the language of cultural relativism. But the assumptions embedded in the statements still determine mainstream conservation discourse, practice, and legislation. The past is seen as a completed development: no longer a vital part of the present, it can be appreciated only at a distance, through the lens of historical knowledge. As such, historical or traditional forms executed in the present are "no longer a genuine expression of either the past or the present."

Certain key concepts in the international charters—concepts like authenticity and significance—derive from various intellectual currents in nineteenth and early twentieth century Europe: historicism, romanticism, and theories about culture that were developing in the emerging social sciences. Both historicism and romanticism posited a radical break between past and present, and the "irreversibility" of the historical time line. Applied in conservation, these ideas mean that an object or monument is to be valued primarily as a record of the past: its material authenticity is thus its primary source of value, and must be preserved indefinitely. In the early twentieth century, the idea of authenticity was approached from another direction: "culture" came to be conceived as an area of inquiry, largely through experience in the European colonies. Culture was understood as a discrete, bounded entity that contained a genuine, unspoiled essence—an essence that could be discerned from the study of material artifacts and settlement patterns.[15] These ideas about culture confirm certain premises of conservation—for example, the notion that a material artifact *represents* the culture that created it, providing evidence of the culture's authenticity.

These historicist and culturalist approaches to authenticity overlapped in complex ways in a number of new disciplines, including archeology, art history, geography, and the social sciences. They would coalesce in theories of environmental determinism, which had a profound effect on the emerging field of urban planning at the turn of the twentieth century. The culturalist approach would be especially important in the colonial context, where it gave rise to the notion of the "traditional city."[16]

14 "Historic Preservation: Criteria, Guidelines," 1976, cited in S.N. Price, *Historical and Philosophical Issues in the Conservation of Cultural Heritage*, pp. 268–70 (italics mine).

15 R. Handler, "Authenticity," *Anthropology Today*, 2:1, 1988; G. Wright, *The Politics of French Design*, Chicago: University of Chicago Press, 1991, Chapter 2.

16 On the traditional city in French historiography and colonial practice, see S. Hamadeh, "The Traditional City: a French Project" in N. Alsayyad, ed., *Forms of Dominance: On the Architecture and Urbanism of the Colonial Experience*, Aldershot: Avebury Press, 1992, pp. 241–59. For the historic city in the wider context of French colonial planning, see especially, P. Rabinow, *French Modern: Norms and Forms of the Social Environment*, Cambridge, MA and London: M.I.T. Press, 1989; and G. Wright, *The Politics of Design in French Colonial Urbanism*, Chicago: University of Chicago Press.

Historic City, Traditional City, Islamic City

As Françoise Choay notes, the idea of an historic city emerged in response to the dramatic transformation of urban space in the nineteenth century.[17] A similar process—modernization and destruction, followed by recognition of historical value—played out, albeit more slowly, in the old towns and cities of the Middle East and North Africa. European-style districts were built by colonial administrators or native rulers. The old cores—for centuries, the heart of vibrant urban cultures—were neglected by colonizers and abandoned by modernizing elites in the nineteenth and twentieth centuries. In some places, notably the French colonies, old towns were conserved as a way to protect native culture, and also to attract tourists. In contrast to historic cities in Europe, old towns in the colonies were rarely provided with modern amenities like water and sewer, which were seen to compromise their charm.[18] They became ghettos: their poverty and backwardness were represented to colonial audiences as picturesque, evidence of some parallel world where time had stopped. They were not just typologically different from modern districts, as in Europe: they were *culturally* different, home to poor people and increasingly, to rural migrants, 'traditional" people who had been left behind by modernity.

An essentialist view of culture persisted alongside modern views of historical development. Certain cultures were seen to be "behind" others, and thus as part of the past—even though they existed in the present. These "others" seemed to be survivals of earlier eras—contradicting the theory of historical development, yet also, paradoxically, providing living evidence for it.[19] The same phenomenon is evident in Rousseau's classic formulation of authenticity.[20] For Rousseau, the authentic self was corrupted by modern society. One could seek for evidence of it within, through personal experience. But one could also seek the antecedents of the authentic self in the outer world: in the toiling peasants of the French countryside, and in exotic natives in lands that were then being conquered and colonized. Portrayed as timeless and pure in picturesque idioms, these others substantiated the modern European self-image by way of contrast: while the latter is individuated, subject to development and progress, the former is anonymous, in the thrall of the collective. The artifacts they produced—crafts, buildings, and towns—were seen to *represent* traditional cultures, even though these cultures existed in the present. They must be protected from modern forms, materials, and values which would compromise their authenticity. In the mid-twentieth century this equation of authentic form and authentic culture would be adopted wholesale into the World Heritage Convention.

The notion of the "traditional city" became a cornerstone of French colonial policy and urban historiography. It had an important variant in the Middle East/North Africa region: the thesis of the "Islamic city." The concept was developed in the early twentieth

17 F. Choay, *The Invention of the Historic Monument*, trans. L.M. O'Connell, Cambridge: Cambridge University Press, 2001, p. 120. See also M. Lamprakos, "The Idea of the Historic City," *Change Over Time*, April 2014, pp. 8–39.

18 P. Rabinow, pp. 238–41; and Hamadeh, "The Traditional City,", p. 253.

19 J. Fabian, *Time and the Other: How Anthropology Makes its Object*, New York: Columbia University Press, 1983, pp. 25–30 and Chapter 2.

20 My discussion of authenticity is indebted to R. Bendix, *In Search of Authenticity: The Formation of Folklore Studies*, Madison: University of Wisconsin Press, 1997, chapters 1 and 2; and to C. Lindholm, *Culture and Authenticity*, Malden and Oxford: Blackwell Publishing, 2008, Chapter 1. On Rousseau see L. Trilling, "The Heroic, The Beautiful, and the Authentic," in *Sincerity and Authenticity*, Cambridge, MA: Harvard University Press, 1972, pp. 81–105.

century by Orientalist scholars studying a small number of cities in North Africa and Syria, controlled by Western powers. On the basis of this sampling, they posited the existence of the "Islamic city" as a general type, which remains constant over time and across space.[21] The assumption was that urban and social forms were shaped primarily by Islam and Islamic law, rather than by specific cultural-historical factors. The city of Fez—which decades later would be one of the first sites added to the World Heritage list—was the focus of many of these studies. "Descriptions of Fez," wrote Janet Abu-Lughod in her classic critique, "are projected back to Fez in the Age of the Marinides [LeTourneau, 1961] and generalizations about all cities of North Africa bear a suspicious resemblance to Fez both in 1900 and in medieval times!"[22]

Since the nineteenth century, European architects had documented and conserved Arab and "Moorish" architecture, and had used these forms in revival styles in both Europe and the colonies. They found that this architecture conformed neither to European aesthetics, nor to evolutionary models of art history. As such, it must exist outside time. As a leading colonial architect in Morocco put it:

> Up until 1909 custom (*cïada*) had not changed [since the fifteenth century]. Mason, sculptor, carpenter, ironworker, each enjoyed himself on his own without breaking the harmony among them, so much had traditions remained stable in these countries situated outside the grand currents of Europe.[23]

The static character of the "traditional" Arab town, the undifferentiated nature of its architecture, and the anonymity of its builders would persist in conservation discourse throughout the twentieth century and beyond: they would become the standard against which authenticity is judged. Here we see the recurrence of certain Romantic tropes. For modern man, authenticity resides deep in the individual self; and in those "noble savages" who predate modernity. For moderns, authenticity requires self-exploration, individuation. But for "traditional" folk, authenticity depends on *anonymity*—that is, on the *absence* of individuation and authorship. Once exposed to modernity, artists lose their authenticity—their connection to primal origins—which can be accessed only in the anonymity of the cultural group. If individual artists do emerge, the products of the group lose value in marketplace.[24] The trope of the anonymous vernacular underwrites the privilege of new experts, who safeguard the authenticity of the cultural object and legitimize its origins.

From Colonialism to World Heritage

In the era of decolonization, modern European-style districts remained the center of government, commerce, and modern life. Native technocrats, schooled in modernization theory and functionalist planning, invited European architects to develop plans for old

21 For the genealogy of the concept see J. Abu-Lughod, "The Islamic City—Historic Myth, Islamic Essence, and Contemporary Relevance," *International Journal of Middle Eastern Studies*, 19, 1987, pp. 155–60. See also A. Raymond, "Islamic City, Arab City: Orientalist Myths and Recent Views," *British Journal of Middle Eastern Studies*, Vol. 21, No. 1, 1994, pp. 3–18; and R. Ilbert, "La ville islamique: réalité et abstraction," *Les cahiers de la recherche architecturale* 10/11, 1982, pp. 6–13.

22 J. Abu-Lughod, "The Islamic City,", p. 159.

23 Albert Laprade, *Croquis*, no date; cited in G. Wright, *The Politics of French Design*, p. 111.

24 See C. Lindholm, *Culture and Authenticity*, pp. 19–20;

cities: examples are André Guitton's plan for Aleppo (1954); Ludovico Quaroni's plan for Tunis (1964); and Michel Ecochard's plan for Damascus (1986).[25] (Quaroni, in association with the Italian consulting firm Bonifica S.p.A., would conduct important studies for the old city of Sanaa in the 1980s.) This was the era of urban renewal in Europe and the U.S., a temporary triumph for high modernist planning. But from the late 1950s it was challenged by public protest, which eventually strengthened the hand of conservation. A similar grass roots movement developed against plans for the old town, or *medina*, of Tunis.

But Tunis was an exception: functionalist planning would leave an even greater mark on the old cities of the region than it did in Europe and the U.S.[26] It served the modernizing aspirations of native elites, who had little interest in conserving old towns. Monuments conservation had been established in many countries in the nineteenth century, by colonial authorities or by local rulers with nationalist aspirations. This patrimony was defined according to the prevailing art historical view—that is, in terms of monuments. Ancient and Islamic antiquities could be prized as signs of civilization. Old cities, in contrast, symbolized backwardness and economic marginality. They were messy tangles of streets with complicated patterns of ownership, not readily amenable to modernization.[27] For Arab elites, old towns were a world apart, inhabited by "traditional" people with whom they could not identify. "In Cairo in the 1970s, the Department of Antiquities was staffed by Egyptologists," said Sa'id Zulficar, who served as UNESCO's Desk Officer for Arab states in those years. "They saw the old city as old buildings and poor people and dirt. Why conserve it? … Some of them had never set foot in the old city."[28] In some socialist states old cities were associated with pre-revolutionary elites, and were actively destroyed in the 1960s and '70s. For the government of South Yemen, for example, Shibam represented the *saada*—the class that had historically dominated the south and which the 1963 revolution had overthrown.[29]

Europeans and Americans, however, were keenly interested in these traditional or "vernacular" settlements and architecture, which seemed to embody an authenticity that the West had lost. This was a time of social upheaval, the revolt against urban renewal, and what has been called the "crisis" of modern architecture. There was a new interest in contextualism, the representational qualities of architecture, and "cultural identity." Around this time, the 1970s, development agencies like the World Bank and the UNDP became interested in conservation as a development strategy, in conjunction with tourism.[30]

25 J. Abdelkafi, "La Médina de Tunis: l'éspace historique face au processus d'urbanisation de la capitale," *Présent et avenir des médinas (de Marrakech à Alep)*, Tours: C.N.R.S n° 706, "Urbanisation au Maghreb," fascicule de recherches n° 10–11, 1982, p. 216. On Aleppo, see S. Bianca, *Urban Urban Form in the Arab World: Past and Present*, London: Thames & Hudson, 2000, pp. 303–24.

26 This point is made by Choay, *Invention of the Historic Monument* (p. 131) and others.

27 R. Lewcock, "Preservation versus Modernization: Realities in Islamic Architecture," in Y.R. Isar, ed., *The Challenge to our Cultural Heritage: Why Preserve the Past?*, Proceedings of a Conference on Cultural Preservation, Washington, DC April 8–10, 1984, Washington, DC: Smithsonian Institution Press and UNESCO, 1986, p. 31.

28 Interview, Paris, May 2005.

29 'Omar 'Abd al-'Aziz Hallaj, Selma al-Radi, personal communication. The *saada* claim descent from the Prophet Muhammad and were a privileged group in both the north and south of Yemen (see below, Chapter 2).

30 M. Cernea, "At the Cutting Edge: Cultural Patrimony Protection through Development Projects," in Serageldin, I., et al., eds, *Historic Cities and Sacred Sites: Cultural Roots and Urban Futures*, Washington, DC: World Bank, 2001, pp. 67–88; and Gianni Brizzi, interview. Brizzi joined the World Bank in the 1970s and directed the Sanaa office from 1998 to 2002.

Until this time, development discourse had been shaped by modernization theory, which saw local models and practices as incompatible with progress. The tone of the discourse shifted in the 1970s. There was a new interest in the link between development and culture; traditional models and the informal sector were now seen as the basis for development.[31]

Influenced by the new thinking in architecture, architects and planners called for the conservation of historic urban fabric and architecture, and the use of traditional building practices and materials. Places like Yemen were seen as repositories of local building skills that could reanimate the lifeless uniformity of modern architecture.[32] On the one hand, this can be seen as a recurrence of the Romantic trope noted by Edward Said, the idea of Asia reanimating Europe.[33] At the same time, it was taken seriously by a new generation of architects in the developing world who, like their peers in the West, were questioning the modernist precepts of their teachers. These young architects believed that regional models could inspire contemporary work; the Egyptian architect Hassan Fathy, whose work had been unappreciated in Egypt, provided a model for many. For some architects this reappraisal led to an interest in conservation.[34]

The World Heritage initiative developed in the same period. In the 1960s there was a growing interest at UNESCO in the "common heritage of mankind"; a project to write a "history of mankind" was also initiated. These interests were related to UNESCO's international safeguarding efforts: the first of these was the operation to rescue the monuments of Nubia, which would have been inundated with the creation of Lake Nasser.[35] The World Heritage Convention was signed in 1972; four years later a committee was established to implement it. It was expected that most applications for World Heritage listing would come from European countries, and this turned out to be the case. The World Heritage Committee compiled lists of prospective sites in other parts of the world, and actively solicited nominations from member states.[36] Initially skeptical, Arab governments began to see the usefulness of a more broadly defined heritage for nationalism. It was also clearly a potential source of tourist revenues. An expanded definition of heritage included vernacular sites and cultural landscapes: countries that did not have "high" artistic traditions could now claim sites of national, even universal,

---

31 This was related to the critique of modernization theory, and alternative models proposed by politicians and intellectuals in the Third World; see J. Rist, *The History of Development*, chapters 8 and 9. I witnessed this shift personally: in the mid-1980s I worked for an international development agency in Cairo. The goals and design of projects emphasized private sector "microenterprises," influenced by a new free market ideology. I managed a project to revive silk production in the Nile Delta, working with producers to provide raw silk cloth to shops that catered to tourists; at the time, only a small minority of educated Egyptians showed an interest in local crafts.

32 See Sa'id Zulficar's thoughtful preface to the proceedings of the 1993 Aga Khan Seminar, held in Sanaa: "Preface," *Development and Urban Metamorphosis*, vol. 1, p. xii.

33 *Orientalism*, Harmondsworth: Penguin Books, 1991, p. 115.

34 For example, 'Abd al-Hakim al-Sayaghi, who worked on the UNESCO Campaign for Sanaa; now with the Social Fund for Development (personal conversation).

35 This Campaign was launched in late 1959 at the request of Egypt and Sudan, and continued until the 1980s. Some 22 monuments and architectural complexes, including the temple of Abu Simbel, were relocated ("International Campaign to save the monuments of Nubia," *Museum*, vol. XIIII, no. 2, 1960, p. 65). For the background and history of the World Heritage Convention, see S. Titchen, "On the construction of 'outstanding universal value'; UNESCO's World Heritage Convention and the identification and assessment of cultural places for inclusion in the World Heritage List," unpublished PhD dissertation, Australian National University, 1995, Chapter 3.

36 Selma al-Radi, an Iraqi archeologist, was asked to compile a list of prospective sites in the Middle East. This discussion is indebted to conversations with her and with 'Omar 'Abd al-'Aziz Hallaj in 2004.

significance. But this meant, paradoxically, that signs of backwardness had to be embraced as "difference" and cultural identity.[37]

From the late 1960s, UNESCO had taken an interest in the cultural heritage of Morocco, particularly the *medina* of Fez. Missions were carried out in the early 1970s, and in 1976, the Fez medina was one of the first sites added to the World Heritage list. It was followed in 1979 by the *medina* of Tunis, the old city of Damascus, and "Islamic Cairo" (distinct from "Old Cairo," which is identified with the Coptic community). Other cities in the region were listed in the following years, including the walled city of Jerusalem (1981); Shibam-Hadhramaut (1982); Tyre (1984); Marrakesh (1985); Aleppo (1986); and Sanaa (1986).

In 1981 UNESCO announced an international safeguarding campaign for Fez; campaigns were subsequently prepared for Cairo, Shibam-Hadhramaut, and Sanaa. According to Ronald Lewcock, a UNESCO consultant for these sites, the Cairo campaign was abandoned because of insufficient political support, but the other campaigns went forward. Until this time most UNESCO campaigns had focused on archeological or monumental sites. (The exceptions were campaigns for Venice and Florence, launched to remediate the effects of catastrophic flooding in 1966.) The campaigns for Fez and other cities in the region were aimed at wider issues: the neglect and development pressures that accompanied modernization. But as Sa'id Zulficar observed, conservation-based development had already been employed in the 1970s at Fez, Algiers, Tunis, and Lahore. The UNDP or the UNOTC (Office of Technical Cooperation) prepared master plans, working with UNESCO on historic areas. "We didn't need campaigns for Fez and Tunis, because we had UNDP projects," said Zulficar. Several former officials said that campaigns were mainly for publicity, a way to attract donor support.

The idea of the international safeguarding campaign seems to have developed separately from the World Heritage Convention, but UNESCO acknowledges that they derive from a common inspiration.[38] The campaigns were seen to provide moral support for the elaboration of the Convention—a way to raise world awareness of special sites in need of financial and technical support. In contrast to the rigorous process of World Heritage listing, the announcement of a campaign required only a request of the state party and a decision by UNESCO's General Director.[39] From the mid-1970s, General Director Mahtar M'bow announced numerous international safeguarding campaigns, especially for sites in the developing world. According a former UNESCO official, the campaigns were politically motivated—a way for M'bow to promote "Third Worldism."[40] In many cases, they were unsupported by historical research. "At the time, no one knew what African art and architecture were," he said. The proliferation of campaigns was problematic in a number of ways: donor states could not be expected to finance countless, costly initiatives; at the same time, the campaigns raised expectations of funding that were unlikely to be met.

37 See A. Arrif "Le paradoxe de la construction du fait patrimonial en situation coloniale. Le Cas du Maroc," *Revue du monde musulman et de la Méditerrannée*, vol. 73–76, 1994/3–4, pp., pp. 153–4.

38 I.C. Clark, et al., "International Campaigns for the Preservation and Safeguarding of the Cultural Heritage of Mankind," report of the special committee appointed by UNESCO's Executive Board, Paris, 9/25/85, p. 4.

39 T. Postel, "L'UNESCO et la sauvegarde du Patrimoine culturel et naturel mondial: Convention du Patrimoine mondial, et campagnes internationals de sauvegarde," unpublished thesis, Université de Droit, d'Economie, et des Sciences d'Aix-Marseilles/Institut d'Etudes Politiques d'Aix en-Provence, 1986, p. 169. Appendix five lists the 29 international campaigns underway as of 1986.

40 This was the time of the North–South conflict within UNESCO, which led to the withdrawal of the United States and the United Kingdom.

In 1985 a special committee was appointed to assess campaigns then underway, and found that few had met their objectives. The committee recommended the rapid conclusion of existing campaigns and a moratorium on news ones.[41]

UNESCO's Campaigns for Islamic Cities

It was under M'bow's direction that campaigns were planned for Fez, "Islamic Cairo," Sanaa, and Shibam, as part of a wider strategy for "Islamic cities."[42] Fez, in particular, became a model for urban conservation in the Arab world. It inspired architects and officials elsewhere in the region to seek World Heritage listing for their own cities, as well as UNESCO safeguarding campaigns. According to officials and architects in Sanaa, Fez was an important model for their own effort.

In the speech that launched the Fez Campaign, M'bow situated Fez and by implication, other Islamic cities, in the narrative of world historical currents and thus, in the new canon of world heritage:

> [S]ituated at the crossroads of the great intellectual and religious itineraries too often overshadowed by the famous routes of trade, Fez formed a nodal point in a vast intellectual network which profoundly influenced the whole fabric of relations between various regions of the African continent, the Islamic East and the European West. It was one of the principal seats of learning from which scientific knowledge and basic philosophical reflection, blossoming under the influence of Islam, were to stimulate and at times even generate an unprecedented development of knowledge on the threshold of the modern world.
>
> A thousand years of history had neither affected the urban fabric, impaired the architectural homogeneity, nor even disturbed the intellectual and artistic activity of the City of Fez. Today, this is no longer the case … The changes that have taken place in recent decades are on such a scale that Fez is in danger, under the pressure of demographic, social, and economic constraints without precedent in its history, of losing the profoundly original quality which makes it one of the purest jewels of Islamic civilization ….

M'bow went on to note that the international campaign for Fez was inspired by the same thinking that led to the Nubia and Venice campaigns. But he notes that the initiative for Fez is "by its very nature a campaign without precedent in the activities of Unesco. It is the first campaign to be undertaken on behalf of an Islamic city …." He concludes with an appeal to the world community to contribute generously to this massive challenge, so as to "preserve, for the happiness of those who inhabit Fez and of those who visit it, its collective soul, the bearer for more than eleven centuries of the most up-to-date messages, that of the solidarity and brotherhood of all mankind."[43] Because Fez was a World Heritage site, ICOMOS—the advisory body to the World Heritage Committee on matters of cultural heritage—was charged with drafting the campaign strategy. The same would be true in Islamic Cairo, Shibam, and Sanaa.

---

41 I.C. Clark, et al., "International Campaigns," pp. 3, 16.

42 I am grateful to Hassan Radoine for calling my attention to M'Bow's strategy for Islamic cities.

43 "Appeal for the Safeguarding of the City of Fez," speech delivered at Fez, April 9, 1980.

In M'Bow's speech we hear clear echoes of the Orientalist discourse of the Islamic city: the Fez medina, and its intellectual and artistic activity, had remain unaffected by "a thousand years of history." All this was now challenged by unprecedented changes, what author and UNESCO consultant Titus Burckhardt called "the irruption of the modern world."[44] The impact of modernization on the old fabric was alarming—much as it had been to cultural elites in Europe and elsewhere. But in the Middle East/North Africa region, the fight to defend old cities against the onslaught of modernization would be colored by notions of the "traditional city" and the "Islamic city." Orientalist sources were used in drafting World Heritage nominations and in developing conservation plans. This is no surprise, since Orientalists wrote some of the best urban histories on the region; LeTourneau's work, for example, was part of the bibliography on Fez.[45] In the process, however, essentialist ideas were absorbed by both Europeans and Arabs working on the conservation and planning of historic cities. These ideas have become tropes, persisting in World Heritage discourse and writing on conservation. In the case of Fez, another variant of Orientalism was at work: so-called perennialist philosophy, an anti-modern movement that saw "traditional" spirituality as an antidote to modern society. Many European adherents of perennialism were drawn to Islam and sufism, and it appealed to some Muslim intellectuals and architects. Burckhardt, a scholar of and convert to Islam, and connoisseur of Islamic art, published the book *Fez, City of Islam* in 1968; he was subsequently hired by UNESCO to consult on the conservation of the Fez medina and traditional crafts. A keen and sympathetic observer, Burckhardt presented an idealized vision of a city shaped by Islam: a "human city" that had become exceptional in "a world more and more mechanized, leveled, and dehumanized."[46]

The tropes of "traditional city" and "Islamic city" are confirmed by the language of the World Heritage Convention and the Operational Guidelines, which equate form and culture. This has led some to argue that World Heritage is a neocolonial discourse, imposed by Western institutions working with, and through, native elites. While there is much to support this argument, the colonial parallel has its limits. For example, colonial planners often resisted the installation of infrastructure and sanitary services in native towns, fearing that this would compromise their picturesque ambience. In contrast, contemporary plans for historic cities have generally emphasized infrastructure and services.[47] A neocolonial mindset is more obvious in attitudes toward the *inhabitants* of historic cities, which may or may not affect what is actually done on the ground.

Attitudes toward the inhabitants of old cities—among both Western and native professionals—seem to oscillate between two poles. Inhabitants may be seen as representatives of a traditional way of life, uncorrupted by modernization; as such, they bear witness to an "authentic" culture that has been lost, or is in the process of disappearing. This has led some contemporary planners (both native and European) to want to "protect" inhabitants from modern ways. The other commonly held view is that inhabitants of old cities are responsible for the decline of the physical fabric. Predominantly rural migrants,

---

44 T. Burckhardt, *Fez City of Islam* (trans. W. Stoddart), Cambridge, England: Islamic Texts Society, 1992, Chapter 8.

45 See G.D. Porter, "Unwitting Actors: the Preservation of Fez's Cultural Heritage, *Radical History Review*, issue 86, spring 2003, p. 143.

46 Lecture quoted in Jean-Louis Michon, "Afterword: Titus Burkhardt in Fez," *Fez: City of Islam*, p. 167. Michon, who also consulted for UNESCO, describes Burckhardt's involvement on pp. 164–70. On perrenialism see M. Sedgwick, *Against the Modern World*, Oxford: Oxford University Press, 2004.

47 See S. Bianca, *Urban Form in the Arab World: Past and Present*, London: Thames & Hudson, 2000, part three.

they are seen to have ways of life that differ from the urban elites who created old cities and their institutions. In essence, this is a new version of the Khaldunian idea of the city as a place where tribal people are "civilized." In the modern era, the civilizing function of the city has broken down: native elites have fled from old districts, traditional institutions have collapsed, and mass migration brings an unmanageable number of newcomers. Houses and other buildings are subdivided to accommodate them, resulting in unacceptable densities and lifestyles that are at variance with urban ways. This line of thinking led conservation planners in Fez, for example, to advocate the "dedensification" of the old city.[48]

The notion of a refined urban civilization decaying under the influence of rural people confirms the prejudices of bureaucrats and planners who are city dwellers themselves, and perhaps descendants of old urban elites. Such prejudices, however, are not unique to conservation or to the Arab world: ideal notions of "urbanity"—aimed at the poor, immigrants, and minorities—have been used to justify a wide spectrum of planning practices across the globe, ranging from urban renewal to preservation and gentrification.[49]

The Question of Authenticity

In the postwar period conservation emerged as a distinct profession, separate from architectural and planning practice. Taught in specialized institutes, implemented through new bureaucracies, and buttressed by new technologies, it was increasingly presented as a "science" of the past. This occurred at precisely the moment that the field faced its greatest challenges. Injunctions against rebuilding in historical styles seemed irrelevant given the massive destruction of war. Architects and planners were leading reconstruction efforts, in a climate shaped by high modernism and urban renewal. At the same time, conservation was being disseminated to other parts of the world, in the context of international cooperation.[50] Many newly independent states were heir to a colonial tradition of conservation; after decolonization, the old antiquities departments continued to operate along much the same lines, with support from organizations like ICOMOS and ICCROM. But some practitioners in these new states started to question the premises of conservation and their relevance for local contexts. Nicholas Stanley Price, former director of ICOMOS, noted that newly independent states approached conservation "with their own sensibilities, starting from traditions completely unlike those of the West and with an altogether different conservation history."[51]

48 G.D. Porter, "Unwitting Actors." For an account of planning in Fez, see S. Bianca, *Urban Form in the Arab World*, pp. 271–302; and H. Radoine, "Conservation-Based Cultural, Environmental, and Economic Development: the Case of the Walled City of Fez," in ed. L.F. Girard, ed., *The Human Sustainable City: Challenges and Perspectives from the Habitat Agenda*, Burlington, VT: Ashgate, 2003, pp. 457–77.

49 The lifestyles of minority and immigrant communities in the U.S., for example, were sometimes cited as justification for urban renewal; see Herbert Gans' classic study, *The Urban Villagers: Group and Class in the Life of Italian-Americans*, New York: Free Press of Glencoe, 1962.

50 See Philippot's assessment of this period in "La restauration dans la perspective des sciences humaines," in C. Périer-D'Ieteren, ed., *Pénétrer l'art, Restaurer l'oeuvre*, 1989, pp. 491–500; excerpted in S.N. Price, *Historical and Philosophical Issues*, pp. 216–29.

51 "The Emergence of Modern Conservation Theory," in S.N. Price, ed., *Historical and Philosophical Issues*, p. 205.

In this debate, which continues to unfold, the Venice Charter of 1964 has been particularly contentious. This document led to the establishment of ICOMOS, and as such laid the basis for a new international practice.[52] It was guided by a faith in universal values, as the preamble suggests:

> People are becoming more and more conscious of the unity of human values and regard ancient monuments as a common heritage. The common responsibility to safeguard them for future generations is recognized. It is our duty to hand them on in the full richness of the authenticity.
>
> It is essential that the principles guiding the preservation and restoration of ancient buildings should be agreed and be laid down on an international basis, with each country being responsible for applying the plan within the framework of its own culture and traditions ....

Intended as an update to the 1931 Athens Charter, the Venice Charter reaffirms the historicist premises of the earlier document, stating them in more categorical terms. It expands the definition of the monument, hitherto understood as a "great work of art," to include "more modest works of the past which have acquired cultural significance with the passing of time" (article 1). Not only the monument itself, but its "traditional setting" must be protected: "no new construction, demolition, or modification which would alter relation of mass and color must be allowed" (article 6). As for restoration, it is

> a highly specialized operation. Its aim is to preserve and reveal the aesthetic and historic value of the monument and is based on respect for original material and authentic documents. It must stop at the point where conjecture begins, and in this case moreover any extra work which is indispensable must be distinct from the architectural composition and must bear a contemporary stamp.

Both supporters and detractors of the charter have noted that the injunction to distinguish new work with "a contemporary stamp" reflects the dominance of modernist ideology in the 1960s.[53]

The concept of authenticity was taken up in the World Heritage Convention, written eight years later. Like the international safeguarding campaigns, the Convention was motivated by the threat of destruction—attributable largely to man-made causes:

> ... [T]he cultural heritage and the natural heritage are increasingly threatened with destruction not only by the traditional causes of decay, but also by changing social and

52 S. Titchen, "On the Construction of Outstanding Universal Value," p. 75. The full name of the Venice Charter is the International Charter for the Conservation and Restoration of Monuments and Sites, adopted by the Second International Congress of Architects and Technicians of Historic Monuments, held in Venice in 1964.

53 See, for example, M. Petzet, "The Venice Charter, Forty Years Later," *World Heritage Review*, vol. 38, December 2004, pp. 20–23. For a comparison of the Athens and Venice Charters and their relationship to architectural modernism, see R. Russell, "Abstraction, Authenticity, and the Abolition of Time," in M. Hardy, ed., *The Venice Charter Revisited: Modernism, Conservation, and Tradition in the 21st Century*, Newcastle upon Tyne: Cambridge Scholars Publishing, 2008, pp. 16–23.

> economic conditions which aggravate the situation with even more formidable phenomena of damage and destruction.
>
> ... [T]he deterioration or disappearance of any item of the cultural or the natural heritage constitutes a harmful impoverishment of the heritage of all nations of the world ....

The Convention calls on the international community "to participate in the protection of the cultural and natural heritage of outstanding universal value." To this end, it establishes a system of protection "organized on a permanent basis and in accordance with modern scientific methods ...."

Under the terms of the Convention, member states may nominate cultural and natural properties to the World Heritage list; these are reviewed by ICOMOS and IUCN (the International Union for the Conservation of Nature) on the basis of expert judgment or "scientific" expertise.[54] To be added to the list, a property must exhibit "outstanding universal value," which may be "intrinsic" to the property, or which may have developed over time. Outstanding universal value is assessed on the basis of criteria in the Operational Guidelines that accompany the Convention. These criteria have evolved over the years to reflect changing, and ever-broadening, conceptions of heritage.[55] They include the old notion of the masterpiece (criterion i), as well as the attribute of significance (criterion vi). The latter, traditionally associated with great men and events, may now apply to "living traditions, with ideas, or with beliefs, with artistic and literary works of outstanding universal significance." Criteria ii–v refer to the notion discussed earlier: that material artifacts and settlements *represent* culture. As such, a site may be listed if it reflects an interchange of human values that has influenced developments in the built environment (criterion ii); or if it bears unique or exceptional testimony of a civilization that is living or has disappeared (criterion iii). The old city of Sanaa was listed based on earlier versions of criteria iv-vi: it was deemed an outstanding example of an architectural ensemble that illustrates a significant stage in human history (criterion iv); and a traditional settlement that represents a culture (or cultures) which "has become vulnerable under the impact of irreversible change" (criterion v). The old city was also listed for its association with events of outstanding universal significance (the spread of Islam; criterion vi).

In addition to demonstrating outstanding universal value, a property must meet the so-called "test of authenticity." Before 2005, the Operational Guidelines stated that a property must "meet the test of authenticity and/or integrity in design, material, workmanship or setting ...." Although several types of authenticity are mentioned, it was widely acknowledged that material authenticity was the prerequisite for other types.[56] As of 2005 the test of authenticity has been amended to read: "a property must also meet the conditions

---

54 ICOMOS and IUCN serve as advisory bodies to UNESCO for cultural and natural sites, respectively. Separate conventions had been envisioned for cultural and natural properties, but it was finally decided to include both in a single document (S. Titchen, "On the construction of 'outstanding universal value'; some comments of the implementation of the 1972 World Heritage Convention," in *Conservation and Management of Archeological Sites*, 1996, vol. 1, pp. 235–42).

55 The full name of the World Heritage Convention is "Convention Concerning the Protection of World Cultural and Natural Heritage." In her unpublished doctoral dissertation, Sarah Titchen carefully documents the evolution of the criteria between 1972 and 1994.

56 *Operational Guidelines*, 2002 24.b.i. See also the widely used manual by B. Fielden and J. Jokilehto, *Management Guidelines for World Heritage Cultural Sites*, Rome: ICCROM, 1993, especially, pp. 16–17.

of *integrity and/or authenticity* and must have an adequate protection and management system to ensure its safeguarding."[57] The change seems to have been a response to the 1994 Nara Document on Authenticity, which was finally annexed to the Operational Guidelines in 2005. But the latter contain numerous qualifications, which suggest that material authenticity remains the overriding concern.

Much of the controversy surrounding the Venice Charter and the World Heritage Convention has revolved around the meaning and intent of the term "authenticity." Until the Venice Charter, the term had not been used in an international convention. But it was an old concept: the Ruskinian emphasis on original materials and intentions had, by the turn of the twentieth century, become a guiding principle of conservation practice. Use of the term "authenticity" in the 1964 document was not accidental: it relates to the culture of the 1960s and '70s, in particular, the disenchantment with established power structures and the search for authentic personal and cultural expression. The term "cultural heritage" (also new at this time) evokes the popular quest to rediscover personal and collective roots.

Attacks on the Venice Charter surfaced in the 1980s, and increased as the fiftieth anniversary of its drafting approached. The critics are an interesting mix: they include architects and conservators from the developing world, and proponents of neo-traditional architecture in Europe and the U.S.[58] They object, in particular, to the modernist bias of the Venice Charter which bans the reconstruction of historic fabric, and requires all new work to "bear a contemporary stamp." Practitioners working in the developing world argue that this modernist bias has the effect of undermining traditional building practice. Master masons, writes A. Krishna Menon, have been marginalized by both the modern construction sector and by the conservation profession: both discourses have rendered local practice "invisible." The creativity of master masons is devalued, but their skills are used when "authentic" techniques are required in restoration. For these masons, the idea of "contrasting" new work with old makes no sense; indeed, it *contradicts* locally held standards of excellence and authenticity. "[T]he master-mason is still able to build and restore in the example of his forefathers," writes Menon. "His pride lies in the fact that it is difficult to distinguish between the two." Strict adherence to historical criteria may make sense in the case of exceptional monuments that are deemed of national importance. But in "hundreds and thousands of less-than-exemplary monuments," writes Menon, "… we could *permit* reconstruction, *encourage* duplication, *value* invention, and *seek* transformation."[59]

---

57 *Operational Guidelines*, 2013, 2.D.78 (emphasis mine). Integrity is defined as "a measure of the wholeness and intactness of the natural and/or cultural heritage and its attributes" (2.E.88). The use of these terms has a complicated genealogy, which can be traced in part to the U.S. National Park Service. See Titchen's dissertation, pp. 102–4, and H. Stovel, "Origins and Influence of the Nara Document on Authenticity," *APT Bulletin*, vol. 39, no. 2/3, 2008," p. 10.

58 Both groups are represented in M. Hardy, ed., *The Venice Charter Revisted* (see, pp. 16–23). This unevenly edited book consists of the proceedings of a conference of the International Network for Traditional Building, Architecture, and Urbanism (INTBAU), held in Venice in 2006. It has a forward by the Prince of Wales, a virulent opponent of modernism. See also N. Gutschow, "Restaurierung und Rekonstruktion Gedanken zur Gültigkeit der Charta von Venedig im Kontext Südasiens," *Deutsche Kunst und Denkmalplege*, vol. 49, no. 2, 1991, pp. 156–60; S. Rab, "Authenticity in Cultural Heritage Management," paper presented at the first International Conference on Architectural Conservation, Dubai, March 14–16, 2004; and L. Smith, *Uses of Heritage*, Abingdon: Routledge, 2006, Chapter 3.

59 A. Krishna Menon, "Rethinking the Venice Charter: the Indian Experience," *TDSR Working Paper Series*, vol. XL, 1992, pp. 49–52. See also "Conservation in India: A search for direction," *Architecture + Design*, 1989, pp. 22–7; and "The Afterlife of the Venice Charter in Postcolonial India," in M. Hardy, ed., *The Venice Charter Revisted*, pp. 16–23.

More recent work by Trevor Marchand echoes these ideas. Trained as an architect and an anthropologist, Marchand has written a number of important works based on apprenticeships with builders in Yemen and Mali. He argues against the "museological view of culture" that underpins modern conservation practice.[60] The expert discourse of masons, writes Marchand, is inseparable from its cultural and social context. This context

> is historically constituted and thus always in flux. It has been influenced by networks of trade, Islam, war, colonialism, globalization, and most recently … incorporation … into the UNESCO framework of World Heritage Sites. The contemporary traditional builder, through his apprenticeship, is inculcated with the technical and social knowledge necessary for innovatively responding to change while simultaneously producing a discourse of locality, continuity, and tradition. In short, it is this expertise that sustains a sense of place. If through legislation and cultural politics (Western-inspired) conservation efforts incapacitate contemporary traditional builders and inhabitants in responding to changing needs and social values, their architectural heritage will inevitably lose its utility and meaning for the living local population, and thus also its authenticity as a valued cultural commodity.[61]

In short, Marchand argues, it is the building *process*—rather than the product—that is "historical." Like Menon, he argues for a new definition of authenticity based on the on-going production of culture.

Critics of the World Heritage Convention fall into two camps: "insider" critics, mainly professionals concerned with the definition of authenticity; and scholars who focus on the discourse itself, and its application at specific sites. The latter see heritage as a globalizing phenomenon: based on transnational processes and networks of expertise, the heritage project depends on, and reinforces, the nation state system and national elites.[62] These critics challenge the universalizing claims of heritage, but few have offered sustained critiques of the concept of outstanding universal value. Michael Herzfeld's work is particularly relevant in this regard. For years anthropologists worked hard to make the case for ethical diversity and relativism, writes Herzfeld. But now some values seem to have become universal after all:

> Notions such as efficiency, fair play, civility, civil society, human rights, transparency, cooperation, and tolerance serve as global yardsticks for particular patterns of interaction.

60 Citing Rowlands; see T. Marchand, "Process over Product: Case Studies of Traditional Building Practices in Djenne, Mali and Sanaa, Yemen," in Teutonico and Matero, eds, *Managing Change: Sustainable Approaches to the Built Environment*, proceedings of the 4th Annual U.S./ICOMOS Symposium, Philadelphia, April 2001, Los Angeles: the Getty Conservation Institute, p. 137. Marchand's key works on this topic are *Minaret Building and Apprenticeship in Yemen*, Richmond, Surrey: Curzon Press, 2001; and *The Masons of Djenné*, Bloomington: Indiana University Press, 2009. See also his edited volume, *Making Knowledge: Explorations of the Indissoluble Relation between Mind, Body, and Environment*, Wiley Blackwell, 2011.

61 T. Marchand, "Process over Product," pp. 155–6.

62 See M. Askew, "The magic list of global status: UNESCO, World Heritage, and the Agenda of States," in S. Labadi and C. Long, eds, *Heritage and Globalization*, New York: Routledge: 2010; and J. Turtinen, "Globalising Heritage—On UNESCO and the Transnational Construction of a World Heritage," Stockholm: SCORE Rapport serie 12, 2000,

> Startlingly, even diversity can become a 'homogenous' product. So, too, can tradition and heritage: the particular itself is universalized ….
>
> The increasing homogenous language of culture and ethics constitutes *a global hierarchy of value*. This hierarchy, which clearly succeeds to the values promulgated worldwide by the erstwhile colonial powers of Europe, is everywhere present but nowhere clearly definable. Its very vagueness constitutes one source of its authority, since, while it often appears as a demand for transparency and accountability, it is itself protected from any demand that it account for itself ….

This global hierarchy of values is based on a reified notion of culture, one that is reproduced in various essentializing discourses and deployed against the "center" ("Asian values," "Islamic values," and so forth). But these discourses, writes Herzfeld, cannot "escape the logic of culture-as-possession that both reinforces the concrete quality of the identity thus promoted and concomitantly favors the commodification of ethics and aesthetics …."[63] Herzfeld's observations confirm a shift that has occurred over the last few decades: diversity is now an integral part of the discourse of the nation state—which is often described as a "mosaic" of regions and cultures. This shift was already apparent in the 1980s in the Yemens, and in other countries in the region.[64]

The response of the conservation establishment to this wide-ranging criticism has been to retrench. Its representatives tend to defend the validity of the international charters on a moral, rather than on a legal, basis. At the same time, they argue that the charters can be interpreted in a variety of cultural contexts.[65] This notion of "cultural relativism" is associated especially with the Nara Document on Authenticity.[66] Alluding to certain post-structuralist concepts, the authors of the Nara Document argued that heritage must be understood within a specific cultural context, and conserved and interpreted according to the values of the culture that produced it. The discussion at Nara was taken up by practitioners and stakeholders in other parts of the world, who have attacked the Venice Charter, in particular, as a Eurocentric document. Many of these critics object to the Venice Charter's emphasis on material form, at the expense of cultural and spiritual meanings associated with form. The Nara Document does not define authenticity, nor does it question the concept of "culture." But in the intervening decades a *de facto* working definition of authenticity has developed: the ability of a property to "communicate" values through various "attributes." In both the World Heritage Convention and Operational Guidelines, as well as in much insider criticism, sites are presumed to "have" or "carry" values—alluding to the positivist idea that value is inherent in the object, rather than projected onto it by the observer.[67]

63 *The Body Impolitic: Artisans and Artifice in the Global Hierarchy of Value*, 2002, pp. 2–3.

64 See below, Chapter 4. On this phenomenon in Morocco, see K. Hoffman and S. Miller, eds, *Berbers and Others: Beyond Tribe and Nation in the Maghrib*, Bloomington: University of Indiana Press, 2010.

65 See M. Petzet, "The Venice Charter, forty years later," *World Heritage Review*, pp. 20–24.

66 Stovel, "Origins and Influence of the Nara Document on Authenticity," *APT Bulletin*, vol. 39, no. 2/3, 2008," p. 9. The meeting at Nara was initiated by Japanese conservators who wanted to expand the concept of authenticity to allow for the periodic dismantling and reassembly of wooden temples, a traditional practice in Japan.

67 See J. Tainter and J. Lucas, "Epistemology of the Significance Concept," *American Antiquity*, vol. 48, no. 4, October 1983, pp. 707–19. In Stovel's words: "the quality of communication of defined

Largely as a result of the Nara Document and the many symposia and conferences that followed it, we have seen a shift toward a "values-based" approach. It takes as its inspiration Alois Riegl's 1903 treatise on the multiple and competing values involved in conservation. Riegl describes "age value"—and hence material authenticity—as the basis of the modern approach to monuments. Following Riegl, the current values-based approach does not do away with older, positivistic notions of value. Rather, it tries to reconcile these with the "postmodern" idea of multiple and shifting values.[68] Relative values are often invoked, but typically with regard to the interpretation and presentation of heritage. Rarely are they allowed to impact the treatment of the object itself. Meanwhile, protection extends to an ever-wider array of artifacts, including those as layered and complex as historic towns and landscapes.[69]

The new category of "intangible heritage" does not help matters. It includes a variety of non-material practices (music, dance, oral and folkloric traditions, social and knowledge-based practices, and so forth) as well "traditional craftsmanship." But the artistic process, which for many is also a spiritual process, involves the creative *interaction* of mind, body, and material. It is precisely this interaction that created the buildings we value today as heritage, and guided their transformation over time. Identifying craftsmanship as "intangible" heritage suggests that it is a distinct *kind* of heritage, separate from the tangible material on (and through) which it acts.[70] The distinction, in effect, allows the treatment of the physical object to remain in the domain of professional expertise. It is perhaps not accidental that the building arts are downplayed in formulations of intangible heritage: in official discourse, "spirit" and "feeling" are seen to compromise authenticity.[71] But in Yemen and other countries where traditional building practices thrive, conservation remains heavily dependent on craftsmen—and this, as we will see, changes the nature of conservation itself.

Scope and Contribution of the Present Work

Professional discourses and practices are often slow to develop reflexivity, which may be threatening to expert niches. If this is true of architecture, it is even truer for

heritage values through the significant attributes carrying these values" ("Origins and Influence of the Nara Document," p. 11). In the 2013 Operational Guidelines, see for example item 2E.82.

68 See M. Lamprakos, "Riegl's Modern Cult of Monuments and the Problem of Value," *Change Over Time*, vol. IV, September 2014, pp. 418–35.

69 On the authenticity of historic towns, see the 2013 Operational Guidelines, item 1.15.ii. Some have hopes that attitudes will change with the adoption of UNESCO's Recommendation on Historic Urban Landscapes (2012); see G. Araoz "World-Heritage Historic Urban Landscapes: Defining and Protecting Authenticity," *APT Bulletin*, 39: 2–3, 2008, pp. 33–7.

70 See Richard Sennett's discussion of material consciousness in *The Craftsman*, London: Allen Lane, 2008. On the problematic distinction between "tangible" and the "intangible," see M. Herzfeld, "Intangible Delicacies: Production and Embarrassment in International Settings," *Ethnologies*, vol. 36, 2014, pp. 41–57.

71 See the 2013 Operational Guidelines, items 2E.83–89. The Convention for the Safeguarding Intangible Cultural Heritage (2003) mentions the "building arts"; but on UNESCO's web site, "traditional craftsmanship" is defined as "craft products" and "motifs." On the exclusion of artisans from conservation projects, see M. Herzfeld, *The Body Impolitic*, p. 19; and his works on Crete (1991, 2004) and Rome (2009). On building and embodied knowledge, see Marchand's studies on Yemen and Mali.

conservation—which only emerged as a distinct profession in the mid-twentieth century. Within the academy, conservation was not seen as an area for serious study until quite recently. Since the 1990s, we have seen a burgeoning literature on discourses of the past, related to, but distinct, from scholarly interest in memory. Of particular relevance are the now numerous works by anthropologists and architectural historians on heritage in a global context. The present work is indebted to this literature, but differs from it in several respects.

Most of these studies portray heritage as a top-down process, controlled and manipulated by state and corporate interests. They see heritage as part of a wider trend to commodify and reify culture, through bureaucratic processes that have become increasingly transnational. Once the heart of communal life and memory, heritage sites are cordoned off by their new monumental and touristic status; they lose their meanings for the very communities whose histories they represent. When entire historic districts are "monumentalized," they are often taken over by shopping and tourism, and many long-time residents and businesses are displaced. In the developing world, where gentrification remains rare, historic areas become stage sets where tourists can observe exotic natives and buy traditional products. This story has played out in countless places across the globe, and shows no signs of abating.

Following Hobsbawm and Ranger (1983), many of these studies dismiss heritage as "invented tradition." But as Herzfeld points out, the notion of "invented tradition" implies that there is a "real" and ultimately knowable historical past. It also tends to privilege elites as the shapers of narratives of the past. As such it risks "complicity with the forces of hegemony to which it is critically addressed."[72] Among these hegemonic forces, we must count the high modernist project itself—which for decades provided the dominant models in architectural and urban practice. In the developing world these models were actively promoted by international consultants and by local elites trained in the West. The universalizing thrust of high modernism, especially urban renewal, led a generation of architects in the 1970s and '80s to look at alternative models rooted in place, history, and cultural identity. Heritage was an important part of that story. The globalizing tendencies of heritage, then, must be considered alongside those of high modernism and the related discourse of modernization.[73]

This book contributes to a growing alternative literature that looks at heritage as a shared discourse.[74] It approaches heritage not as a top-down process—overdetermined by the state and official agencies—but rather as a process that unfolds on the ground, through the work of multiple actors. In Sanaa and elsewhere in Yemen, heritage initiatives have encountered strong local cultures with a strong sense of the past and its meaning for the present. This book looks at heritage through the eyes of those who build it and manage it: officials, architects, builders, and residents. These groups claim to represent the past, based on inherited or acquired criteria of authority, expertise, and belonging. Their agency

72 *A Place in History: Social and Monumental Time in a Cretan Town*, Princeton: Princeton University Press, 1991, pp. 12–13.

73 While modernity, modernization, and modernism are distinct terms, they share certain ideological premises. If modernization consists of a set of quasi-scientific practices and institutions that aim to establish certain indices of progress, then "high modernism" can be described as a faith in these practices and institutions, and their ability to transform society; on this point see J. Scott, *Seeing Like a State*, pp. 88–9. See also D. Tipps, "Modernization Theory and the Comparative Study of Societies: A Critical Perspective," in *Comparative Studies in Society and History*, vol.15, 1973, pp.199–226.

74 Herzfeld's notion of "cultural intimacy" will be a touchstone in this work (*Cultural Intimacy: Social Poetics of the Nation State*, New York: Routledge, 2005). On heritage as a shared discourse in Yemen, see T. Marchand, *Minaret Building*, p. 21.

resides in their ability to make various choices, and to frame these choices in terms of a discourse that is at once global and local. Aspects of international theory—which may be seen by outsiders as "Eurocentric"—are sometimes employed by local actors to support their notions of the "historic" and the "traditional." This interplay of local and global discourses of the past have produced a fascinating synthesis—one that holds lessons for those practicing heritage outside Yemen's borders.

Since the 1970s, heritage has been an aspect of development assistance. International agencies like UNESCO-UNDP, the World Bank, and the Aga Khan Trust for Culture have pursued various heritage initiatives in the old towns of the developing world under the rubric of "conservation in development." Working with national and local governments and counterparts, these projects have included building restoration, urban upgrading, the promotion of heritage tourism, and economic and community development. Despite decades of experience and the transnational networks of expertise that have emerged to support it, we know little about the reception and outcomes of these efforts. Once the projects are over and the last reports duly filed, the residents of old towns are left to inhabit these upgraded and restored environments, and local architects, officials, and builders are left to manage them. What do locals make of these projects? What difference do they make in their lives? To what extent are these old towns "owned"—and how are they transformed in the process? This is one of the few scholarly studies to review a major international effort through a close study of the local context. As such, it contributes to the growing critical literature on the history and ethnography of heritage.[75] I pay close attention to what various actors say, what they build, and how they build it—drawing on my training as an architect and historian, and on years of practice in architecture, construction management, and third world development.

It is important to clarify what the book does *not* do. It does not enter into the scholarly debate concerning the concepts of "modernity" and "tradition"; rather, it looks at how these terms are used locally. Apart from the preceding discussion of the regional context and a brief discussion of Shibam in Chapter 6, the book does not compare Sanaa to other cases of conservation in the region. Rather, it looks at how conservation emerged and developed in a specific locale, the old city of Sanaa. I focus on the historic core of the city, the only area where conservation has been applied at the urban scale; and on the elements that make up the fabric, rather than on individual monuments.[76]

The first part of this book deals with the development of the city of Sanaa (Chapter 2) and the history of the UNESCO Campaign (Chapter 3.) In chapters 4 and 5 we move to the mid-2000s, the moment of my fieldwork; they can be considered as a pair. Chapter 4 looks at the ways in which conservation has been understood and applied by residents and officials. Chapter 5 looks at those charged with caring for the old city: the new heritage experts (architects and other professionals); and the "traditional" experts, builders and craftsmen (*usta*s, literally "masters"). In Chapter 6, I trace the evolution of local discourse and practice through a series of projects: several early projects conducted under the auspices

75 In addition to studies by Herzfeld and Marchand, see Peutz (2008); Labadi and Long (2010); and Joy (2012). Of the numerous historical studies that might be cited, see Hall (2011); Toxey (2011); Kavuri-Bauer (2011); Grigor (2009); Alsayyad (2001, 2005); Mason and Page (2004); El-Habashi (1998, 2001); and ʻAbdelkafi (1982). See also "Conserving the City," special issue of *Change Over Time* (April 2014), which I guest edited; and my article "The Idea of the Historic City" in the same volume (pp. 8–39).

76 The boundary of the World Heritage site excluded several historic suburbs adjacent to and near the old core; this issue, which has been subject to much debate, is discussed in Chapter 3.

of the Campaign, and later projects led by the Social Fund for Development (a quasi-governmental agency supported by the World Bank). The concluding chapter assesses the successes and the failures of the Campaign, and its legacy for current work.

Yemen can be a difficult place to conduct research—and the Sanaa Campaign has a complicated history, involving multiple agencies, donor nations, and consultants. Some of the agencies and offices are defunct; written records of the Campaign are incomplete, both in Sanaa and at UNESCO/World Heritage offices in Paris. In the pages that follow, I weave together oral accounts, written records, and assessments of the built work. Most of the research presented here thus combines history and ethnography. This is fitting since, as Paul Dresch notes, the distinction between history and anthropology does not readily apply in Yemen.[77] But the same may be true for other places: academic and professional disciplines have been shaped, and constrained, by the distinction between change and atemporal pattern. If we blur that line, we may better understand architecture and its role in lived societies, past and present.

I conducted the research for this book during the years 2004–2005, and during a short reconnaissance trip in fall 2003. It was a fortuitous moment, in ways that were only partially clear at the time. The Campaign ended in the early 1990s, and major projects were brought to a close shortly thereafter. Little conservation work happened in the mid-1990s, during the economic crisis that followed the first Gulf War. Work resumed later in the decade under the Social Fund, in partnership with the government conservation office; the latter's mandate, originally restricted to Sanaa, was now expanded to include all historic cities in Yemen. A generation of architects and other professionals trained under the Campaign were able to shape this second wave of projects in accordance with lessons learned in the 1980s and early '90s. Through these architects and the builders with whom they worked, I was able to track the development of local discourse and practice. I also had the good fortune to speak with individuals who initiated and supported the Campaign, several of whom have since passed away. The moment of my fieldwork was significant in another way. It followed 9/11 and the U.S. invasion of Iraq. There were tensions around the Yemeni government's involvement in the U.S.-led "war on terror," but not to the degree felt later in the decade. Since the 2011 uprising, access to the country has been difficult for foreign researchers. Donor aid for cultural heritage has dwindled, new projects have been shelved, and government oversight of existing projects is lax. The future of the heritage enterprise is unclear; and the same is true of the country itself.

This situation provides an opportunity to take stock of the UNESCO Campaign and similar large-scale initiatives in Yemen and beyond. When foreign funding resumes, what priorities should donor agencies pursue? What are the appropriate mechanisms to employ? Which stakeholders should be engaged, and how? I have argued that "values-based" conservation has not yet transformed institutional strategies: the concept itself is compromised by a discursive history that remains largely unexamined. But in Yemen, multiple "values" and claims have exerted pressure from below, creating new and local approaches to heritage. There are creative and potentially empowering opportunities which have so far been missed. How should international agencies and donors deal with this new reality? In the conclusion of this work, I offer some concrete proposals in this regard.

77 See *Tribes, Government, and History in Yemen*, New York: Oxford Press, 1989, p. 31.

Chapter 2
The Old City of Sanaa: An Overview

This chapter provides background for the conservation of the old city of Sanaa, which began in the early 1980s. I will first briefly discuss the wider historical and political context, and then look more closely at Sanaa itself. The outlines of the city's historical development until 1962 have been well charted. Serjeant and Lewcock's edited volume remains the definitive source on pre-revolutionary history and society. Written during preparations for the Sanaa Campaign, it informed planning for the old city and thus can be seen as a kind of primary source.[1] It has been supplemented and enhanced by later works, especially Kopp and Wirth's study which is important for the Ottoman period and later developments.[2] Drawing on these sources, I will focus on certain elements of city form that will be important for subsequent chapters. The last section of the chapter looks at the development of the city after the 1962 revolution, setting the stage for the conservation effort.

Situating Sanaa

The area that became modern Yemen managed to preserve a relative degree of autonomy throughout most of its history. With the exception of the British presence at Aden, Yemen was never occupied by a European power; nor did two Ottoman occupations (1538–1635 and 1872–1918) bring about the kind of domination associated with European colonialism. At the same time, Yemen's modest petroleum reserves did not attract the kind of foreign intervention that occurred in neighboring states. This has produced what some observers see as a unique cultural confidence among Yemenis.[3] It is reinforced by the sense of Yemen as a natural unit—an idea that is rooted in Islamic history and local literature, and also

1 R.B. Serjeant and R. Lewcock, eds, *San'a, an Arabian-Islamic City*, 1983. See also R. Lewcock, *The Old Walled City of San'a*, Paris: UNESCO, 1986; and P. Bonnenfant, *Sanaa: Architecture Domestique et Société*, Paris: CNRS Editions, 1995.

2 H. Kopp and E. Wirth's *Sanaa: Développement et organisation de l'espace de la ville arabe*, Cahiers de l'Institut de Recherches et d'Etudes sur le Monde Arabe et Musulman (IREMAM), CNRS-Universités d'Aix Marseilles, 1994. Kopp and Wirth's book, as well as P. Costa's article "Sana" in Serjeant, R.B., ed., *The Islamic City*, Paris: UNESCO, 1980, include discussions of architecture and urbanism after 1962. For a more extensive treatment based on an eye witness account, see F. Varanda, "Tradition and Change in the Built Space of Yemen," unpublished dissertation, Department of Geography, University of Durham, UK, 1994 (now available on www.archnet.org, document no. DPC0927). See also G. Grandguillaume, et al., eds, *Sanaa hors les murs*: *une ville arabe contemporaine*, Collection Villes du Monde Arabe, vol. 1, Tours: URBAMA, 1995; R. Stadnicki and J. Touber, "Le Grand Sanaa. Multipolarité et nouvelles forms d'urbanité dans la capital du Yemen," *Annales de géographie*, no. 659, 2008, pp. 32–53; and N. al-Kawkabani, "Al-ittijahat al-mi'mariyya al-haditha fi madinat Sana'a ma bayn 1962–90" ("Modern architectural trends in the city of Sanaa from 1962–90"), unpublished Master's thesis, University of Sanaa, 2000).

3 These points are made by G. Grandguillaume in "Sanaa, Ville d'Arabie," in Grandguillaume, et al., eds, *Sanaa hors les murs*, pp. 4–6. Even the British occupation of Aden (1839–1967), he notes, did not radically transform the tribal structure, but allowed for the emergence of a modernizing elite.

in the popular imagination. In contrast to other states in the region, notes Paul Dresch, Yemen's history is somehow "real"—although the idea of Yemen as a single *political* unit is a modern idea, that developed in a context created by outside powers.[4]

Yemen is the only fertile region of the Arabian peninsula, which earned it the name *Arabia Felix* in ancient times (Figure 2.1). It has also always been much more heavily populated than the rest of the peninsula: there are a few large cities, numerous towns, and thousands of villages and hamlets.[5] Many of these towns are striking and varied in their architectural form. They are characterized by multi-story houses of either stone and fired brick, as at Sanaa and Hajjara (Figure 2.2), or of earth, as at 'Amran and Sa'da. The oasis towns of the Hadhramaut Valley, in the former PDRY, are known for their exceptional and varied earthen architecture.[6] The tiny settlement of Shibam, in particular, is known for its multi-story tower houses. Shibam was listed as a World Heritage Site in 1982 and like Sanaa, was the site of a UNESCO safeguarding campaign. Over the course of the last decade, conservation of the town was supported by an internationally recognized project (to be discussed in Chapter 6). Tarim has come to be appreciated for its remarkable palaces built by merchants in the first half of the twentieth century, which incorporate South Asian and European neoclassical features. The fortified town of Zabid, in the Tihama, was an important center of Islamic learning and commerce; in later centuries, merchants who made their fortunes in the indigo trade built houses of fired brick that have much in common with Persian models. Added to the World Heritage list in 1993, it was placed on UNESCO's list of endangered sites seven years later. Attempts to enforce conservation regulations have been embroiled in controversy—perhaps reminiscent of the situation in Sanaa in the 1980s.[7]

Sanaa was historically the largest and most spectacular of the highland towns. Located at a strategic point in the northern plateau at the intersection of ancient trade routes, the city has always been an important political center.[8] One of two capitals of the ancient Sabaean empire, it was the governor's seat under the early Islamic caliphs, an important administrative center during the later medieval and Ottoman periods, and from the

4 P. Dresch, *A History of Modern Yemen*, p. 1. The following discussion draws on this work and on Dresch's *Tribes, Government, and History in Yemen*, New York: Oxford Press, 1989; R. Burrowes, *The Yemen Arab Republic: the Politics of Development, 1962–1986*, Boulder, CO: Westview Press, 1987, and *Historical Dictionary of Yemen*, 2nd edition, Lanham and Toronto: Scarecrow Press, 2010; S. Day, "Power-Sharing and Hegemony: a Case Study of the United Arab Republic of Yemen," Ph.D. dissertation, Georgetown University, 2001; and S. Philips, *Yemen's Democracy Experiment in Regional Perspective: Patronage and Pluralized Authoritarianism*, London: Palgrave Macmillan, 2008.

5 R. Burrowes, *The Yemen Arab Republic*, p. 6.

6 On the Hadhramaut see especially S. Damluji, *The Valley of Mud Brick Architecture: Shibam, Tarim, and Wadi Hadramut*, Reading, UK: Garnet Publishing Limited, 1992. On highland architecture, see F. Varanda, *The Art of Building in Yemen*, 2nd edition, Lisbon: Argumentum, 2009; on Zabid, P. Bonnenfant, *Zabid au Yémen: archéologie du vivant*, Aix-en-Provence: Edisud, 2004; "Zabid, anti-développement et potentialités," *Peuples Méditerranéens*, nos 72–3, July–Dec. 1995, pp. 219–42.

7 Local resistance to conservation intensified after an incident in June 2010, following a new UNESCO warning that the city might be removed from the World Heritage list. Informants report that demolition crews, sent to remove illegal structures, were resisted by protesters. A security officer fired on the crowd, killing one man; others were injured in clashes. Demonstrators attacked GOPHCY's office, which was forced to close; GIZ suspended its project there, but resumed work in 2012. In the English language press see especially I. al-Mawseemi, "Is Zabid destined to lose its place among historical heritage cities?" Arab Reporters for Investigative Journalism/*Al-Thawra*, July 21, 2013, http://arij.net/en/zabid-destined-lose-its-place-among-historical-heritage-cities (accessed 7/30/14).

8 *Sanaa* probably derives from an ancient South Arabian word that means "well-fortified"—suggesting the origin of the city as a fortified settlement.

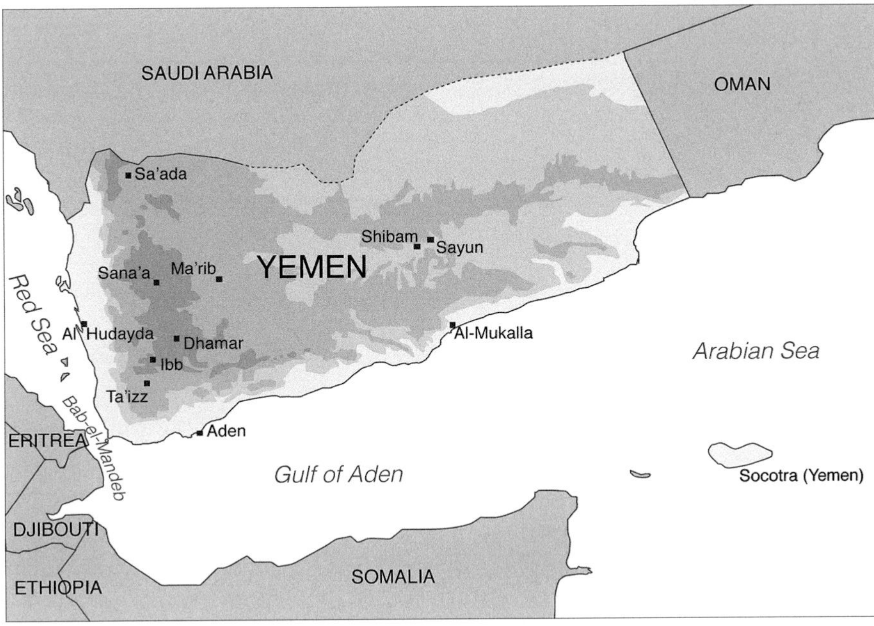

Figure 2.1 Map of Yemen

eighteenth century, the capital of the Qasimi imams. In 1962 their successors, the Hamid al-Din imams, were overthrown and Sanaa became the capital of North Yemen (Yemen Arab Republic/YAR). A socialist revolution in the south the following year created the state of South Yemen, with its capital at Aden (People's Democratic Republic of Yemen/PDRY).[9] In May 1990 the two countries unified, with Sanaa as capital of a united Yemen Republic.

Although scholars have emphasized the division between the two former states, the more salient division is that between Upper Yemen and Lower Yemen, or the upper and lower highlands. The division is not so much religious as socioeconomic, cultural, and political.[10] A poor area that supports only subsistence agriculture, Upper Yemen is home to the country's two main tribal confederations, the Hashid and the Bakil. For these groups the tribe is still the primary source of group identity, in many cases more important than the state; they follow the sect of Zaydi Shi'ism espoused by the former imams.[11] Lower Yemen, in contrast, is an extremely fertile region; most inhabitants are Sunni of the Shafi'i school, with affiliations based on the village and kinship group. The tribes of Upper Yemen formed the power base of the Zaydi imams: often acting as the imams' mercenaries, they depended on raids of the rich agricultural lands of Lower Yemen for their livelihood.

9 This name was adopted in 1970, replacing "People's Republic of South Yemen."

10 S. Philips, *Yemen's Democracy Experiment*, p. 42. The division between Upper and Lower Yemen is usually placed at the Samarra Pass, to the north of Ibb.

11 Zaydism is a moderate form of Shi'i Islam that does not differ significantly from from Shafi'i Sunnism on matters of doctrine: its main difference is the principle of imamic rule. As Dresch notes, most tribes are settled and agricultural, not nomadic.

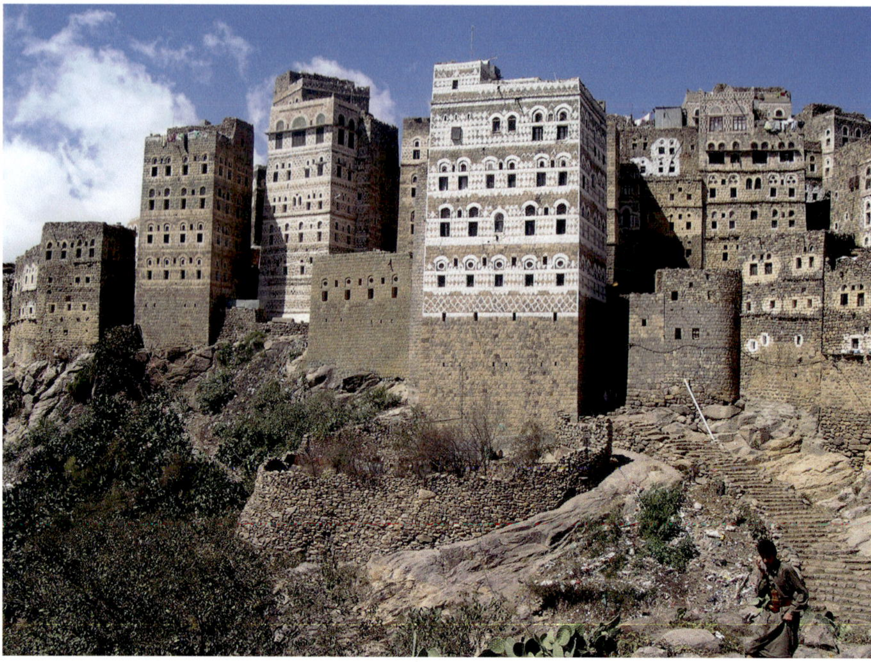

Figure 2.2 The town of Hajjara in the Harraz mountains to the west of Sanaa

According to Zaydi political theory, the imam had to be a *sayyid* (pl. *saada*), that is, a descendant of the Prophet. Zaydi rulers relied on *saada* in their administration, and on a class of learned men called *qudah* (sing. *qadi*, "judge").[12] The *saada* were resented for their privileged status by other groups, including tribesmen; tribal sheikhs often competed with the *saada* for regional control. The 1962 revolution, inspired by Nasserism, sought to end the old socio-political order based on descent. A new elite emerged, made up of military and tribal leaders, businessmen, and technocrats.[13] The latter were primarily from Lower Yemen, individuals who fled the socialist regime in the south. They represented the majority of the country's managers and intellectuals, but their opportunities for advancement were limited by northern political patronage. Since unification, in particular, the country has been dominated by the political culture and elite of former North Yemen. Made up of tribal and military leaders (sometimes one and the same), and from the 1980s, a new commercial element, this elite has monopolized the fruits of economic liberalization. This is sometimes portrayed as a new kind of "tribal raiding" by people who do not respect the rules of civilized society.[14]

12 On the "estates" of Yemeni society see P. Dresch, *Tribes, Government, and History in Yemen*, New York: Oxford Press, 1989, pp. 118–23, 136–45.

13 G. Vom Bruck, "A House Turned Inside Out: Inhabiting Space in a Yemeni City," *Journal of Material Culture*, vol. 2, no. 2, July 1997, p. 144; S. Caton, *"Peaks of Yemen I Summon": Poetry as Cultural Practice in a North Yemeni Tribe*, Berkeley: University of California Press, 1990, p. 7.

14 See S. Philips, *Yemen's Democracy Experiment*, pp. 47–57.

'Ali 'Abdullah Saleh (himself of the Sanhan tribe, Hashid confederation) was president for over three decades, from 1978 to 2011. During this time, he managed to govern by taking into account the relatively weak powers of the center and the strong centifugal forces of tribes, regions, social and religious factions. He balanced coercion and negotiation, and used patronage to win support. The patronage "pyramid" grew more pronounced after 1994, fueled by oil revenues and development aid; Saleh himself was the greatest beneficiary of the system.[15] Saleh's political balancing act became more difficult after 9/11. The U.S. pressured him to crack down on terrorism; he now had to balance a powerful external actor against domestic forces. From the mid-2000s, the rebellion of the Houthis in the north and calls for secession in the south challenged the regime. A revitalized al-Qa'ida emerged; U.S. drone strikes enraged Yemenis, and undermined Saleh's legitimacy. In 2011, inspired by the Tunisian revolution and mirroring events in Egypt and Syria, democracy protesters took to the streets in cities across Yemen. Other factions saw the opportunity to attain power and the situation turned violent—although it did not degenerate into civil war, as in Libya or Syria. Saleh was severely injured in an attack on the palace compound in 2011; he finally stepped down in favor of his Vice President, 'Abdu Rabbu Mansour Hadi. A national dialogue ensued which many saw as hopeful, but elements of the former regime remained powerful. By early 2015, the situation had deteriorated. Houthi insurgents took over the north and seized Sanaa; they forced Hadi to resign, but he fled to Aden and withdrew his resignation. With al-Qa'ida still a powerful force in the south, peace and stability seemed unlikely.

This political turmoil has played out against a backdrop of poverty, failing state systems, and severe environmental challenges. Yemen's oil reserves are nearly exhausted, the country faces grave water shortages, and the birth rate—over five children per woman—is one of the highest in the world. The situation is especially acute in the capital city, which has experienced spectacular growth over the last four decades. Before 1970, development did not extend beyond the *intramuros* core and suburbs; taken together, these areas covered around 300 hectares (741 acres) and were home to 55,000 people in 1962. Fifteen years later, in 1977, the city covered nine times that area. By the early 1990s, the city had expanded to 25 times the original surface area, with a resident population of 500,000 and an additional 250,000 daily commuters. Today the old city represents only about 3 percent of the total urbanized area.[16] Sanaa is now home to more than two million people—the largest city in a country that has the highest rate of urban growth in the Middle East/North Africa region.[17]

In this context, it is perhaps not surprising that cultural heritage is low on the agenda of both policymakers and donors. At the time of this writing, cultural heritage projects have stopped in Sanaa; projects that had been carefully managed suffer from lack of resources and oversight. From the early years, the conservation effort was driven by foreign aid—in a

15 R. Burrowes, *Historical Dictionary of Yemen*, pp. 327–8. In early 2015 a panel of UN experts found that Saleh had amassed between USD 32–60 billion during his time in office, transferring much of it abroad under other names (reported by the BBC, February 25, 2015; see http://www.bbc.com/news/world-middle-east-31632502, accessed 3/1/15).

16 R. Stadnicki and J. Touber, "Le Grand Sanaa." The earlier figures were estimated by F. Varanda, "Tradition and Change," p. 198; and 'A. Saqqaf, "Sana'a: A profile of a changing city," in 'A. Saqqaf, ed., *The Middle East City: Ancient Traditions Confront a Modern World*, New York: Paragon House Publishers, 1987, pp. 120–21.

17 For a sobering look at the figures, see the 2009 CDS Report, pp. 10–12. 2004 figures from the Ministry of Planning and International Cooperation report a rate of urban growth that hovered around 10 percent per year (cited in Stadnicki and Touber, "Le grand Sanaa," p. 37).

country that was, and remains, heavily dependent on it.[18] It was also driven by the promise of tourist dollars. In the 1970s and '80s, policymakers hoped that the awe-inspiring heritage sites throughout Yemen would make tourism a major driver of the economy. Yet tourists never arrived in expected numbers—and they have all but disappeared in the wake of the 2011 uprising. In the following pages, I argue that the relevance of heritage to everyday life, and livelihood, is much greater, and more profound, than tourism-driven heritage: it has become an integral part of what is normally called "development." This is acutely appreciated by local professionals and others, but less often by policymakers who, despite evidence to the contrary, continue to pin their hopes on a tourism bonanza.[19]

Sanaa, "Mother of Yemen"

The mythic origins and noble history of Sanaa have been related by historians since medieval times, and are often evoked in conservation literature. The nomination form submitted to the World Heritage Committee in 1985, drafted by the Yemeni staff of the conservation office, stated:

> Sanaa is considered as one of the most ancient cities of the world. Al-Hassan Bin Ahmad al-Hamdany (10th c. Yemeni geographer and historian) describes Sanaa as the oldest city on earth and 'Shem'—son of Noah—as the Founder of the City of Sanaa.

An earlier name for the city, 'Azal, links it to the line of the Prophet Nuh (Noah) and his son Sam (Shem). This established the place of the city in an ancient and sacred genealogy, and also had political undertones. Founded by the descendants of Sam—the "true" Yemenis, as opposed to the *saada*—the city was portrayed as the origin and symbol of the Yemeni Republic.[20]

Histories of Sanaa generally rely on a limited number of sources, most of which refer back to the tenth century descriptions of Ibn Rustah and al-Hamdani.[21] Working within a well-established tradition of hortatory description, al-Hamdani called Yemen "the best of the world's lands," the center of the Islamic world; and the city of Sanaa "the mother of

18 Like democratization (Philips, *Yemen's Democracy Experiment*, p. 49), commitment to cultural heritage is invoked in part to please foreign donors.

19 See for example the CDS report.

20 S. al-Suleihi, personal communication. Qahtan—a descendant of Nuh and Sam—was a key figure in revolutionary populism. As the putative ancestor of "southern Arabs," he was a symbol of native rebellion against the *saada*, who are "northern Arabs" (S. Caton, *Peaks of Yemen*, p. 34; G. Vom Bruck, *Islam, Memory, and Morality*, p. 10; R. Lewcock, *The Old Walled City of San'a*, p. 19; S. Day, "Power-Sharing and Hegemony," pp. 87–94, 110). It is interesting that in conservation literature, biblical genealogy has eclipsed Qahtan—perhaps because the latter is less recognizable to Western audiences.

21 For the medieval period the primary sources are descriptions of the city by Ibn Rustah who wrote after 903 CE, and al-Hamdani who wrote slightly later; as well as al-Razi's history of the city, written in the eleventh century, which draws heavily on al-Hamdani. (Al-Razi is said to have constructed a map of the medieval city, but it has not been made available to scholars.) Jamal al-Din 'Ali's *Wasf Sanaa* (*Description of Sanaa*), written in the eighteenth century, has been a standard reference; it is based largely on al-Razi. *Wasf Sanaa* was edited and published in Arabic by 'Abdullah ibn Muhammad al-Hibshi and translated into English by Tim MacIntosh Smith under the title *City of Divine and Earthly Joys: the Description of Sanaa*, American Institute of Yemeni Studies, 2001.

Yemen and its axis, for it is at the centre."[22] According to al-Hamdani, Sam chose the site of Sanaa to found the first city after the great flood. His founding act was the construction of the palace of Ghumdan, legendary for its height and rich ornamentation—and often cited as the prototype of the Sanaani house. As Ghumdan grew over time, through successive additions by rulers, so the Sanaani house has been added to by successive generations of patrons and builders. Hence the founding of the *bayt*—a word that means both house and family line—replicates the founding act of the city.

These descriptions of the city and its origins are often cited in contemporary texts with little critical reflection. At the same time, archeological excavations in the city have been extremely limited. Ronald Lewcock and others have supplemented these accounts with analysis of urban form and visual evidence, working to reconcile conflicting details and to locate iconic sites in the contemporary urban fabric.[23] The following discussion does not repeat the well-known narrative that Lewcock and others have constructed, and which is widely accepted. Rather, it focuses on some aspects of urban morphology that will be important for the chapters that follow. The reader is referred to the 1929 map by Von Rathjens and Von Wissmann (Figure 2.3), as well as an aerial view from 1970 (Figure 2.4).[24]

Sanaa is believed to have developed around two nodes, the palace and the citadel. The palace, located at the site of a present-day tell to the east of the Great Mosque, was built in the third century CE; it was probably part of a larger fortified complex that was destroyed, rebuilt, and refortified over the centuries. Following al-Hamdani, the palace has been identified by historians as the legendary Ghumdan. The Great Mosque—one of the oldest in the Islamic world, now undergoing a major restoration—was built in the garden of the palace at the order of the Prophet Muhammad, when Yemen was converted to Islam in the seventh century. The citadel or *qasr* was located at the eastern edge of the present walled city, at the foot of Jabal Nuqum. It consisted of two parts: an upper fort, and a lower fortified area—called the *qasaba* in medieval texts—which may have been the original settlement.[25] The land slopes westward from the *qasr* to the Wadi Sa'ila, a dry river bed that bisects the old city from north to south; used as a roadway through the old city, it was paved during the conservation effort. A second wadi, the floodcourse of Jabal Nuqum, extended along the edge of the present market. Located in the heart of the old core, the market (*suq*) is made up largely of single story buildings, probably some of the oldest in the city. These are interspersed with large caravansarays (*samsara*, pl. *samasir*) that served the whole-sale trade; the oldest probably date to the seventeenth century, when the coffee trade was at its height. The core of the city was surrounded by walls from the ninth century or earlier, pierced by monumental gates.[26]

---

22 Cited in Dresch, *Modern Yemen*, p. 205.

23 See especially R. Lewcock, et al., "The Urban Development of *San'a*," in Serjeant and Lewcock, *San'a*, pp. 122–43; and R. Lewcock, *The Old Walled City of San'a*. Lewcock's unpublished paper, "Early and Medieval Sanaa: the Evidence on the Ground," presented at a 2004 conference at the University of Sanaa, elaborates and revises the earlier accounts. The following discussion is based on these sources, as well as P. Costa, "Sana'" in Serjeant, ed., *The Islamic City*.

24 Von Rathjens and Von Wissmann's map has provided the basis for most scholarly studies. See G. Gillot, "Généalogie des cartes de Sanaa," in M. Barcelo, ed., *Les jardins de la vielle ville de Sanaa*, Yemen, Bellaterra: Centre Français d'Archéologie et de Sciences Sociales, 2004, pp. 15–16.

25 R. Lewock, *The Old Walled City of Sana'a*, p. 35. The *qasr* remained an important military installation through Ottoman times, and is still controlled by the Yemeni military.

26 Al-Razi writes that Sanaa did not have a wall around it until the ninth century, but reference to city gates in earlier periods suggest that walls existed and had been destroyed. Lewcock presents

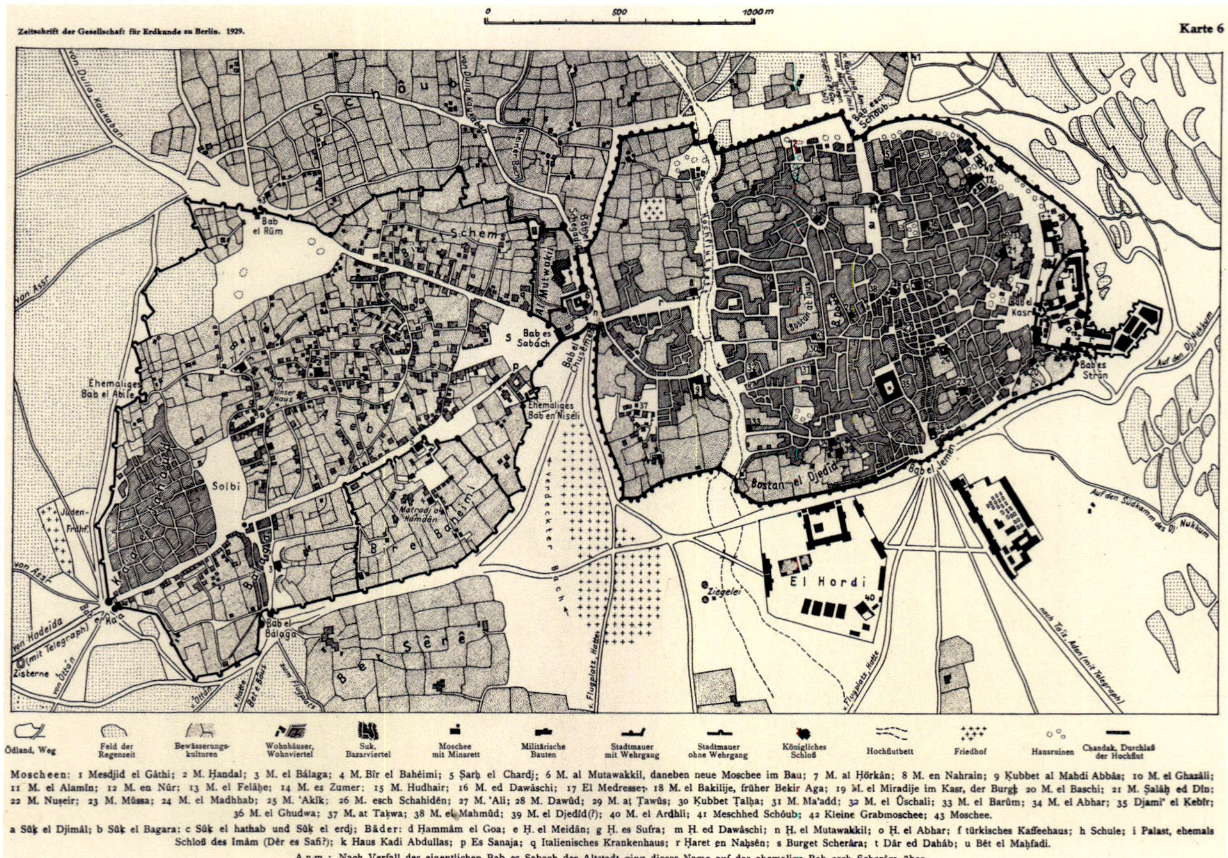

Figure 2.3 **Map of Sanaa, 1929**

Source: Carl Rathjens und Hermann von Wissmann, Sanaa. Eine südarabische Stadtlandschaft, Berlin: Zeitschrift der Gesellschaft für Erdkunde, 1929, S.329–353, Map 6. Reprinted with permission of Horst Kopp.

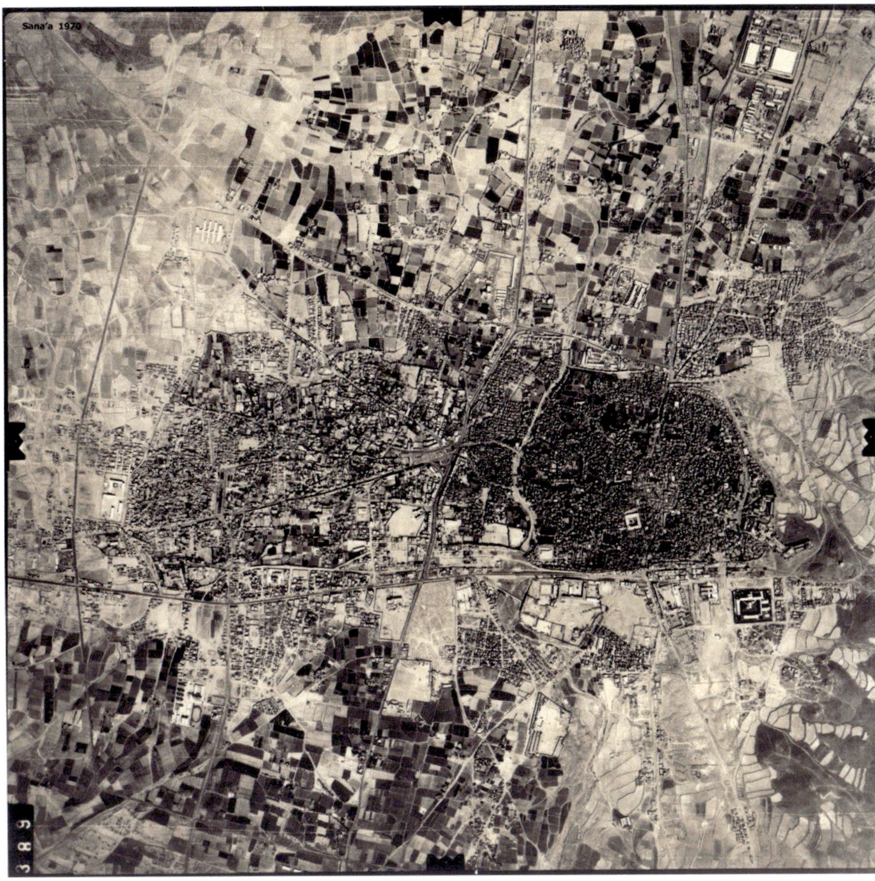

Figure 2.4 Aerial view of Sanaa, 1970

Source: Horst Kopp und Eugen Wirth, Beiträge zur Stadtgeographie von Sanaa, Wiesbaden 1990 (Beihefte zum Tübinger Atlas des Vorderen Orients, Reihe B, No.95), Figure 1. Reprinted with permission of Horst Kopp.

The distinctive urban pattern of Sanaa probably emerged in the centuries following the Islamic conquest, as the city expanded up to the banks of the Sa'ila.[27] The basic social and administrative unit of the city was the quarter (*hara*), which had at its heart an endowed complex of mosque, bathhouse (*hammam*; pl. *hammamat*), and agricultural garden (*maqshama*; pl. *maqashim*). The fabric of the *hara* consisted of tall, densely packed houses which, however, rarely shared common walls. Houses were designed for extended patrilineal families: as sons married additional stories were added, or new structures were built on adjacent land. Houses are often grouped around a *maqshama*, and some houses

---

schematic drawings of the probable positions of walls and gates in various periods in "Early and Medieval Sanaa."

27 P. Costa, "Sana," in Serjeant, ed., *The Islamic City*, p. 156.

can only be reached through it (Figure 2.5).[28] The *maqshama* was endowed as income-producing property for a mosque or, less frequently, for a bathhouse.[29] In contrast to the small squares and streets that wind through the city, the edges of the gardens are somewhat regular, if not orthogonal—attesting to their origin in endowment deeds, which must have specified their boundaries. In contrast the small squares, or *sarhat*, seem to be left-over spaces. Paul Bonnenfant suggests that the *sarhat* were originally the courtyards of family compounds: over time, as houses were sold to individuals outside the family, the private court (*hush*) became public but retained the name of the original owner.[30] The boundaries of quarters in Sanaa are not precise, and are not generally known by residents. Perhaps because of Sanaa's long subservience to central governments, quarters lost the kind of autonomy they enjoyed in other settlements.[31]

The endowed complexes (*awqaf*; sing. *waqf*) of Sanaa—consisting of mosque, bathhouse, lavatory, agricultural garden, and water supply system—must date to the early centuries of Islamic rule in Sanaa. They form a unique urban morphology, inserted into what may have originally been a quasi-rural settlement pattern.[32] The location of the complexes, and the development of the city in general, were closely tied to a complex system of water supply and drainage. Water for drinking and irrigation was drawn primarily from underground wells, which are often linked by man-made water channels called *ghayl*s, similar to the Persian *qanawat*. In parts of Yemen wadis are used to collect and distribute rainwater, but in Sanaa they seem to have been used primarily for drainage. During the rainy season the dry river beds, or wadis, flood and carry rainwater away from the city. The position of the *ghayl*s and the wadis, then, responded to and influenced the development of the city.[33]

The *waqf* complex, together with surrounding houses, created a kind of eco-system based on the recycling of human waste and water, while providing for the production of perishable agricultural products within the walls. Because fresh and waste water had to circulate through the various parts of the complex, its design is conditioned by topography. Gardens were thus constructed at a low elevation, sometimes two to three meters below street level.[34] While the street level has risen over the centuries, the garden had to remain at a low elevation relative to the mosque, its ablutions pool (*birka*), and the canals that

28 G. Vom Bruck, "A House Turned Inside Out," p. 147.

29 In the 43 gardens surveyed by the UNESCO mission directed by M. Barcelo in 2000–2001, 34 were attached to mosques, and nine to *hammam*s (M. Theresa Marquez, "Les jardins potagers urbains de Sanaa: étude hydraulique," in Barcelo, ed., *Les Jardins de la Vielle Ville de Sanaa*, p. 18).

30 "Maisons. Voisins, et Extérieur," in P. Bonnenfant, ed., *Sanaa, Architecture Domestique et Société*, pp. 65–6. I found evidence for this during my fieldwork: residents sometimes identified a *hush* with a particular family.

31 R.B. Serjeant and H. al-'Amri, "Administrative Organization," in Serjeant and Lewcock, eds, *San'a*, p. 146). In 1995 GOPHCY attempted to map boundaries of quarters (N. Monasser, D. Heepen, "The Boundaries of the Quarters in the Old City of Sana'a," unpublished report).

32 See R.B. Serjeant and H. al-æAmri, "Administrative Organization," in Serjeant and Lewcock, eds, *San'a*, pp. 151–4. Gardens are more numerous in the western part of the city, suggesting its origin as an area of farmsteads.

33 Heymeyer notes, for example, that two mosques were built along the course of the Ghayl al-Aswad so as to have access to this water source ("Mosque, Bath, and Garden," p. 109). For a discussion of the *ghayl*s, see R.B. Serjeant, et al., "The Ghayls of San'a," in Serjeant and Lewcock, eds, *San'a*, pp. 19–31.

34 Some were constructed at the bottom of clay quarries, which were excavated to make bricks (R.B. Serjeant and R. Lewcock, "The Houses of *San'a*," in Serjeant and Lewcock, *San'a*, eds, p. 461).

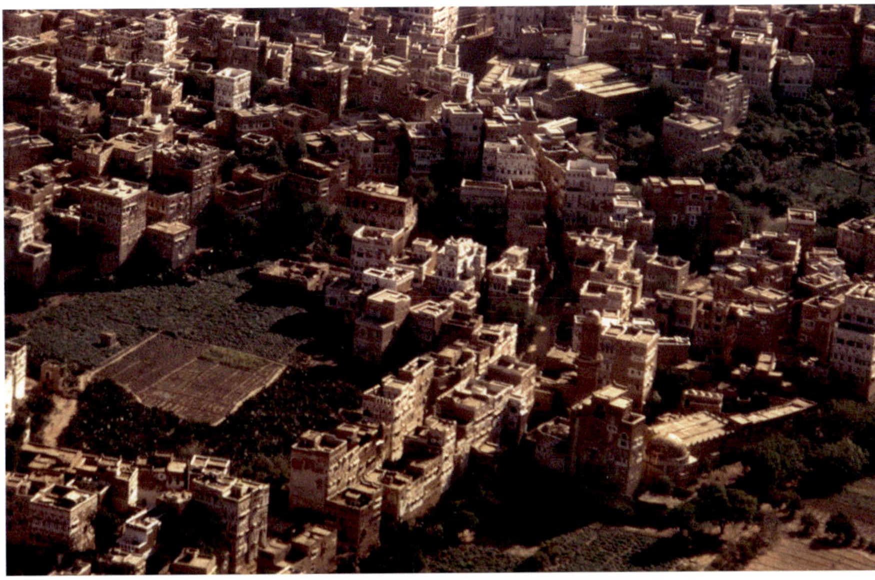

Figure 2.5 View of the old city circa 1973, showing houses grouped around the urban gardens

Source: Photograph by Ronald Lewcock.

distributed water to it.[35] Grey water from the *hammam* was also distributed to the garden, so it too had to be at a higher elevation. Other key relationships can be found in surviving wells and enclosures (*maran'i*, sing. *marna'*) which have been abandoned in favor of electric pumps. Water lifting devices were operated by animals that walked and back and forth along long ramps, which were often roofed. The length of the ramp is directly proportional to the depth of the well—both of which were modified over time, in response to changing groundwater conditions.[36]

Since *waqf* land cannot be divided or sold, the urban endowments ensured that the land, and the facilities to which it was linked, would be kept intact. A close physical study of the *awqaf*, along with the endowment deeds, would provide the basis for an environmental and social history of the city, while revealing much about its physical development. After the establishment of the republic, the endowment system was reorganized: most endowments were placed under the direction of a special government ministry. Among other things, this meant that rents and income were no longer necessarily associated with a specific complex, but were rather redistributed according to the ministry's schedule of priorities. R.B. Serjeant and H. al-'Amri estimated that some two thirds of the old city of Sanaa is endowed as one of several types of *waqf*, including most of the public buildings and much of the market. Representatives of the conservation office (the General Organization for the Preservation

35 T. Marquez, "Les jardins potagers de Sanaa, étude hydraulique," in M. Barcelo, ed., *Les jardins de la vielle ville de Sanaa*, p. 21.

36 I. Hehmeyer, "Mosque, Bath, and Garden: Symbiosis in the Urban Landscape of San'a," *Seminar for Arabian Studies*, Brepols, vol. 28, 1998, p. 109.

of the Historic Cities of Yemen, or GOPHCY) and the Ministry of Endowments believe the number is closer to 90 percent.[37] While the Ministry's jurisdiction does not extend to all types of *waqf*, the property under its control is enormous. The modern transformation of the *waqf* system, and the consequences for urban administration and conservation, is a tremendously important but under-studied issue not only in Sanaa, but in other countries in the region.[38] At the same time, the tendency to inventory *individual* structures tends to obscure the functional and historical relationships that bound them together into an integrated urban whole.

By the twelfth century or earlier, the city had expanded across the banks of the Sa'ila. The Ayyubids built their palace complex on the west bank of the wadi. The area is still called Bustan al-Sultan—"Orchard of the Sultan"—although most of the garden has disappeared: surrounded by development, it is now a *suq* for the sale of *qat*, the mildly intoxicating leaf chewed daily by Yemenis. The Ayyubids repaired the old city wall, and extended it around the new districts. Where the new wall crossed the wadi, arched openings (*khanadiq*, sing. *khandaq*) allowed floodwaters to pass through it.[39] In subsequent centuries large houses with gardens appeared farther to the west, beyond the Ayyubid walls. This area—which came to be called Bir al-'Azab after one of its quarters—would be favored by the Ottomans as a garden residential quarter when they occupied the city in the sixteenth century. During this first of two occupations (1547–1623), the Ottomans shifted the locus of power away from the Ayyubid precinct to the site of the *qasr*, at the eastern edge of the city. They formalized the square below the citadel, building the Bakiriyya mosque in the imperial style and a *hammam* to support it. The Ottomans laid out a wide, tree-lined avenue from the square to the northern gate of the city, Bab al-Shu'ub. They also upgraded the fortifications during this period, in accordance with the new requirements of heavy artillery.[40]

In 1636 the Zaydi imams drove out the Ottomans and managed for the first time to bring much of Yemen, including the Hadhramaut, under the rule of a single dynasty. The unification of territories under the Qasimis is sometimes viewed anachronistically as the first Yemeni state.[41] Sanaa became their capital in the eighteenth century: it enjoyed a period of prosperity under the so-called "Zaydi renaissance," due in large part to the role of Yemen in the coffee trade. There was a surge in the building of *samsara*s (caravansarays) in this period.[42] In the early eighteenth century the Imam al-Mutawakkil al-Qasim (r. 1716–1727) initiated a building program outside the western wall near Bab as-Sabaha, creating a new

37 During my fieldwork in 2004–2005, the Ministry was conducting a national survey of *waqf* properties. For the earlier estimate, see R.B. Serjeant and H. al-'Amri, "Administrative organization," in Serjeant and Lewcock, eds, *San'a*, pp. 151–2.

38 The effects are in some sense parallel to the expropriation of Church properties in Catholic countries of Europe during the late eighteenth-nineteenth centuries, but are probably more far-reaching.

39 The *khanadiq* were destroyed by floods and rebuilt several times during the sixteenth and seventeenth centuries. See R.B. Serjeant, et al., "The Ghayls of Sanaa," p. 26; R. Lewcock, *The Old Walled City of San'a*, pp. 44–6; P. Costa, "*San'a*" in Serjeant, ed., *The Islamic City*, p. 156.

40 P. Costa, "San'a," p. 156; R; Lewcock, *The Old Walled City of San'a*, pp. 46–7; Kopp and Wirth, *Sanaa*, p. 22.

41 P. Dresch, *Modern Yemen*, p. 5 and, p. 223, fn. 5.

42 Nancy Um, personal communication. She believes that the majority of *samsara*s date from the first Ottoman occupation and later, based on their *waqf* deeds. See N. Um, *The Merchant Houses of Mocha: Trade and Architecture in an Indian Ocean Port*, Seattle: University of Washington Press, 2009, p. 227; and Serjeant and Lewcock, eds, *San'a*, pp. 277–8.

center of power distinct from the Ottoman imperial precinct around the *qasr*.[43] The complex included royal residences, gardens, a mausoleum, and public baths. Several open spaces, enclosed within walls, fronted this complex on the east and west, including a ceremonial parade ground. The mausoleum is the only surviving part of the original complex, which was transformed in the following centuries. In the last decades of the seventeenth century, a new Jewish quarter was created to the west of Bir al-'Azab, which came to be known as Qa'a al-Yahud. This area and Bir al-'Azab were absorbed into the city when the walls were extended in the early nineteenth century.[44]

Origins of Modern Sanaa

The advent of modernity in Sanaa is usually identified with the second Ottoman occupation (1872–1918), but some scholars point to earlier origins. The coffee trade enabled the Qasimis to consolidate their rule; they introduced new formalized practices and bureaucratic structures, which tended to supersede the old form of rule based on the charisma of the imam.[45] Deteriorating conditions in the nineteenth century—called the "time of corruption" by Yemeni historians—are attributed to the decline of the coffee trade and the breakdown of Zaydi rule. As a result, the second Ottoman conquest met with a degree of local support in Sanaa.[46] The city was again made capital of the *vilayet* of Yemen, and remained so until the dissolution of the empire in 1918.

The Ottomans reestablished their presence in the area of the *qasr*, lining the avenue with new administrative buildings and schools, and restoring the mosque. At one end of the square—presumably *al-Maydan*, the square below the *qasr*—was a thoroughfare that led to the old *suq*, lined with cafes and shops frequented by Turkish officers. Called "el-Casino" by the Arabs, it was the site of a new kind of urban life during the period. The area of the old imam's palaces was reused for official purposes; the palace of al-Mutawakkil was destroyed, and a military hospital built on the site.[47] Bab al-Sabah, the old western gate of the city, was demolished presumably to facilitate circulation with Bir al-'Azab.[48] Outside the southern gate of the city, Bab al-Yaman, new barracks were built in the late 1880s. The gate itself was later rebuilt to the design of an Ottoman military engineer, along with a stretch of wall on either side of it. Despite its Ottoman origins, Bab al-Yaman—substantially rebuilt

43 N. Um, "Eighteenth century patronage in Sanaa: building for the new capitol during the second century of the Qasimi imamate," *Proceedings of the Seminar for Arabian Studies*, vol. 34, 2004, p. 6.

44 Lewcock, et al., "The Urban Development of *San'a*," p. 137. The Jewish community had been exiled to the Tihama in 1678 for refusing to convert to Islam; they were allowed to return two years later but were made to settle two kilometers to the west of the old city. See A. Shivtiel, et al., "The Jews of San'a," in Serjeant and Lewcock, eds, *San'a*, pp. 391–431.

45 B. Haykel, *Revival and Reform in Islam*, Cambridge: Cambridge University Press, 2003, pp. 16–17.

46 P. Dresch, *A History of Modern Yemen*, p. 4.

47 Skovgaard-Peterson "The Appearance and Disappearance of Public Space," in H-C K. Nielson and J.S. Peterson, eds, *Middle Eastern Cities: 1900–1950*, Proceedings of the Danish Institute in Damascus I, Aarhus: Aarhus University Press, 2001, pp. 152–4. See also T. Kühn, "Ordering Urban Space in Ottoman Yemen, 1872–1914," in J. Hanssen, et al., eds, *The Empire in the City*.

48 The name Bab al-Sabah was attributed to Bab al-Shiarara, which survived until the mid-1960s at the northern edge of the square called Maydan al-Tahrir (R. Lewcock, "The Urban Development of *San'a*,"p. 131, fn. 125 and, p. 138). On Manzoni's map (1879) the name of the gate appears as Bab al-Shiarara; in Von Wissman's map (1929) it is called Bab al-Sabah.

during the conservation effort—has become a kind of symbol of the city of Sanaa and of Yemeni nationalism.

Although historians have portrayed the second Ottoman occupation as a period of decline, Kopp and Wirth maintain that it was a time of prosperity for the city, the result of its new position in a wider imperial and world market. During this time, they argue, many buildings were built and renovated in the old city, and the distinctive architectural style of Sanaa reached its apogee—even as the agricultural hinterland and its villages fell into ruin. A new trade in European products, destined for the Ottoman elite, enriched merchants in the markets of Sanaa and the port of al-Hudayda.[49] The use of glass appears to have expanded greatly from the end of the nineteenth century: the opening of the Suez Canal in 1869 made it easier to obtain, and the Ottoman preference for wide expanses of glass began to affect local tastes. Garden villas with large glass bays appeared on the outskirts of Sanaa, in Bir al-'Azab, al-Rawdha, Wadi Dhahr, and Sanhan, built by princely and merchant families desiring a second residence for summering. Around the same time, the windows of the tower house underwent a significant change. The old vertical type—consisting of a small window surmounted by circular openings (originally alabaster, later glass) gave way to a wider window with arched lite above, filled with colored glass. This window type appeared in new floors added to older buildings—and perhaps suggest an attempt to emulate the garden villa in the *mafraj* of the tower house.[50]

After the collapse of the Ottoman Empire in 1918, Imam Yahya Muhammad Hamid al-Din assumed control of the north and extended his control to Lower Yemen. In the political rhetoric of the time, Yahya had re-established the millennial rule of the Zaydi dynasty and also founded the modern Yemeni state—since this was the first time in history that polity and territory were coterminous.[51] The new Yemeni nation was defined by borders and underpinned by historical and geographical studies, sponsored by the Imam. Yahya is often portrayed as a "traditional" ruler who sought to insulate the country from foreign influence and nationalist ideologies. Yet the term "progress" was often heard in official discourse.[52] The trappings of a modern state were introduced, including a standing army and a ministerial system, yet power remained concentrated in the person of the Imam. Yahya did not allow travel outside the country, but did send some young men to Iraq and Egypt, to learn skills required for government. Those who went to Egypt in 1947—later known as the "Famous Forty"—were to become the country's first generation modernists.[53] The imam undertook hesitant, modernizing projects, built public buildings, including schools, and restored and expanded mosques.[54] His efforts focused largely on the capital. But the

49 Kopp and Wirth, *Sanaa*, pp. 30–42.

50 P. Bonnenfant, et al., "Le Livre des Jours et des Modes," in P. Bonnenfant, ed., *Sanaa*, pp. 484–6, The old ratio of window opening to the lite (*qamariyya*) above, typically 2:1, was reduced to 1:1; see Bonnenfant's illustration reprinted below, Figure 5.3.

51 P. Dresch, *Modern Yemen*, pp. 7–9. The following account of the later imams and the post-revolutionary period draws from Dresch's book and R. Burrowes, *The Yemen Arab Republic*, chapters 2 and 3; Kopp and Wirth, *Sanaa*, pp. 42–68; and R.B. Serjeant, "The Post-Medieval and Modern History of San'a and the Yemen," in Serjeant and Lewcock, eds, *San'a*, pp. 68–107.

52 P. Dresch, *Modern Yemen*, pp. 50–51.

53 R. Burrowes, "The Famous Forty and their Companions: North Yemen's First-Generation Modernists and Educational Emigrants," *Middle East Journal*, vol. 59, no. 1, Winter 2005, pp. 81–97; *The Yemen Arab Republic*, pp. 20–22.

54 Yahya's projects in Sanaa included a new library at the Great Mosque (1925–1926); the mosque and palace at the complex of al-Mutawakkil (1936); the near-by palace of Dar al-Shukr (1933); and the palace at Wadi Dhahr (1941–1946). The al-Mutawakkil complex has been restored to house

expansion of the city was limited by tribal and endowed agricultural land, which surrounded the old core and suburbs.

Yahya was assassinated in 1948 in a plot by a group of progressive intellectuals. After ordering tribes to sack the city—a brutal undertaking that still burns in popular memory—Yahya's son and successor, Ahmad, transferred the capital to Ta'izz and never set foot in Sanaa again. There is very little information on development of the city during this period, but because the capital was located elsewhere, it has been assumed that little change occurred.[55] The Jews left Yemen in 1950, doubtlessly impacting the city's economy; the abandoned quarter of al-Qa'a would be appropriated in the following decade by rural migrants. Ahmad played the politics of the time, opening the door to foreign assistance and intervention. An anti-British pact with Egypt and Saudi Arabia (1956) led to closer ties with Nasser's Egypt, the supply of Soviet arms, and Egyptian military advisors. Ahmad began a road-building program which would, write Burrowes, "come to symbolize great power competition for influence in less developed countries: the Soviets, Americans, and Chinese each working to build a highway on a different side of the triangle formed by Sanaa, Taiz, and al-Hudayda." After the revolution these asphalted roadways would become the major agents of integration and change.[56] But often state power did not extend beyond the triangle they formed.

The 1962 Revolution

From the late 1950s through the 1960s, Yemen became a theater of regional and international politics, resulting in a series of overlapping wars.[57] Ahmad and his son, Crown Prince al-Badr, were ousted in 1962 and a republican government was established under the sponsorship of Egypt. The civil war that ensued became a proxy war between Egypt and Saudi Arabia, which supported the republicans and the royalists, respectively. In the south, the new federation of Aden and the hinterland opposed the British. When the Egyptians and the British withdrew in 1967 Yemen remained divided as two states: North Yemen and South Yemen, the latter a socialist state allied with the Eastern bloc.

By 1963 a full-fledged occupation authority was in place, with some 30,000 Egyptian soldiers in North Yemen. As a new government was created; each office had an Egyptian advisor without whom no decision could be taken. "To Egyptians, Yemen was a 'primitive' country," writes Dresch. "They seemed convinced that Yemenis were incapable of conducting their own affairs and, as journalists noticed, spoke of them in the racist fashion the British once used of Egyptian peasants."[58] Sanaa seemed to embody this primitive state. Foreshadowing the language of conservation, Cairo newspapers portrayed the city

---

the new National Museum; the palace at Wadi Dhahr is a major tourist destination. Both restorations were directed by architect 'Abdullah al-Hadrami, who began his career in the UNESCO campaign. On Yahya's modernizing projects see Kopp and Wirth, *Sanaa*, pp. 43–4; for buildings, see R.B. Serjeant, "The Post-Medieval and Modern History of San'a and the Yemen," p. 100.

55 Kopp and Wirth, *Sanaa*, pp. 47–8.

56 B. Burrowes, *The Yemen Arab Republic*, pp. 6–7, pp. 20–21.

57 The following account draws from Dresch, *Modern Yemen*, Chapter 4; P. Costa, "Sana," in R.B. Serjeant, ed., *The Islamic City*; Kopp and Wirth, *Sanaa*; F. Varanda, "Tradition and Change," Chapter 6. It also draws on interviews with officials who served in the Ministry of Public Works, who asked not to be named; and with Fernando Varanda, Paolo Costa, 'Abdo 'Ali 'Othman, and others.

58 *Modern Yemen*, p. 93.

as a relic of the Middle Ages, in an attempt to gain support of the Egyptian public for an unpopular occupation and war.[59] The idea of Yemen as backward and "medieval" was internalized by many Yemenis, especially republicans, who aspired to the modernity they saw in Nasser's Egypt.

Soon after the revolution, the leaders of the Egyptian-backed government undertook to remake Sanaa as the capital of the new Republic. They created a modern administrative and commercial center at the eastern end of Bir al-'Azab, just to the west of the old city. In the process, they demolished all of the walls around the old suburbs, and portions of the wall around the old core. The planning of this period is rarely discussed—presumably because, in the era of conservation, the demolition of the walls is seen by many as a fateful mistake. Two of the individuals who directed the demolitions were among the "Famous Forty" sent for education abroad by Imam Yahya in the 1940s. 'Abd al-Latif Dayfullah was one of the five army officers who conspired to overthrow Imam Ahmad; he held a number of high-level posts after the revolution. According to a former official at the Ministry of Public Works, Dayfullah directed the first round of demolitions with the encouragement of Egyptian advisors. "For the Egyptians, the only antiquities (*athar*) were in Egypt," said the official. "The *athar* of Yemen were worth little."[60] 'Abdullah al-Kurshimi, another of the Famous Forty and one of Yemen's first engineers, served as Minister of Public Works from 1967–1968 and from 1975 until the mid-1990s.[61] A senior official who worked under him in the 1970s said that al-Kurshimi's policies were informed by his training in Egypt. "He knew only Egypt," said the official, who asked not to be named. "He destroyed everything."

The western wall of the old city was demolished first, along with part of the southern wall and several gates, around 1964. Later, the northern and southern *khandaq*s (flood gates) over the wadi were destroyed. Bab al-Khuzaymah and Bab al-Saqadif, which had enclosed the precinct of al-Mutawakkil, were demolished to open up a new street along the path of the old western wall, 'Abd al-Mughni Street. In 1966 the gate known as Bab al-Sabah, which stood at the corner of the present Yemen Arab Bank, was demolished and the open space outside the old palace precinct was transformed into a public square, Maydan al-Tahrir (Liberation Square; Figure 2.6).[62] 'Abd al-Mughni Street, and the street that ran along the southern edge of the old city, al-Zubayri, generated a grid that would guide subsequent development. 'Abd al-Mughni Street was lined with buildings in reinforced concrete intended for commercial and multi-family residential use. Designed after Egyptian models, the buildings were imposed by the state and supported by funds from the United

59 'Abd al-Karim al-Iryani, former Prime Minister, interview.

60 He added that the Ministry of Public Works owned all the earth-moving equipment in Sanaa, which had been given to them by the Soviets. Varanda confirms that planning in the 1960s was directed by the Ministry of Public Works with the assistance of foreign, mainly Egyptian, advisors ("Tradition and Change," p. 204).

61 Al-Kurshimi spent a decade in Egypt (1948–1959) and was one of three engineers to return to Yemen before the revolution; between 1968 and 1975 he held a number of cabinet posts and served briefly as prime minister (R. Burrowes, "The Famous Forty," p. 4, 10, 12; *The Yemen Arab Republic*, p. 36; *Historical Dictionary*, p. 218). Dayfullah served as Prime Minister and Deputy Prime Minister in 1963–1964; he served as Minister of Public Works in 1974 and perhaps earlier. According to Burrowes, Dayfullah could have directed the demolition of the walls in any of these capacities (personal communication).

62 The square was designed by British-trained engineer Muhammad Ahmad al-Junayd, founder of the Ministry of Finance (R. Burrowes, personal communication).

Arab Republic (the short-lived union of Egypt and Syria; Figure 2.7).[63] They were never accepted for residential use; most floors came to be occupied by retail and offices. The contractors were from Aden, part of the first great wave of migration from the south: they had learned new construction techniques under the British and sometimes used Egyptian architects.[64] ‘Abd al-Mughni Street and the near-by Jamal ‘Abd al-Nasser School are still seen as emblematic of both the revolution and the Egyptian occupation.

Although the city did not begin to expand rapidly until the mid-1970s, significant building took place during the years of civil conflict: by 1970, the built-up area covered an area three times the size of the city as it was in 1962 (Figure 2.8).[65] The development was generally low density and scattered, lacking in public utilities. In some areas, as in the zone to the west of the Sa’ila, development followed the kind of gridded layout proposed in an Egyptian plan drafted during these years.[66] Much of the building activity in and around Bir al-‘Azab and in the area of Bab al-Shu‘ub provided new government infrastructure. Jamal Street—named after the Egyptian president—was widened and developed into a secondary commercial center: while many of the buildings were typologically new, they retained a local style and detailing.[67] Palaces that had belonged to the imam were adopted for government purposes: the Ministry of Finance, for example, was located in Dar al-Sa‘ada in the 1970s.

The attention that has been given to the walls of Sanaa both after the revolution and in the era of conservation—roughly, from 1980 onward—reveals their potent, and mixed, symbolism. In the medieval period, as now, the walls defined the image and identity of Sanaa, whose protective function was not only physical, but symbolic and even magical.[68] During the civil war, the city’s vulnerability from the air demonstrated that these land defenses were outdated and futile. At the same time, the new republican ideal envisioned the end of tribal conflict, making the walls unnecessary.[69] In the years following the revolution, the walls of Sanaa recalled an era that republicans wanted to forget: an era of theocracy, isolationism, and sieges.

63 J.-L. Arnaud, “Formation de l’architecture contemporaine à Sana‘a (1965–1990),” in Grandguillaume, ed., *Sana‘a Hors les Murs*, pp. 72–4. Arnaud notes that these projects had more to do with political expediency rather than a need for housing.

64 The late ‘Abd al-‘Aziz ‘Abd al-Ghani, former Prime Minister and chairman of the Shura Council, interview. ‘Abd al-Ghani worked briefly for one of these contractors in 1967, Sharikat al-Nil, which built four of the early buildings on ‘Abd al-Mughni Street. The company used Egyptian architects, he told me, and the workers and clerks had worked with the British in Aden. ‘Abd al-Ghani later directed the Board of Trustees for the Preservation of Old Sanaa. He was fatally wounded in the 2011 attack on the presidential compound, which also wounded Saleh. On ‘Abd al-Ghani see Burrowes, *An Historical Dictionary of Yemen*, pp. 2–3.

65 A. Bertaud, “Report on planning, building, and related fields in the municipality of Sana‘a,” Information Paper no. 1, UNDP, July 1970. Costa says that development outside the city walls, at least on the northern side, did not affect the aspect of the city until 1969 (“Sanaa” p. 161).

66 A. Bertaud, “Report on Planning,” p. 4; Kopp and Wirth, *Sanaa*, p. 53; F. Varanda, “Tradition and Change,” pp. 203–4. When Varanda arrived in Yemen in 1973 the Egyptian plan was still being used by the surveyors at the Ministry of Works.

67 F. Varanda, interview; Kopp and Wirth, *Sanaa*, p. 53.

68 Al-Razi, writing in the eleventh century, related the magical ability of the walls to ward off attacks. Two talismans on the gates, he wrote, prevented snakes from entering the city (F. Mermier, “Les Fondations Mythiques de Sanaa et d’Aden,” *Revue du monde musulman et de la Méditerranée*, vol. 67, special issue, “Le Yemen, présent et passé de l’unité,” 1994, p. 134).

69 Varanda, “Tradition and Change,” p. 194.

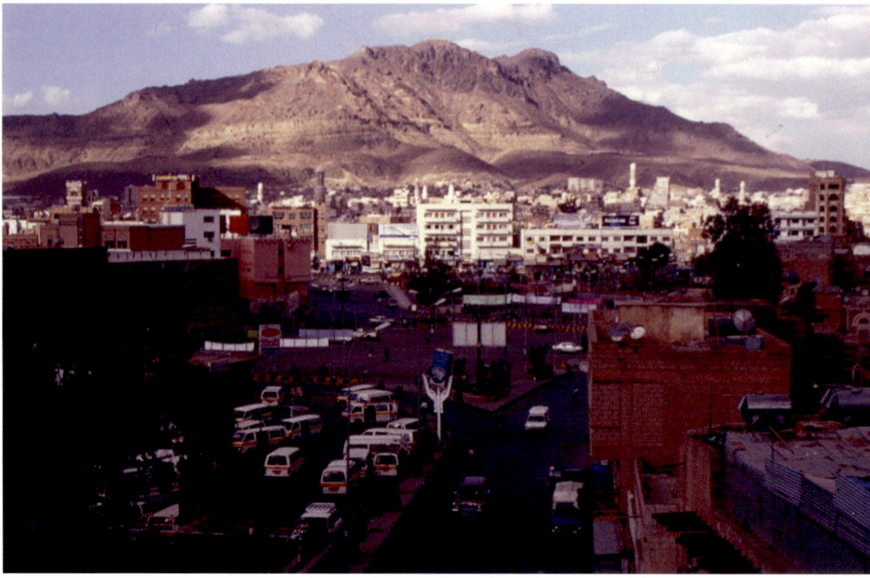

Figure 2.6 View across Tahrir Square toward ‘Abd al-Mughni Street and the old city

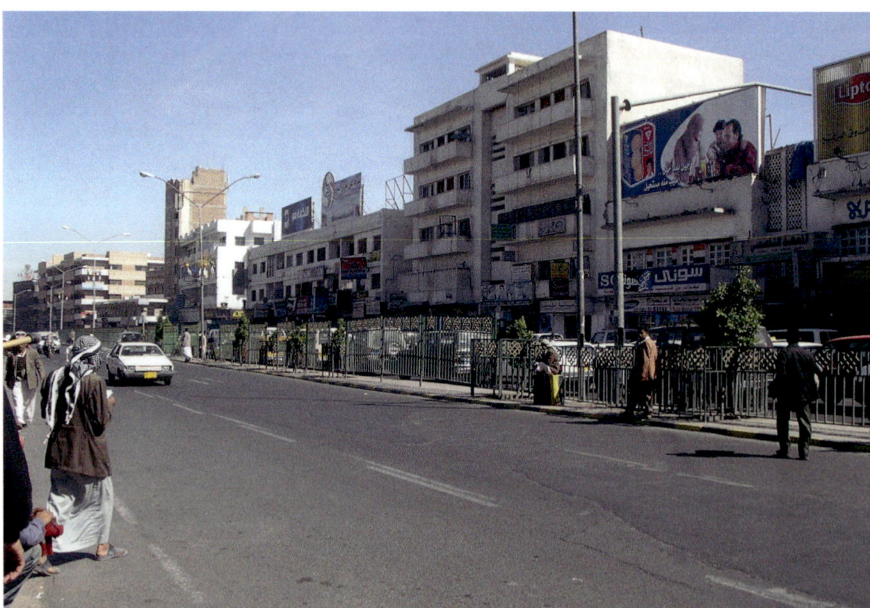

Figure 2.7 ‘Abd al-Mughni Street, which runs along the line of the old western wall

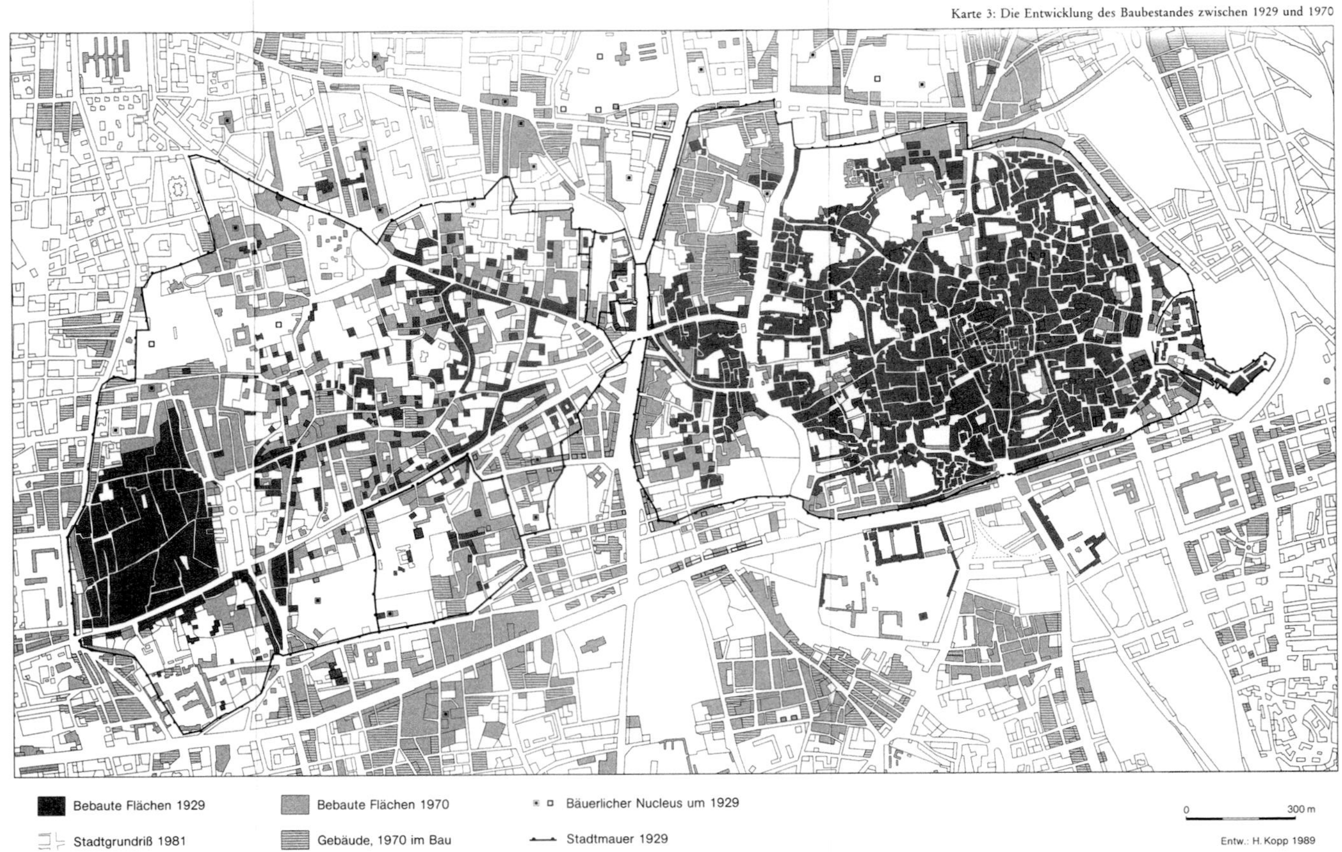

Figure 2.8 Growth of the city, 1929–1970

Source: Horst Kopp and Eugen Wirth, *Sanaa: Développement et organisation de l'espace d'une ville arabe, 1994* (Les Cahiers de l'IREMAM 5), p. 115. Reprinted with permission of Horst Kopp.

Other than the western and southwestern walls, the walls around the old core and a stretch of wall near Bab al-Balaqa were more or less intact in the early 1970s, but were slated for demolition. Several foreign consultants tried to convince the Minister of Public Works to save the remaining walls, but he refused. "We do not want things that remind us of a time when Sanaa was backward—the imam's time, when Sanaa was surrounded by mud walls," he reportedly said. "We do not want these symbols."[70] The Ministry was finally persuaded—how or by whom, it is not clear—to leave the remaining walls around the central core intact. But the walls around Bir al-'Azab and al-Qa'a, along with their gates, were gradually demolished by encroaching development.[71]

In many European countries, city walls were demolished in the nineteenth and early twentieth centuries to facilitate traffic and commerce. Over time, surviving city walls and gates came to be seen as historic "monuments," and were saved in some places. But this kind of historical distance was lacking in Sanaa. As Trevor Marchand puts it:

> Sieges of 1904 and 1911 against Sanaa, the sacking of the city by tribes in 1948, the revolution in 1962 and hostilities which continued until late that decade, and the civil war in 1994 are significant events of this century which hardly afford the temporal distance necessary for viewing such structures as the city walls and fortifications as defused 'historic monuments' set in 'Victorian pastoral landscapes.' The city walls are still a potent reminder of a turbulent past and a not-so-stable present.

In the era of conservation, local professionals and foreign consultants came to see the walls of Sanaa not only as important historical features of the city, but as a new kind of defensive structure: one that safeguards the historic character of the old core from modern encroachments. But for some the walls evoke, and inscribe, a sense of "traditional" identity—as city walls did in French settlements during the colonial era.[72]

Urban Expansion in the 1970s

The Egyptians withdrew in 1967. The regime that had ruled under the occupation was replaced by a conservative coalition of moderate republicans and tribal leaders. The coalition was strengthened by the National Reconciliation Pact of 1970, which ended the civil war between republicans and royalists; moderate royalists (including *saada*) were

70 This was recounted to me by F. Varanda, one of the consultants who tried to convince the Minister to save the walls. A former employee of the Ministry noted the irony of the association, since most of the existing walls had been built by the Ottomans, not the imams.

71 By the early 1970s the earthen towers at Bab al-Qa'a, at the far end of Jamal Street, had been destroyed (F. Varanda, interview). In 1976 Paolo Costa called for the protection of the surviving stretch of wall near Bab al-Balaqa, whose southern (exterior) face was particularly impressive. A new road along that stretch prevented development alongside the wall, but intense building activity was taking place along the northern side "without any plan to assure appropriate access. As each house is completed, the owners open a breach in the walls according to their personal requirements or haphazardly, without the least consideration for preserving the old wall. In one case the aperture has been dug right through a tower, which consequently was completely destroyed" (Costa, "Sana," in R.B. Serjeant, ed., *The Islamic City*, p. 64).

72 T. Marchand, "Walling Old Sanaa," pp. 49–51. See also O. Creighton, "Contested townscapes: the walled city as world heritage," *World Archaeology*, Vol. 39, No. 3: *The Archaeology of World Heritage*, September 2007, pp. 339–54.

brought into the fold of the regime. The new coalition, notes Burrowes, "contained a strong restorative undercurrent." It sought to return control over Yemen's destiny to Yemeni hands, and also to "restore certain traditional Yemeni values and practices that had been challenged by the rhetoric of Arab socialism and by other ideas during the Egyptian interlude."[73] This restorative current was also evident in architecture. In the years following the revolution many in Sanaa had been attracted by the new way of life that the Egyptians brought, and wanted to live in modern houses. They associated the old architecture with the imam's regime.[74] But they also "yemenized" the Egyptian-style apartment buildings—enclosing balconies and installing decorative, multi-lite windows (*qamariyya*s). In institutional architecture, too, Egyptian models were rejected in favor of a more domesticated modernism that incorporated local features.[75]

The new government faced the task of creating an institutional structure and providing basic infrastructure for much of the country, in conditions of ongoing financial and political crisis (including a war with South Yemen in 1972). The new state was entirely dependent on foreign aid, which was shaped by Cold War politics. State institutions, as such, were not truly operational before 1975. Nevertheless, a government agency for antiquities and libraries was set up and a National Museum established in 1969. In 1972 an Antiquities Law was passed, which provided for the surveillance of sites and monuments of historic value. A small architecture studio was installed in the new National Museum, charged with initiating an inventory and survey of the principal monuments.[76]

In the 1960s tribal and *waqf* land surrounding the city had been purchased or illegally seized, ending traditional barriers to development outside the walls.[77] As civil conflict abated, pressure for development became intense. The major impetus in this regard were remittances from emigrant workers, which fueled private speculation and building, and contributed to sky-rocketing inflation. The massive influx of foreign aid also contributed to inflation.[78] There were no mechanisms in place to guide this growth. Although the Ministry of Public Works had been created soon after the revolution, as of 1970 there were no cadastral records or aerial surveys, no land use or zoning guidelines, land development laws, or safety and health regulations. A building permit process had recently been implemented but municipalities, which did not provide utilities and infrastructure, had little authority to

73 Burrowes, *The Yemen Arab Republic*, pp. 28–9.

74 ‘Abd al-‘Aziz ‘Abd al-Ghani, interview.

75 S. al-Suleihi, personal conversation; J.-L. Arnaud, "Formation de l'architecture contemporaine à Sanaa," pp. 72–4. The "yemenized" apartment buildings can be seen to the east of ‘Abd al-Mughni Street, at the edge of the old city in the area around Bustan al-Sultan.

76 A. Daolatli, mission report for ICOMOS, September 1982, 2.1. It is uncertain if this studio was the archeological section that opened in 1971 in Dar al-Shukr. Costa was appointed to catalogue the archeological collection at the museum in 1971 (P. Costa, interview; "Sana," in R.B. Serjeant, ed., *The Islamic City*, p. 162; Varanda, "Tradition and Change," p. 226).

77 Tribal custom had prevented the sale of land to individuals outside the tribe. When a capitalist land market emerged after the end of the civil conflict (1970), this custom began to lose its force, and tribesmen began to engage in speculation (H. Kopp and E. Wirth, *Sanaa*, pp. 43–4, 52–5). Much of the agricultural land outside the walls was endowed as *waqf*, although Kopp and Wirth do not mention this. According to several informants, the illegal seizure and sale of *waqf* land was widespread throughout the 1960s.

78 J.E. Peterson, *Yemen: the Search for a Modern State*, Baltimore and London: John Hopkins University Press, 1982, pp. 18–19; Philips, *Yemen's Democracy Experiment*, p. 44.

enforce it. Moreover, local staff lacked training in surveying, planning, and architecture.[79] The government did not have the funds or capacity to deal with urban improvements such as the installation of utilities, roads, and solid waste collection. Four decades later, many of these institutional and financial challenges remain as urban expansion continues, largely uncontrolled.[80]

In 1975 Sheikh Sinan Abu Luhum—a leader of the revolution and perhaps the most important tribal politician of the period—formed a Committee for the Improvement of the Capital (*lajnat tahsin al-'asima*), composed of officials and prominent citizens. By lobbying well-to-do businessmen (including migrants to the Gulf), the Committee was able to raise money to improve streets and install utilities. But the streets improved were outside the walled city, some along its perimeter.[81] In the 1980s Sheikh Sinan would serve on the Board of Trustees for the UNESCO campaign to save the "old city." The area had been hooked up to an electrical grid in the mid-1960s, but there was no running water until 1975. Sewers were not installed until the late 1970s.[82]

Over the course of the decade a series of plans were commissioned by the Ministry of Public Works and the Ministry of Municipalities.[83] But these plans were not implemented, or were implemented only in part—primarily because there was no cadre of trained technicians, and no legal or planning framework to guide the work. The Ministry of Municipalities and Housing lacked legal authority, budgetary control, technical capacity, and an effective system of land registration. Most importantly, perhaps, the planning

---

79 Varanda was a consultant to the Ministry of Public Works from 1973–1975, charged with implementing Bertaud's Master Plan at the level of the neighborhood. He and several other European consultants undertook to train the staff of the surveying office in their spare time in English, geometry, and the rudiments of planning; Varanda also developed a manual of land-use planning. Some of the trainees earned scholarships to study abroad and returned to assume responsible positions in government (Varanda, "Tradition and Change," pp. 204–9, and interview; see also A. Bertaud, "Report on planning, building, and related fields", Information Paper no. 1, UNDP?, July 1970).

80 See the assessment of planning capacity in the 2009 CDS Report, pp. 19–20, 35–6.

81 The streets were al-Zubayri, al-Shu'ub, al-Qasr, and al-Qiyada (S. Abu Luhum, *Haqa'iq wa watha'iq 'ishti, Sana'a: mu'assasat 'afif al-thiqafiyya*, 2005, pp. 20–21; 'Abdo 'Ali 'Othman, Professor of Sociology, Sanaa University, interview). 'Othman was then Minister of Municipalities and served on the committee, which included Muhammad al-Ruba'i, Minister of Social Affairs; Muhammad 'Abd al-Malik al-Mutawakkil, head of the People's Cooperative for the Development of Sanaa; Qadi 'Ali 'Abdullah al-'Amri, province of Sanaa; 'Ali Lutf al-Thawr, President of the Yemen Bank; and others. Qadi 'Ali Abu Rijal told me that a "committee of notables" (*majlis al-'ayan*) paved the area around Bab al-Sabah, but others said this did not happen. "We wanted to introduce the benefits of modern life into the city," said Abu Rijal, "while preserving its special qualities." He served as Deputy Minister of Public Works, but says that he assumed the position after the walls had been demolished.

82 This according to Muhammad Sa'd, Professor of Urban Geography at the University of Sanaa. "Our house was the fourteenth house to get electricity," he recalls. "I remember that children were dancing in the streets!" Water and sewerage were still provided through on-site wells and cesspits, both in the old districts and in new development. As water and sewer lines were laid, hookups were usually made by individual owners on an *ad hoc* basis; the situation is much the same today.

83 These include plans by a Japanese mission (1968); the UNDP (1972); the World Bank (1975); and by Chinese and Saudi teams (circa 1975, according to 'Othman who was Minister at the time). Plans for utilities had begun earlier: in 1970, Italconsult (Italy) had drawn up utility plans for Sanaa and al-Hudayda. For a partial list of planning studies that had been done by 1982, see report carried out for the Ministry of Municipalities and Housing by Berger Louis, International and Kampsax, International, A/s, "Urban Planning Study of San'a, al-Hudaydah, Ta'izz, Ibb, Dhamar," 1979, volume one; and J. 'Abdelkafi, ICOMOS mission report, September 1982, appendix one and bibliography.

initiatives were opposed by powerful interests.[84] Following current urbanistic thinking, few of the plans were concerned with local context or the historic core. In his opening address to the 1983 Aga Khan Seminar in Sanaa, Prime Minister Yahya al-Iryani would look back on these plans with a new perspective:

> Our ancient tradition of building, developed over centuries, reflects the experience of generations. But many characteristic features of our buildings have been ignored by planners and those international consultants responsible for the implementation of development projects.
>
> It is worth stressing that in general terms received technology and methods are better and more suitable than modern technological means because they originate in the environment. Because of developed communications, technology transfer has become easier and faster in the twentieth century. But it has led to the adoption and literal imitation of technologies and methods of development that are alien to the need of many societies. Many countries have adopted means of development that are in contradiction to their environments.[85]

The Egyptian plan of the 1960s, for example, would have extended one or more arteries of a proposed grid through the old city to Jabal Nuqum, destroying the houses that stood in the way. In the 1970s Saudi consultants proposed a high-speed road above the Sa'ila, in order to make the corridor usable during seasonal flooding. The proposal would have destroyed houses on both sides of the wadi, driving a wedge through the old core and attracting development along the banks.[86] Voices rose in opposition to these proposals. European and American consultants at the UNDP (the Sanaa office opened in 1970) charged that the proposals treated the walled city as little more than a slum. They cited the unique architectural, urban, and environmental qualities of Sanaa, and called for their conservation.[87]

Architect Alain Bertaud, consultant to the Ministry of Public Works, was charged with developing a Master Plan for Sanaa (1972), which was primarily concerned with guiding and controlling future growth. But it differed in several important respects not only from the Egyptian and Saudi plans, but from plans proposed by Europeans for other cities in the region. Bertaud insisted, for example, that new development in Sanaa should be informed by local models, and that local materials and building skills should be used in new construction. This would reduce capital expenditures. But more importantly, money

84 'A. al-Shawish, "Conservation Policies in a Historic City: A Case Study of Sanaa," unpublished Master's thesis, Department of Town Planning, University of Wales Institute of Science and Technology, CARDIFF, October 1987, pp. 52–3.

85 "Opening Remarks," *Development and Urban Metamorphosis*, vol. 1, p. xvii.

86 F. Varanda, interview; R. Lewcock, "Restoration and preservation of the historic architecture of the old city of San'a," mission report to UNESCO, 1982, p. 55. In 1982 two proposals for building a road above the Sa'ila were still under consideration; one of these was the so-called "Chinese plan" (designed circa 1975) which channeled floodwaters in a culvert below a raised roadway; the other was that by Quaroni Bonifica, which would have kept the roadway at the level of the wadi floor and place the culvert underneath it (Yoshida, "Study Report on the Wadi al-Sa'ila Improvement Project," 1990, pp. 12–13; Quaroni Bonifica, "Safeguarding of Sanaa Historical Center in relation to the urban development of the city," Rome, 1981, Section PII). Lewcock pointed out that the wadi was only impassable a few days each year; this could be solved by providing drainage.

87 F. Varanda, "Tradition and Change," p. 226. The following accounts of plans for Sanaa rely on Chapter 6 of this work.

spent on stone construction would stay in the community.[88] His plan also called for the protection of the city's cultural heritage from surrounding development; to underline this point, he undertook the first improvement of public space in the walled city by paving al-Abhar Square in 1972.[89] Some interesting features of Bertaud's proposal were rejected, like collective urban gardens for new districts based on the model of the *maqshama.* But he managed to convince decision-makers to use stone rather than concrete in institutional buildings. Also at his suggestion several of the imams' houses, including the palaces at Dar al-Hamd and al-Rawdha, were restored and adapted for use as hotels. Another architect, Derek Matthews, was sent to Sanaa by the UNDP in 1972: as director of institutional projects, he designed a number of important buildings in a local idiom.[90]

When civil conflict ended in 1970, aid from Western countries poured into North Yemen, and foreigners (mostly Europeans) arrived in increasing numbers. Embassies were transferred from Ta'izz to Sanaa, and some set up operations in old houses. Yemeni officials were surprised at this: they thought that foreigners would want to set up embassies in modern buildings. Instead, they expressed awe and admiration for the local architecture, and this prompted Yemeni officials and intellectuals to look at it with new eyes.[91] In October 1970, Italian filmmaker Pier Paolo Pasolini visited Yemen to shoot footage for the film *Arabian Nights* (1974)—a film infused with outrageous Orientalist tropes. He was moved by the poverty and underdevelopment of the country, and by the beauty of its architecture. The demolition of the walls, he warned, was a prelude to the destruction of the historic core: he feared that Sanaa would follow the course of Italian cities, only to regret it later. Pasolini used some extra film to shoot a short documentary, *Le Mura di Sana*, framed as an impassioned plea to UNESCO to save the city. The prayer-like incantation, addressed to an international agency, evokes the quasi-religious role of heritage in the modern era:

> We appeal to UNESCO—Help Yemen save itself from destruction, which began with the destruction of the walls of Sana.
> We appeal to UNESCO—Help Yemen to become aware of its identity and of what a precious country it is.
> We appeal to UNESCO—Help stop this pitiful destruction of national heritage in a country where no one denounces it.
> We appeal to UNESCO—Find a way of giving this nation the awareness of being a common good for mankind, one which must protect itself to remain so.
> We appeal to UNESCO—Intervene, while there is still time, to convince an ingenuous ruling class that Yemen's only wealth is its beauty, and that preserving that beauty means possessing an economic resource that costs nothing. Yemen is still in time to avoid the errors of other countries.
> We appeal to UNESCO—In the name of the true, unexpressed wish of the Yemeni people. In the name of simple men whom poverty has kept pure. In the name of the grace of the Dark Ages,
> In the name of the scandalous, revolutionary force of the past.[92]

88 F. Varanda, "Tradition and Change," p. 205 and fn. 18–19.

89 F. Varanda, "Tradition and Change," p. 226. After Bertaud left in late 1972, recalled Varanda, "the only people talking about conservation were Costa, Lewcock, Serjeant, and myself" (interview).

90 For projects by Bertaud and Matthews, see Varanda, "Tradition and Change," pp. 194, 205–9.

91 'Abd al-Karim al-Iryani, interview; P. Costa, "Sana," p. 160.

92 *Le mura di Sana*, 1974 (shot on October 18, 1970), documentary; translation by S. Iovino, "The Ashes of Italy: Pier Paolo Pasolini's Ethics of Place," in J. Gifford and G. Zezulka-Maillous,

Pasolini appears to have submitted the film to UNESCO in 1973—a year after the idea of heritage as "a common good for mankind" was canonized in the World Heritage Convention. In what may have been an accompanying statement, the filmmaker wrote:

> Nowadays there is no longer any such thing as a Western or an Eastern culture. For some time now, the boundaries have been blurred. For Europeans, the old city walls of Sanaa are just as important as the lagoon city of Venice.[93]

There is no record of the receipt of Pasoloni's film by UNESCO, or the impact it had on decision-makers there. But it is often cited in Yemen, and by Italian consultants who worked on the project, as the initial spark that led to UNESCO's interest in Sanaa.

Over the next few years, rising oil prices transformed the regional situation. From 1974, hundreds of thousands of Yemeni men migrated to the Gulf to take advantage of booming oil economies, sending back their earnings in the form of remittances.[94] Through the 1970s and '80s, labor migration was the major catalyst of economic and social change: the shortage of labor and the influx of cash artificially inflated the economy, which had little to offer in the way of industry and services. The rapid growth had a tremendous impact on a variety of sectors, such as water and other natural resources and the form of towns and cities, while hastening the decline of the agricultural sector in a country that was, and is still, overwhelmingly rural. Yet growth was kept in check to some extent by the labor shortage.[95]

Capital in the form of remittances was in private hands and was generally not channeled into the development of industry or commerce: rather, migrants preferred to invest in land and later, in the construction of family homes.[96] By the mid-1970s, speculation extended through outlying farmlands in a wave of development that was largely unplanned. Buildings were executed partly or wholly in new, imported materials like concrete, iron, plastics, and paint, which were diffused via an expanding road network. The result was, for European observers, a great breach between this new construction and the simple beauty of local forms: even the occasional "discords" of the Ottoman period, noted Costa, had been softened through the use of local materials.[97]

Little infrastructure was in place, nor could the city afford to provide it. The opening of the eastern portion of the first ring road in 1975—intended to set the limits

---

eds, *Culture and the State: Landscape and Ecology*, Edmonton: CRC Humanities Studio/University of Alberta, 2003, pp. 70–91. For the transcript of the film see L. Betti and M. Gulinucci, eds, *Le regole dell'illusione*, Roma: Associazione Pier Paolo Pasolini, 1974.

93 1973; quoted in *Programme 13: Venise, Sanaa*, UNESCO/Transtel, M0365, no date; video film on file at UNESCO Visual Archives, Place Fontenoy, Paris.

94 By the end of the 1970s it was estimated that some 800,000 North Yemeni men had migrated for work (Dresch, *Modern Yemen*, p. 131).

95 Troin reports that building activity slowed in Sanaa in the mid-1970s, due to the shortage of labor ("Sana'a: géographie d'une explosion urbaine," in Grandguillaume, et al., eds, *Sana'a hors les murs*, pp. 15–30). For an account of the far-reaching economic effects that accompanied the remittance boom and the introduction of an economy based on wage labor, see Dresch, *Modern Yemen*, pp. 137–40; for the impact on the construction sector through the early 1990s, see F. Varanda, "Tradition and Change," pp. 198–202.

96 Dresch, *Modern Yemen*, pp. 137–8. Philips notes, however, that some remittances were channeled through local development associations (LDA's) for projects like schools, roads, and basic infrastructure in areas that were beyond the reach of government (*Yemen's Democracy Experiment*, p. 44 and fn. 23).

97 "Sana," p. 160.

of expansion—served rather to increase speculation along it; at the same time radial connections to the center of the city put increasing pressure on the old districts.[98] In the coming years development would spread westward to meet a second ring road, and later, along the north-south axis of Airport and Ta'izz Roads (Figure 2.9).[99] Implementation of the 1972 Master Plan for Sana and other provincial centers was handicapped by pressure to develop more and more land. It was also limited by the absence of a cadastral survey, the looseness of building permit procedures, and the absence of proper equipment and trained personnel. Laying out roads was dangerous work: surveying teams were often fired upon.[100]

The Decline of the Old City

Meanwhile, conditions in the old districts were worsening as new circumstances overwhelmed established mechanisms of urban management—rupturing what local architects describe as a self-sustaining "environmental cycle." There was little room for expansion within the city walls, and in any case, the vast majority of land was *waqf*. The population was growing, due to a rising birth rate and in-migration from the countryside. In some cases, relatives were brought from villages to live in family homes, as prosperous residents moved to the new districts.[101]

Old mechanisms could not keep up with changes, or were replaced by new ones that functioned less effectively. Manufactured products had been introduced on a relatively large scale, but there was no solid waste disposal system in place. By the early 1970s garbage was accumulating in the streets; the municipal street-sweeping system that had been set up by the Ottomans could no longer keep the city clean. Changes in the local authority structure released the *'aqil*s (heads of quarters; Arabic plural: *'uqal*) from maintenance responsibilities, yet new municipal structures were not in place or were ineffective.[102] Most streets had probably been paved from the medieval period onward, but in recent decades paving had been torn up by motorized vehicles. The practice of tamping down the earth with water was gradually abandoned; dust was exacerbated by vehicles, which also rutted the streets during the rainy season. Wastewater from house lavatories (containing urine, but not feces) flowed in pipes on the house facades into the unpaved streets, making them humid and unsanitary. Conditions were especially bad in the *suq*, where the muddy roads were congested and strewn with garbage. The state of the streets were believed to be the single greatest cause of outmigration, which accelerated over the course of the decade (Figure 2.10).[103]

---

98 R. Lewcock, interview; P. Costa, "Sana."

99 Stadnicki and Touber call the later north-south expansion a "second urban explosion," following the return of migrant workers after 1990 ("Le Grand Sanaa," p. 37).

100 Varanda recalls numerous threats to his surveying team, and on one occasion a surveyor was shot with a machine gun. In roadwork executed by the Committee for the Improvement of Sanaa, Sheikh Sinan Abu Luhum was asked to intervene when there were disputes (A-A Othman, interview). Despite numerous plans in the subsequent two decades, draft planning legislation was not finalized until 1989.

101 M. Sa'd, interview.

102 Varanda, "Tradition and Change," pp. 201–2; A. 'Oshaish, "Essential description of the historical city of Sanaa for the purpose on inclusion in the World Heritage List," TOPOCS (no date), p. 10.

103 Bertaud, "Report on planning, building, and related fields in the municipality of Sanaa"; R. Lewcock, interview and mission reports, 1984; 1987, para. 26.

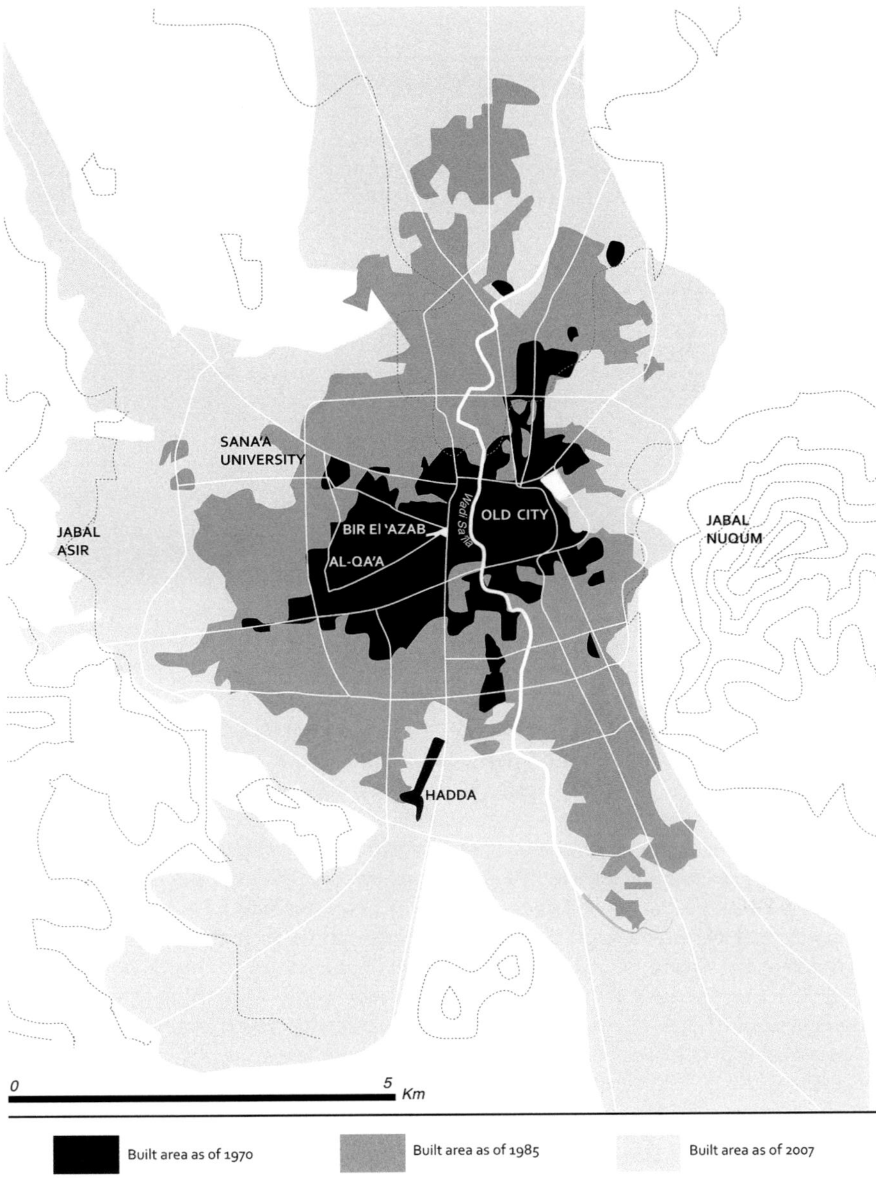

Figure 2.9 Growth of Sanaa since 1970

Source: After Kopp and Wirth (1994); Troin (1995); and Troin and Stadnicki (2007).

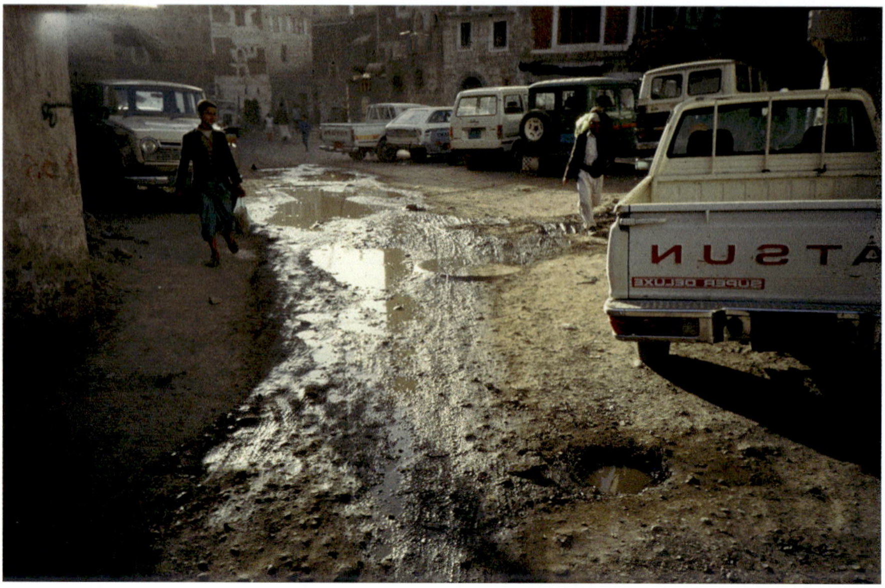

Figure 2.10 View of a muddy street in the old city circa, 1982
Source: Photograph by Ronald Lewcock.

One of the main reasons for deteriorating conditions was the change in supply and consumption of water. Household water in Sanaa had been obtained from on-site wells, drawn by bucket to upper floors; it was stored in masonry jars, often placed in a *shubbak*—a projecting element of stone or brick—where it would be cooled by air currents. Water was used sparingly, and recycled within the house for various uses.[104] Lavatories within the house provided for the separation of liquid and solid wastes. Wastewater from bathrooms and kitchens flowed down a recessed channel on the house facade; from there it was conducted into a cesspit and soaked into the surrounding earth.[105] Human feces dropped down a chute within the house that connected toilets on various levels; it quickly dried in the mountain air and left little smell. This "night soil" was collected via small doors at ground level and sold to *hammams*, where it was burned as fuel; the ashes were subsequently used as fertilizer on the *maqshamas*. The cleanliness of Sanaa, and the absence of smells, had been noted over the centuries by travelers. The nineteenth century ethnographer Joseph Halevy, for

104 See Varanda, "Tradition and Change," p. 86. The following account of water supply and sewerage disposal before and after the revolution draws from this work, pp. 86–8 and 199–202; from P. Helmore, "Water Supply, Sewerage, and Stormwater Drainage," ICOMOS mission report, September 1982; and from Howard Humphreys, Ltd., "Improvements to water supply, sewerage, and stormwater drainage systems," September 1982.

105 The cesspit was located in the street adjacent to the building, or less often, within the property boundary. There was no provision for emptying the cesspit: rather when a pit filled it essentially sealed itself, and a new pit was dug (R. Lewcock, interview).

example, considered Sanaa the most beautiful and cleanest town in Arabia, with straight, wide streets that were usually paved.[106]

By the 1970s, due to increasing water consumption and a decreasing water table, wells in both private homes and in well-houses attached to mosques were fitted with motor-driven pumps. From the mid-1970s a new water supply system was introduced in phases: water was now retained in reservoirs, piped into the city, and stored in tanks on the roofs of buildings. Early phases of the system, installed first in the old city and al-Qa'a, were flawed in design and installation.[107] With the introduction of piped water, on-site water usage increased dramatically. But because a sewer system was not installed at the same time, there was no means to carry the water away. By 1982 flush toilets had been installed in some houses at the recommendation of the World Health Organization, adding human feces to the waste collected in the cesspits.[108] Cesspits overflowed under the increased load: sewage soaked into the ground and seeped into the streets and wells. Consultants feared that areas of water supply under low pressure might become contaminated; and that the mixing of sewage with wastewater would have adverse effects. "The quantity of water now available for consumption in the old City and al-Qa'a [sic] areas," wrote Helmore

106 Bidwell, "Western Accounts of *San'a*, 1510–1962," in Serjeant and Lewcock, eds, *San'a*, p. 112. On the dry toilets, which are characteristic of highland towns, see R. Lewcock, "Three Problems in Conservation: Egypt, Oman, and Yemen," pp. 72–3; and drawing in Lewcock and Serjeant, "The Houses of *San'a*," p. 440. In the late eighteenth century or early nineteenth century, the duty of collecting night soil was imposed on Jews (they were also charged with removing dead animals from the city). Prior to this time, the collection of night soil appears to have been done by the Jews within al-Qa'a, and by Muslim bathkeepers in other areas (A. Shivtiel, W. Lockwood, and R.B. Serjeant, "The Jews of *San'a*," in Serjeant and Lewcock, *San'a*, p. 394). The mass exodus of the Jews after 1948 doubtlessly affected the system. According to middle-aged residents I interviewed, night soil collection continued at least through the mid-1970s. But it was sold outside the city, they said, because people objected to the use of its ashes in food-producing gardens.

107 The southern portion of the old city, the area south of Tahrir Square, and al-Qa'a were the first areas to receive an "emergency water supply system" in 1975, funded by USAID. But design and installation were faulty; pipes leaked, affecting building foundations which are typically shallow and of dry joint masonry. In 1975–1979, phase one of a water project for Greater Sanaa was carried out by Annem and Thabet/Lilly, a Lebanese contractor, under the direction of the National Water and Sewer Authority. This phase served the remainder of the old city and its immediate suburbs to the west, north, and south. In this case, design seems to have been sound, but there were leaks due to inferior pipes and workmanship. The phase one system was tied into the old emergency system, which served most of the old city and was still in extensive use in 1982. As parts of the phase one system were brought into operation pipes burst, causing severe damage to some buildings in the old city (Helmore, ICOMOS mission report, 1982, pp. 15–20; Howard Humphreys, Ltd., "Improvements to water supply, sewerage, and stormwater drainage systems," 2.1). A second phase of the water supply system was designed by Howard Humphreys Engineers and completed by contractor SOBEA in 1982. Phase two extended water supply to newer areas of the city and upgraded service to the earlier systems. Pressure and water use increased yet again: a new round of structural issues emerged in the old city, and a number of buildings collapsed. Fewer problems associated with the water supply and wastewater disposal were observed in al-Qa'a, perhaps because the houses there are smaller (Helmore, ICOMOS mission report, p. 29).

108 Lewcock and others questioned the WHO recommendation, arguing that the traditional dry chute system was effective and ecologically sound. See R. Lewcock, "Three Problems in Conservation: Egypt, Oman, and Yemen," in R. Holod, ed., *Conservation as Cultural Survival*, Geneva: Aga Khan Award, 1980, p. 72, and UNESCO mission report, September 1982, 145. In 1982 most houses appeared to be using the old lavatories, but it was expected that increasing numbers of households would install flush toilets; the load could not be supported by cesspits, so the installation of a sewer system was seen as highest priority.

in 1982, "is greater than can be sensibly permitted until adequate sewerage facilities have been constructed and put into operation."[109] Faulty pipes and workmanship had resulted in extensive leaking; when pressure within the system improved after 1979, the leaking increased. Trapped below a perched water table, the groundwater destabilized building foundations, which were not designed for water-laden soil. By 1982 some 30 houses had collapsed (Figure 2.11).[110]

As a result of these conditions, as well as changing lifestyles and preferences, those with means left the old districts. This outmigration—and the concurrent inmmigration from the countryside—has been underestimated in conservation literature, which has painted a portrait of old Sanaa as an unchanging place. Although the numbers are uncertain, the scope of the transformation was dramatic. "Don't believe it when they tell you that *some* people left the old city," said Amal al-Shaiba, native of Sanaa and widow of one of the architects who was a moving force behind the conservation effort. "*Everyone* left, except for very few."

The attraction of the new districts were many: they had clean streets and modern amenities, including the possibility of owning a car. New houses types—modeled on rural precedents, or western-style "villas" that migrants had seen in the Gulf—allowed for new comforts, and independence from extended families. Even middle income families could now afford to build new houses, thanks to remittances.[111] The family houses in the old districts were usually left in the hands of elderly or less prosperous relatives, who did not have the means to maintain them. In some cases houses were rented to migrants from the countryside, who had limited financial resources and were not invested in the maintenance of the houses and the city streets. New construction within the walls was limited by various factors. There was little unbuilt land, and vehicular access was difficult. But there was increasing pressure for development, especially with the opening of the Sa'ila as a through road. The dearth of land in the walled city led to demolitions and building on open spaces including the *maqshama*s, where cultivation had declined in recent years.[112]

In the course of only a decade the historic core had been marginalized both physically and economically: in 1971 it was still the dominant physical and social space of the city; by 1982 it occupied a twelfth of the total urbanized area, and housed only one fifth of the population.[113] The new metropolis was organized by new principles, those of consumer culture. Until the end of the 1960s, wrote historian 'Umar al-Jawi, "every house in Sanaa

109 Helmore, ICOMOS mission report, 1982, p. 21. By the early 1980s there was evidence of contamination of the aquifers by sewage (Varanda, "Tradition and Change," p. 200). Howard Humphreys had designed a sewage system for the city in 1978; scheduled for construction in 1980, it was delayed. In September 1982, Prime Minister al-'Iryani assured UNESCO consultants that work on the sewer system already begun (R. Lewcock, UNESCO mission report, Dec. 1983).

110 R. Lewcock, interview; P. Helmore, "Water Supply, Sewerage, and Drainage." For an explanation of the perched water table, see R. Lewcock, "The problems of subterranean water in the old urban areas of Arab cities," in I. Serageldin, I. and S. El-Sadek, *The Arab City: its Character and Islamic Heritage*, Proceedings of a Symposium held in Medina, Saudi Arabia, 28 February–5 March, 1981.

111 Building permits—introduced after 1968, but unevenly enforced—rose steadily after 1975 and reached a peak in 1984 (see chart in Varanda, "Tradition and Change," p. 213). Shortly after joining the UNDP in 1970, Alain Bertaud did a survey of traditional and new house types which has been reproduced in various publications (including Kopp and Wirth, *Sanaa*, pp. 55–7). The original report is entitled "Housing in Sana'a," UNDP, no date.

112 M. Welbank and J. Figueiras/Shankland Cox Partnership, "San'a: a Report to UNESCO," London, 1978, 3.3.1.

113 R. Lewcock, mission report, 1982, paragraph 2.

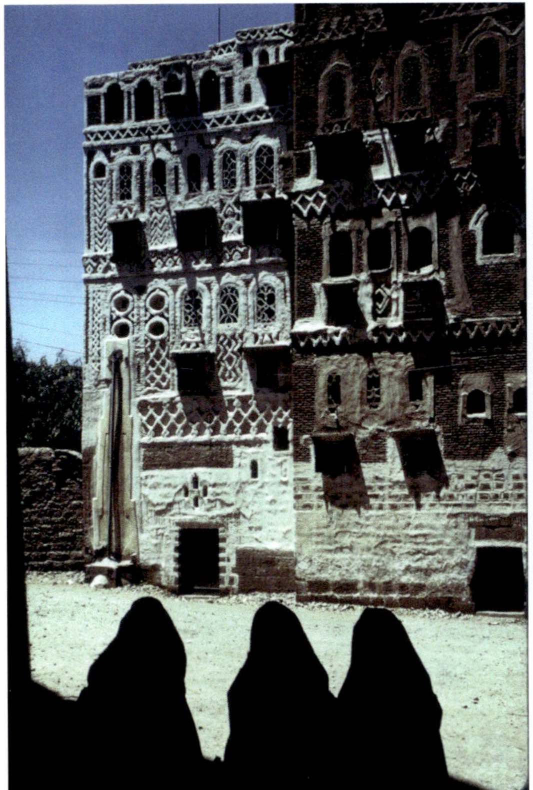

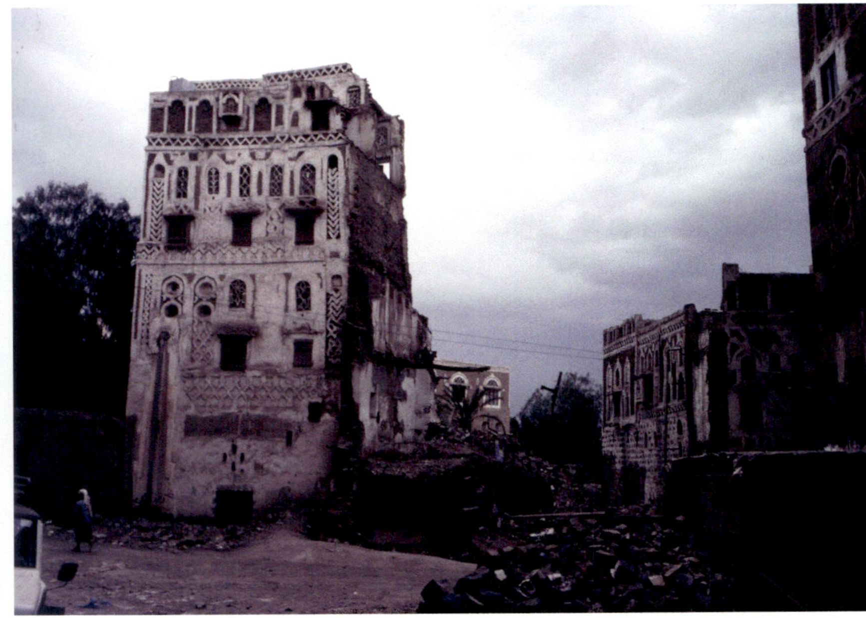

Figure 2.11 **Ronald Lewcock took these photos circa 1982. By the time he took the second photo (above), the house on the right had collapsed**

was a self-contained economic unit, and the original inhabitants depended on what was in their store rooms … and in their own wells." Over the course of the following decade, Dresch observed, economics replaced ecology.[114]

114 *Modern Yemen*, pp. 167–8. The quotation is from al-Jawi's *Hisar San'a*, 1975, cited by Dresch.

Chapter 3
The Campaign to Save Old Sanaa

In the early 1970s, the idea of Yemeni heritage was largely restricted to antiquities (*athar*). By the end of the decade, this notion had expanded to include a wider conception of heritage, including settlements and customs. This reflected the international discourse of world heritage, and also a new political era in the YAR. In 1978 ʻAli ʻAbdullah Saleh of the Hashid—one of two major tribal confederations in Yemen—seized power in a military coup, after years of political upheaval. State institutions were growing, with heavy infusions of foreign aid and advisors. But state authority was weak, at times limited to the environs of the capital. North Yemen was often described as a nation, but not a state.[1] The idea of Yemen as an historical and natural unit easily translated, in the minds of Yemenis, into nationalist sentiments. But the nation meant a unified Yemen—and leaders of both North and South Yemen were constitutionally bound to work toward unity. It was through unification that Yemen would become a nation-state, where territory, history, and polity coincide.[2]

It was in this context that images of popular heritage (*turath*) started to circulate in the media. Television shows broadcast images of landscapes and towns, and the "customs and traditions" (*ʻadat wa taqalid*) of both north and south.[3] The phrase is commonly heard today in Yemen, among individuals of various ages and educations. Another factor was mass education: textbooks that "make available, in compact form, a set of discrete state histories (Tahirid, Rasulid, Zaydi) as part of a national history."[4] For revolutionaries the period of the imamate represented darkness and oppression, and some saw the walls of the old city as reminders of this. An increasingly self-confident state would assimilate the Zaydi past, and the walls of the old city would become part of the national, historical narrative.

The idea of the "old city" as a protected area developed over the course of the 1970s, as scholars began to study its history. At the same time, there was growing interest in Sanaa at UNESCO—perhaps sparked by Pasolini's film. By 1975 Sanaa had been listed as an endangered site by the World Heritage Committee. As such, it must have been on the Committee's list of prospective World Heritage sites in the region. But the associations with the old order were still fresh: ex-royalists, including many who had worked under the last imam, were brought into the government after 1970. In 2005, I asked Qadi ʻAli Abu Rijal, a prominent *qadi* historian (and former republican) about this; he had been an official in the Ministry of Public Works and later became a staunch supporter of conservation. "Old Sanaa

1 Dresch, "A Daily Plebiscite," *REMMM*, vol. 67, no. 1, p. 67; *Modern Yemen*, p. 159.

2 Here I am extrapolating from Dresch's point: "Since unity in 1990 … there seems to be little doubt in Yemenis' minds, wherever they are from, that the state now is properly a nation-state" ("A Daily Plebiscite," p. 67). On the prior point, see F.G. Gause for the situation in the 1980s ("The Idea of Yemeni Unity," *Journal of Arab Affairs*, vol. 6, no. 1, 1987, pp. 55–81).

3 "In the context of progressive, developmental rhetoric promoted by the State and parties," wrote Dresch in 1993, "tribal ʻcustoms and traditions' (*ʻadat wa taqalid*) have become a marker of Yemeni cultural authenticity" ("A Daily Plebiscite," p. 73). For the earlier period, see *Modern Yemen*, p. 150.

4 Dresch, "A Daily Plebiscite," p. 75.

cannot be seen as a symbol for the return to the old order," he said. "Anyway, most of the revolutionaries were from the old city."

Imam Yahya had encouraged historical studies as part of a protonationalist program; later, the pre-Islamic past played an important role in revolutionary ideology. At the end of the 1960s, the new Antiquities Department began to document important monuments and artifacts. Art historian Paolo Costa oversaw this work and published a number of studies, including one of the Great Mosque of Sanaa in 1974.[5] R.B. Serjeant of Cambridge University, a noted scholar of Islamic architecture and urbanism and former director of SOAS, had long dreamt of studying North Yemen. Serjeant was a specialist in south Arabia, and as such had broadened the canon of "Islamic architecture" that had focused on a handful of Mediterranean Arab cities.[6] Under the imams northern Yemen had been relatively inaccessible to European scholars and archeologists. With the end of civil conflict, Serjeant saw an opportunity. He sought and obtained funding for an expedition, which he carried out with a team of foreign and local experts in 1972, including Ronald Lewcock. The results of what came to be called the Cambridge Expedition were exhibited the following year. In 1976, a World of Islam Festival was organized in London, with the support of Arabists and officers at the British Foreign Office and funding from Saudi Arabia. As part of the festival an exhibition called "Nomad and City" was installed at the British Museum, with a section on Sanaa conceived by the Cambridge Expedition team. Intended as a brief exhibit, it remained open for several years due to overwhelming interest and attendance. The "City of San'a" was one of 23 exhibits at the festival on the art, music, and culture of the Arabian peninsula. The exhibit featured full-sized replicas of part of the *suq*, a *mafraj* (sitting room) and a typical Sanaani kitchen (*dayma*).[7]

During the mid-1970s, Serjeant and Lewcock spent extended periods in Sanaa, researching and coordinating the articles that would form the book, *San'a, An Arabian Islamic City* (1983). In 1976 Lewcock was asked to undertake studies of the northern Gulf, Oman, and Muscat for a variety of sponsors, and worked with UNESCO on the preparation of an international campaign for Islamic Cairo. He was also approached by UNESCO to undertake missions to Sanaa and Shibam, and was later appointed coordinator for the campaigns to safeguard those cities. Many of the individuals who worked in Sanaa had also worked in Cairo, including Michael Welbank, who would direct UNESCO's first major study of Sanaa in 1978.[8]

This study, by the British firm Shankland Cox, was carried out in the framework of a master plan for Sanaa and four other cities by the German firm Berger Kampsax

5 *La Moschea Grande di Sana'a*, Naples: Istituto Orientale di Napoli, 1974.

6 Serjeant died in 1993. This account is based on interviews with Ronald Lewcock, who collaborated with Serjeant in his work on Sanaa.

7 The exhibit was conceived and created by Serjeant and Lewcock, in association with Paolo Costa, Walter Dostal, Martha Mundy, Rex Smith and Robert Wilson; interpreted and curated by Margaret Hall and her team at the British Museum (Serjeant and Lewcock, eds, *Sanaa*, p. 12). See the exhibition catalogue edited by J. Kirkman, *City of Sanaa: Nomad and City Exhibition*, Museum of Mankind, Department of Ethnography of the British Museum, London: World of Islam Festival Publishing Co. Ltd., 1976.

8 S. Zulficar, personal conversation. The study directed by Welbank for Shankland Cox (to be discussed below) was preceded by a brief study by G. Ehrbach, "Cultural Town Planning in Sana'a," dated March 1977 (on file at the Office of Arab States, UNESCO). UNESCO's first work in North Yemen was the conservation of the Zafar Dhi Bin Mosque, overseen by Zulficar in the early 1970s.

International.[9] The Berger Kampsax plan identified an historic core, composed of the walled city and the old Jewish quarter of al-Qaʿa (but not the old suburbs that were collectively called Bir al-ʿAzab). The plan called for a specialized study of these areas and the creation of an administrative framework for their protection.[10] Like the Berger Kampsax plan, the Shankland Cox study was prepared for what was now the Ministry of Public Works and Municipalities (Municipalities would separate from Public Works shortly thereafter). The firm was asked to advise the ministry on matters related to "cultural planning" and conservation problems in "Old Sanaa … [which] has recently come to be regarded as a key element in the world's cultural heritage."[11]

Michael Welbank and Julio Figueiras carried out a number of missions to Sanaa for Shankland Cox that year, and submitted their report in June. The Shankland Cox proposal would have created *three* conservation districts—the walled core, al-Qaʿa, and Bir al-ʿAzab. The consultants stressed the need to preserve these areas as living entities, and at the same time to protect them from development pressures. In order to accomplish this, the existing housing stock and the mix of socio-economic groups should be maintained; accessibility and services should be improved; and commercial viability should be enhanced. The plan called for sensitive and selective "surgery," distinguishing it from the approach used for monuments.

> In accordance with the general strategic objective of maintaining the city as a busy living entity, the concept of conservation here cannot be an absolute one, as it would be for instance in the case of the Pyramids of Giza or other similar monuments. If changes are required to keep the city a living entity they must be accepted even at the price of some destruction of the existing physical fabric. How the changes are actually effected is a far more critical issue.
>
> In other words, it is better to accept minor surgery now to maintain a healthy living city than to reject all surgery at the expense of the patient's death. The conservation task is therefore as much concerned with the sensitive handling of change as with defining the exent of acceptable change ….

The authors proposed a degree of "dedensification," a strategy that was proposed around the same time for Fez—although this does not seem to have been based on a close study of current and historic conditions.[12] Specific proposals for change in the physical

---

9 Berger Louis, International and Kampsax, International, A/s, "Urban Planning Study of San'a, al-Hudaydah, Taʿizz, Ibb, Dhamar," 1979 (for the Ministry of Municipalities and Housing). It is not clear whether the Shankland Cox study was initiated by UNESCO or Berger Kampsax.

10 The Berger Kampsax plan was concerned mainly with zoning and road networks, and was thus largely passive in its attitude to directing growth. Most observers and consultants saw the plan as lacking sensitivity to local models and conditions (Varanda, "Tradition and Change," pp. 210–11, and A. al-Shawish, "Conservation Policies in a Historic City," p. 52; see also Ludovico Quaroni's proposal to the government of North Yemen on file at the Office of Arab States, UNESCO, Paris; no date, circa 1981). Nevertheless the Berger Kampsax plan remained the main planning tool for the city. Stadnicki notes that its growth projections were based on the model of an industrializing city. In reality, the consumer-remittance economy produced much more rapid growth (Stadnicki, "Le Grand Sanaa," pp. 36–7).

11 Shankland Cox report, 1978, 1.3, 5.4.2.

12 M. Welbank, M. and J. Figueiras, "Sanʿa: a Report to UNESCO," Shankland Cox Partnership,

fabric—for example, the transformation of the largely unused ground floor of the tower houses into car parks—were rejected by later UNESCO consultants, as well as by members of the government.[13]

Nevertheless, the main strategic objective of the plan was adopted, which in some ways sanctioned the pattern of development up to that point. The functional core of the city—in essence, a kind of central business district—was displaced to the west, around Tahrir Square. The authors of the plan wanted to protect the adjacent historic core from development, but keep it integral to the life of the city.[14] Later consultants, however, would see the central position of the historic core as a problem: the confluence of important arteries and networks exerted too much pressure on it.[15]

UNESCO normally works through Antiquities Departments, but in this case, their counterpart was the Ministry of Public Works and Municipalities—presumably because the study was carried out in the context of the Berger Kampsax plan. The historic districts identified by the team would be administered by a conservation unit within this ministry.[16] The Department of Antiquities was also interested in conservation.[17] The director of the department at the time, Qadi Isma'il al-Akwa', was a distinguished historian who had worked closely with Serjeant and Lewcock's team. He was also a friend of Mounir Bouchenaki, who would later become Desk Officer for Arab States at UNESCO.

Bouchenaki recalls that the idea of protecting the "old city" came up at a conference of culture ministers in Sanaa: al-Akwa' and Youssuf 'Abdullah (who later directed the Department of Antiquities) were frantic about the state of the old city. This was probably the Advisory Committee on Arab Culture, held in Sanaa in June 1979; this was the first of a number of regional conferences in the field of culture held in Sanaa during these years.[18] The following year, at UNESCO's General Conference in Belgrade, North Yemen would refer to the recommendations of the June 1979 meeting in its Draft Resolution to the UNESCO Secretariat. The resolution, submitted on October 3, 1980, requested international assistance to save the "Islamic monuments of Yemen"; the preamble of the resolution called special attention to "the ancient city of Sanaa."[19] Three days earlier, at

London, 1978, 3.3, 5.1. Population density in the old Sanaa is not mentioned by Lewcock and other consultants as a matter of concern in the early 1980s.

13 R. Lewcock, mission report to UNESCO, April 9–12, 1982, pp. 22–3.

14 M. Welbank, "Conservation and Development," in *Development and Urban Metamorphosis*, Proceedings of Seminar Eight in the Series "Architectural Transformations in the Islamic World," Aga Khan Award for Architecture, Singapore: Concept Media, 1983, vol. 1, p. 9.

15 For example H. Eckert,"The Chief Urban Functions of the Old City of San'a Today," Technical Report, Paris: UNDP, 1992; B-M Nardella, "Mission Report," prepared for UNESCO—Culture, Tangible Heritage Department, January 2004; D. Pini, et al., *Inventory of the Historic City of Sanaa: a tool for urban conservation*, Paris: UNESCO Press, 2008.

16 M. Welbank, M. and J. Figueiras, "San'a: a Report to UNESCO," 5.4.

17 This is the predecessor to the General Organization of Antiquities, Museums and Manuscripts of Yemen (GOAMM).

18 Other conferences held in Sanaa were the Arab Conference in Architecture, organized by the Arab Organization for Education Culture and Science (Ninth Session); and a seminar sponsored by the Aga Khan Award, entitled "Modernity and Tradition." The proceedings of the latter conference were published in two volumes under the title *Development and Urban Metamorphosis* (Proceedings of Seminar Eight in the Series "Architectural Transformations in the Islamic World," 1983). The conference was attended by many of the individuals, both Yemeni and foreign, who had been involved in conservation and conservation policy in the country.

19 Draft Resolution 323, General Conference, Twenty-First Session, Belgrade 1980. Two goals of the Recommendations of the Third Session of the Advisory Committee on Arab Culture were cited:

the thirteenth plenary meeting, North Yemen's representative to UNESCO had called for an international campaign to save the country's heritage, which "we consider a unique and rich example of human heritage, that is the common property of mankind (*juz' asilan wa mukmilan lil turath al-insani alladhi huwa mulk lil bashariya jam'an*)."[20]

South Yemen had prepared a similar initiative earlier that year, and North Yemen must have been aware of it. Several months before the Belgrade conference, South Yemen had submitted a draft resolution for consideration, which called for the protection of "monuments and sites of historic, cultural, and natural value in Wadi Hadramaut and particularly the architectural heritage of the city of Shibam." The resolution notes that UNESCO had already supported a general survey of historic sites and monuments in the Wadi Hadhramaut in 1979–1980.[21] South Yemen called for an international safeguarding campaign at the Belgrade plenary meeting of October 2, referring to its draft resolution.[22] This was two days after North Yemen made its request for assistance. The General Assembly authorized preparatory studies for campaigns for the two cities.

The competition between the two countries would continue in the following years. Shibam was listed as a World Heritage site in 1982. Sanaa was not listed until 1986; the delay seems to have been on the North Yemeni side. Lewcock recalls, however, that not all UNESCO officials were persuaded of the "outstanding universal value" of Sanaa. In contrast to Shibam-Hadhramaut which had enchanted many at UNESCO, Sanaa was seen by some as simply "vernacular." One suspects that Sanaa may have been less appealing because of its highly ornamented architecture: in his 1974 film Pasolini called the beauty of Sanaa "unreal … almost excessive." A UNESCO film, produced in 1994, referred to the "flamboyant, harmonious, amusing architecture" of Sanaa that was reminiscent of the *Thousand and One Nights*, "like a child's dream come true."[23] In contrast, the restrained earthen tower houses of Shibam, with their planar surfaces and regular openings, were easily assimilated to the modernist aesthetic. Such culture-specific preferences would color the assessments and recommendations of consultants in the years to come—an often unacknowledged factor in the search for "outstanding universal value."

Preparations for the Campaign

Before an international safeguarding campaign can be launched, the state party must submit an application, based on a technical report.[24] It is also recommended that the state party prepare a nomination form for the World Heritage list. In 1980 North Yemen ratified the World Heritage Convention, so it was now a state party and could submit an application. In the same year, a

---

the "appreciation and respect for cultural identity"; and the "preservation and presentation of the cultural and natural heritage."

20 Proceedings of the 13th Plenary Meeting, 30 September, 1980, p. 675. North Yemen's representative was Muhammad al-Khadim al-Wajih, Minister of Education and President of the National Commission for UNESCO.

21 Draft Resolution 88, General Conference, Twenty-First Session, Belgrade 1980, submitted on July 16, 1980.

22 Proceedings of the 13th Plenary Meeting, 2 October, 1980, p. 972.

23 *Sanaa and Shibam, the Towns of Sand*, directed by Descamps, 3+ MDM Productions in cooperation with UNESCO; on file at the UNESCO Visual Archives, Place Fontenoy, Paris.

24 R. Lewcock, report on mission conducted 9–16 April, 1982, items 135–6; the terms of reference above are from the same report, p. 3.

presidential decree called for the preservation of the Yemeni architectural style and the use of traditional materials.[25] The nomination form was not submitted until 1985, after the Campaign had been announced, but all parties expected Sanaa to be added to the World Heritage list.[26]

After the General Assembly adopted North Yemen's resolution, UNESCO's General Director Mahtar M'bow traveled to Sanaa to meet with government authorities. He also organized preliminary missions by 'Abd al-'Aziz Daolatli, then president of ICOMOS, and by Ronald Lewcock.[27] In the spring of 1982, when Bouchenaki joined UNESCO as Desk Officer for Arab States, he recalls that M'bow was concerned: preparations for the Sanaa campaign were moving slowly. He sent Bouchenaki to Sanaa later that year to obtain a commitment from the government. Bouchenaki returned with a letter outlining a preliminary plan of action, approved by the authorities.[28] The next step was for ICOMOS to develop the plan of action with state authorities, along with a preliminary cost estimate; on the basis of this plan, the executive board of UNESCO would authorize the General Director to launch a campaign. ICOMOS organized a mission of five experts for this purpose, carried out in September 1982. This mission will be discussed in some detail below.

The support of North Yemeni policymakers was not self-evident: after years of political and financial crisis, there were more pressing matters at hand. After the revolution Yemenis across the political spectrum saw modernization as a panacea that would undo the effects of decades of isolation and oppression. Many in Sanaa embraced the modern lifestyle brought by the Egyptians. At the same time, the grand old houses of the walled city retained a certain status—and semiotic meaning—that had developed over centuries, in part due to their association with dominant social groups.[29] Many of the post-revolutionary political elite came from these groups: they had been born in the walled city, and still had relatives living in their family homes. Their sense of obligation to the walled city must have been as much personal as political. Yet the conditions there were an embarrassment, a sign of the country's backwardness. It may be for this reason that the walled core and al-Qa'a were the first parts of the city to receive a water supply system, even before UNESCO turned its interest to the city. The ruling elite was also aware of the economic potential of the new "cultural tourism" that was developing in the 1970s. Tourism development was seen as a potential source of revenue to supplement the declining agricultural sector, and by 1977 it was a stated policy of the government.[30] Decision-makers were gradually persuaded of the

25 Presidential decree no. 50 (A. 'Oshaish, "Toward a strategy for the preservation of the urban cultural heritage of Yemen and the experience of the preservation of the old city of Sanaa," Urban Cultural Heritage Project, 2002).

26 The nomination file was received by UNESCO in July 1985; "Essential description of the Historical City of Sanaa for the Purpose of Inclusion in the World Heritage List," Technical Office for the Preservation of the Old City of Sanaa (TOPOCS), no date.

27 A-A Daolatli, ICOMOS mission report, 1982, pp. 2–3. The dates of the missions were December 1981 and April 1982. Both Daolatli and Lewcock participated in the ICOMOS mission later that year which drafted a campaign strategy (see below).

28 In Sanaa Bouchenaki, formerly director of Antiquities in Algeria, reestablished contact with his colleague Qadi Isma'il al-Akwa', and was able to leave with a letter from Dr. Abd al-Karim al-Iryani, then Foreign Minister (M. Bouchenaki, interview).

29 See S. al-Suleihi, "The Lyrical Facades of Sanaa," unpublished Master's thesis, Department of Architecture, M.I.T., 1992.

30 J. Binous, ICOMOS mission report, 1982, p. 17. The first Five Year Plan, published in 1977, listed among its objectives the development of "the tourist industry by taking advantage of the country's climate and historic landmarks" (cited in Shankland Cox, 1978, 2.5). The responsible agency was the National Tourist Corporation. Cultural heritage was an important part of the tourism development

usefulness of UNESCO's interest—because it carried a certain prestige, affirming the value of Yemeni civilization before the world community; and because it carried a promise of funds to deal with living conditions in the old districts. "When we saw that they [foreigners] were interested," said a prominent official, somewhat shyly, "we became interested."

The lobbying efforts of resident foreigners had been important in persuading policymakers, and several embassies had promised to sponsor projects.[31] Two young Yemenis who worked at the Ministry of Municipalities—Ahmad al-Ibbi, an architect trained in Syria, and 'Ali 'Oshaish, an engineer trained in the U.S.—were also instrumental. They persuaded Muhammad al-Tayyib, the Deputy Minister, to secund them to work full-time on the preparation of the international campaign. This allowed work to begin on the ground, making the campaign a practical reality.[32] In 1984 they ran what was to become the Campaign office from the Ministry of Municipalities; after the official announcement of the campaign, the office got its own headquarters. In the early 1980s, Ahmad al-Ibbi had written his Master's thesis on the old city of Sanaa with a fellow student at the University of Damascus. They knew that Tunis and Islamic Cairo had just been listed as World Heritage sites, and were aware of the international safeguarding campaign for Fez. They believed that Sanaa deserved the same attention. "The Campaign was their idea," says Amal al-Shaiba, al-Ibbi's widow. "My husband was upset at the neglect of the old city of Sanaa … he thought that it should be world heritage (*turath 'alami*)."

The Politics of Conservation

Many in Sanaa associate the beginnings of conservation with Italy. The government of Italy had made urban planning and cultural heritage a focus of its foreign aid program. Between 1979 and 1985 some 300 Italian experts were sent to Sanaa in the fields of architecture, planning, engineering, and archeology.[33] At the 1983 Aga Khan Award Seminar in Sanaa, Qadi Isma'il al-Akwa' noted that Italy had been the first country "to express its willingness to participate in the campaign to save Sanaa."[34]

The Italian effort seems to have begun independently from that of UNESCO. Around 1981 the Department of Development Cooperation of the Italian Ministry of Foreign Affairs initiated a study for the conservation of the old city; its counterpart was the Department of Antiquities, which was also engaged in preparations for the UNESCO campaign.[35] The study was carried out by the partnership of Studio Quaroni and Bonifica

---

program envisioned by Binous in her mission report (pp. 23–30). Foreigners in the city had emphasized the tourist potential of Yemeni towns and villages since the early 1970s in an effort to encourage conservation (for example, Pasolini's reference to the old city as an economic asset).

31 In the late 1970s and early 1980s these resident foreigners included Paolo Costa; Rémy Audoin, archeologist and director of the French Institute for Yemeni Studies; Selma al-Radi, an Iraqi archeologist working on the restoration of the 'Amiriyya *madrasa* with the Department of Antiquities; Werner Ligenau, German attaché to the Department of Antiquities, and others.

32 R. Lewcock, interview.

33 See details in M. Yoshida, "A Concept on 'Conservation-Based Development' for Historic Cities."

34 "Opening Remarks," *Development and Urban Metamorphosis*, vol. 1, "Yemen at the Crossroads," p. xiv.

35 Studio Quaroni S.r.l./Bonifica S.p.A, "Action Proposal: Implementation of a Conservation Reinstatement Measure for a Sample Nucleus of the Historical Centre of Sanaa and the Establishment of a Restoration Workshop/School," January 1986, p. 19.

S.p.A.; it was one of several master plans for foreign cities executed by the partnership as part of Italy's program of foreign aid. Studio Quaroni was directed by Ludovico Quaroni, a master of modernism and a scholar of Arab-Islamic architecture, in partnership with civil engineer Adolfo DeCarlo (cousin of the architect Giancarlo DeCarlo). They conducted the study in three phases over the course of the 1980s. Work was carried out by three teams within Quaroni's office: architecture and conservation, led by Quaroni; urbanism, led by DeCarlo; and sociology, led by Corrado LaRocca.[36] Building on certain studies, including several that would become part of Serjeant and Lewcock's book, Studio Quaroni and Bonifica S.p.A. produced reports and drawings that would inform much of the subsequent work in the old city.[37] Echoing current thinking about historic centers in Italy, Quaroni and his team focused on the role of space as a constant in Islamic culture, in terms of the city's physical structure and symbolism. The morphology of the city, animated by the "spirit of Sanaa," should guide its conservation and development. In the words of Maria Luisa Neri, an architect and historian who worked on the team:

> Profoundly convinced of the inevitability of transformation, we believed that the more the city could reinvent itself and reaffirm its centrality with respect to its surroundings, the more sensible and constructive [the intervention] would be. Starting from the 'message of nobility' that Sana'a al-Qadima—the old center of the city—could spread to the outside world, we sought to protect it as a 'triggering factor' to activate changes.[38]

The team's proposals, then, tended to emphasize revitalization over conservation. The "authenticity" of the building elements should be protected, but balanced with services and infrastructure that would make the old city a desirable place to live. In some cases these elements would be retained and modernized: the tower house, for example, would be reconceived as a multi-family dwelling. In other cases, historic elements might be replaced with modern ones that were typologically similar to the old. Perhaps most controversial was the team's proposal to demolish the existing one-story shops of the *suq*—replacing them with a modern, two-story facility and underground parking (Figure 3.1).

All of these efforts were embroiled in politics. First, there was the competition between the departments of Antiquities and Municipalities. The 1978 Shankland Cox study, sponsored by UNESCO, had as its counterpart the Ministry of Public Works and Municipalities; the study had called for a conservation unit to be established in that ministry. Since then Municipalities had become a separate ministry. It saw itself as sponsor of the Campaign; this would be reinforced in the following years, as the ministry's employees

36 Maria Luisa Neri, personal communication. The preliminary or "Reconnaisance Phase" was conducted in 1981; the "Proposal Phase," in 1983; and the "Implementation Phase" in 1989–1990. The Proposal Phase consisted of an introduction and five volumes (PO-PIV) and included folios of large-format maps and drawings. Part I, "Internal and External Traffic Schemes and General Framework for Services," was led by DeCarlo; Part II, "The 'Border' of the Old City: the Wadi Sa'ila and the Walls," and Part III, "The Urban Texture," were led by Quaroni; and Part IV, "The Central Suq," was led by LaRocca. There are two sets of drawings for the "Implementation Phase": "Preliminary Design" (1989) and "Final Design" (1990); only some of the drawings, and none of the accompanying reports, are at CATS.

37 The authors of the Quaroni Bonifica study acknowledge their indebtedness in C. La Rocca, "Studio per la salvaguardia del centro storico di Sanaa," *Urbanistica*, vol. 92, 1988, p. 58.

38 M.L. Neri, "Occidente/Oriente. Un corto circuito per la salvaguardia di Sana," *Innumerevoli paesaggi: Scritti di storia economica per Roberta Morelli*, in D. Felisini, et al., Bari: Cacucci Editore, 2013, p. 166.

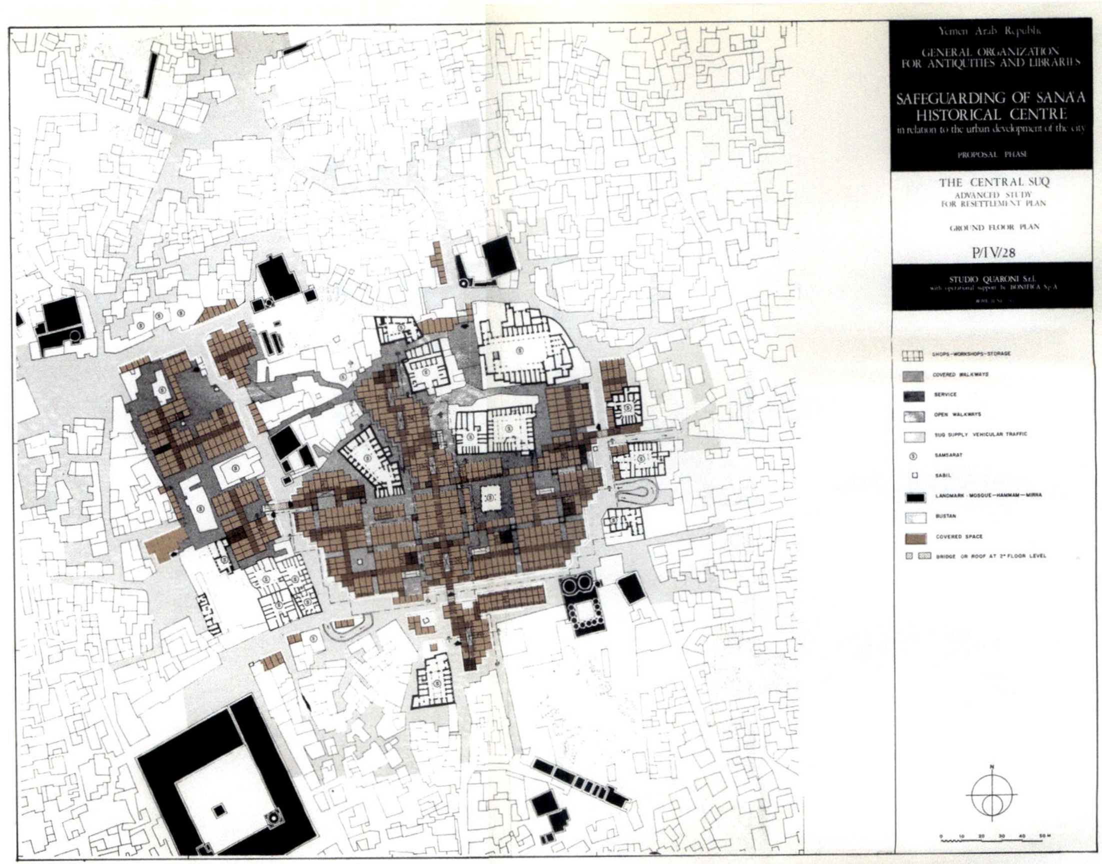

Figure 3.1 **Quaroni Bonifica's proposal for the *suq*, which would have demolished the existing shops**

Source: Studio Quaroni S.r.l./Bonifica S.p.A, "Safeguarding of Sanaa Historical Center in Relation to the Urban Development of the City," Proposal Phase, 1981; reprinted with permission of Maria Luisa Neri.

took charge of the Campaign office. There was also competition between UNESCO and Quaroni Bonifica. While Quaroni Bonfica was well-financed by the Italian government, UNESCO had no money—only moral authority. A high-level UNESCO official recalled that the Campaign was intended in part to counterbalance Quaroni Bonifica's influence. It was a way "to help Yemenis refuse [Quaroni Bonifica's] urban strategies and proposals." UNESCO perhaps feared that the firm would replicate Quaroni and DeCarlo's approach in Tunis, which was opposed by some professionals there—including 'Abdel-'Aziz Daolatli and Jelal 'Abdelkafi, both of whom were consultants for ICOMOS on the Sanaa Campaign.[39] According to Lewcock, UNESCO sought to bring the Italian project within the framework of the international campaign, and eventually succeeded. By the mid-1980s Italian cooperation was presented as part of campaign activities.[40] In 1986 North Yemen signed a second agreement with Italy: a pilot project for the conservation and upgrading of one of 30 "islands" or "action areas" in the city.[41] Both Quaroni and DeCarlo died in 1987, just as the Implementation Phase was starting up; direction of Studio Quaroni was assumed by Quaroni's wife.

In September 1982 ICOMOS sent a team of five experts to Sanaa to help develop the Campaign strategy. There they met with Dr. 'Abd al-Karim al-'Iryani, then Minister of Foreign Affairs. Al-'Iryani had been one of the architects of development and domestic and foreign policy since the 1970s; he became one of the main supporters of conservation among the political elite. Al-'Iryani told the ICOMOS mission that no official structure was yet in place to manage the Campaign. But a committee composed of representatives of concerned ministries had been formed to coordinate conservation work.[42] He assured the consultants that the installation of a sewage system in the old city was about to begin, with financing guaranteed by the World Bank. Al-'Iryani shared the mission's concern with infrastructure and especially street paving, emphasizing the importance of making the old city an attractive and desirable place to live. But what was expected of the international campaign, he said, was preserving the old city as it is: it should not be transformed into a modern city.[43] Here he may have been referring to other proposals under consideration, including the Quaroni Bonifica studies. At the time, two proposals were still on the table to transform the Wadi Sa'ila into a high-speed thoroughfare.

---

39 The differences between UNESCO-ICOMOS and the Italian team are clear in the mission reports, particularly the report of J. 'Abdelkafi who was asked to review the Italian project, then in the Proposal Phase. A founder of the Association pour la Sauvegarde de la Medina de Tunis, 'Abdelkafi had opposed the work of Quaroni and DeCarlo in Tunis.

40 See, for example, the UNESCO-UNDP promotional brochure and catalogue, which presents the Quaroni Bonifica studies as work accomplished under the Campaign: M.B. Lane, *Pilot Restoration Projects for the International Campaign to Safeguard the Old City of San'a*, General Organization for the Protection of the Historic Cities of Yemen, UNDP-UNESCO Report YEM/88/06, p. 18.

41 R. Lewcock, UNESCO mission report #4, 1987. The preliminary studies by Quaroni Bonifica divided the neighborhoods of the old city into "islands"; La Rocca sets the number at 50 ("Studio per la salvaguardia del centro storico di Sanaa," p. 59; and Quaroni Bonifica, "Safeguarding of Sanaa Historical Center," Reconaissance Phase, part III, "Sanaa al-Qadima: Approach to the Building Texture," p. 80).

42 This ministerial committee, chaired by the Minister of Housing and Municipalities, was created by Cabinet Decree no. 5; the first meeting of the committee was held in January 1983 (A. 'Oshaish, "Toward a strategy for the preservation of the urban cultural heritage of Yemen").

43 Minutes of meeting recorded by J. 'Abdelkafi and R. Lewcock, mission reports, September 1982.

The five consultants' reports from the ICOMOS mission treated a variety of issues: the restoration and development of monuments and sites; urban planning; regulations for development and protection; promotion of the crafts and tourism (which were, and would continue to be, considered jointly); and infrastructure, communications, and sanitation.[44] The conservation area was defined as the area within the walls of the old city as they existed at the time Von Wissman drew his map (1929), as well as the area of al-Qa'a. This contrasts with the plan that Quaroni Bonifica was preparing, which would define the "old city" as the area *east* of the Sa'ila: the quarters to the west of the Sa'ila, the authors would argue, had been compromised by modern development.[45] Such changes were seen to compromise the integrity and authenticity of the site, required for admission to the World Heritage List. Similar reasoning led the ICOMOS consultants to exclude Bir al-'Azab: the area had been "so intruded upon by modern buildings that it is difficult to include it in a conservation project."[46] As for al-Qa'a, consultants suggested that if funding were limited priority should be given to the walled city. In the end, al- Qa'a was not included in the conservation area: it was not included in the application for the World Heritage List submitted by North Yemen, and there is no mention of it in subsequent mission reports. There may have been other reasons for the exclusion of both al-Qa'a and Bir al-'Azab, since they were not included in North Yemen's application. Some observers note that they were associated with Jews and Turks, respectively, and so perhaps did not fit as easily into the new national narrative. In the mid-2000s, they were inventoried and declared as protected zones, but by this time they had been radically transformed by new construction.[47]

---

44 The consultants were 'Abdelaziz Daolatli (team leader); Ronald Lewcock; Jelal 'Abdelkafi; Jamila Binous; and, P. Helmore. See "Consultant Reports on the Missions Executed from 12 September–1 October 1982," ICOMOS.

45 "It is necessary to maintain the 'break' of the wadi, this constituting the natural border between the historic part of the city and the modern part, making it possible at the same time to enjoy the beauty of the houses overlooking the wadi from the wadi bed structure … Its value and potential reside in its capacity—increasingly endangered by total lack of control—of becoming one of the urban 'nodes' able to reconcile the two realities, the old and the modern, of Sanaa" (Quaroni Bonifica, "Safeguarding of Sanaa Historical Center," PII, pp. 2–4).

46 R. Lewcock, ICOMOS/UNESCO mission report no.1, 1982, p. 1. The walls around Bir al-'Azab and al-Qa'a had been largely destroyed by the mid-1970s (P. Costa, "Sana," in Serjeant, R.B., ed., *The Islamic City*, Paris: UNESCO, 1980).

47 In 2004 the Prime Minister declared that various areas belonging to the capital authority (*amanat al-'asima*) should be considered historic and protected: Bir al-'Azab, al-Rawdha, the villages of al-Qabil, al-Hadda, Sanaa, 'Attan, Bayt Baws. These areas should be studied and eventually proposed for inclusion in the World Heritage List (*qirar majlis al-wuzara'*, no number, 2004). Another proposal drafted by GOPHCY that year included al-Qa'a as part of the historic surroundings of old Sanaa. In 2006, as the inventory of the old city was underway, a government decree extended protection to Bir al-'Azab, Al Gabal, Hadda village, Sana (village), al-Rawdha, 'Attan, and Bayt Bows as heritage sites of national interest. As a result the inventory included these areas. Al-Qa'a was considered part of Bir al-'Azab and included in the inventory.

Changes in Bir al-'Azab stem mainly from commercial development that has occurred along Jamal Abd al-Nasser Street, which borders it on the south. Real estate pressures led to infill of the district and the gradual demolition of many old houses; existing roads have been widened and new ones created. Nevertheless, the neighborhood preserves a low-density character, with gardens and a few agricultural areas. Al-Qa'a has been less affected, largely because of poverty. Many houses have been reconstructed and open areas filled in, but in the late 2000s the character of the old fabric was more or less intact (D. Pini, et al., *Inventory of the Historic City of Sanaa*, pp 73–5, 83).

According to Lewock, there was a tension in the ICOMOS mission between two different views of conservation. He and 'Abdelkafi, both architects, argued that "like Venice, the value of Sanaa derives from the whole fabric, not just monuments. We insisted upon this." The whole city should be considered a monument, and also a living entity: in order to keep it alive, residents must be persuaded to stay. Infrastructure works and street paving had to urgently be carried out, to improve living conditions and also to prevent further damage to buildings. "We always said that individual monuments should be conserved," Lewcock told me. "But this should only happen after the problem of groundwater was addressed—otherwise the restored buildings would collapse." Other members of the team, Lewcock recalled, argued for a more conservative approach. Invoking the Venice Charter, they argued that it went against the prevailing practice of ICOMOS, whose mandate was the careful conservation of historic monuments. The alleged disagreement reflects the different interests of UNESCO and ICOMOS. Many UNESCO officials and consultants realize that conservation involves rehabilitation and revitalization; ICOMOS follows a conservative approach, focusing on building restoration. When asked about this many years later, however, Sa'id Zulficar was surprised. Daolatli had been director of the Association pour la Sauvegarde de la Médina de Tunis, a town-planning body that did roads and infrastructure as well as conservation. "He was used to this kind of work," said Zulficar.

Certain key individuals in the Yemeni government and at UNESCO supported the idea of conserving the entire historic city:

> [I]t would be dangerous to focus too much attention … on one section of the city or another, or on the significance of a few monuments at the expense of the whole. (Indeed, the Prime Minister has specifically indicated that the government does not favor such an approach).[48]

It was clear, however, that the cost of infrastructure work would have to be borne by North Yemen. UNESCO had approached European and Arab embassies in Yemen, asking them to participate in the Campaign by sponsoring projects. With a few exceptions, those that agreed preferred to sponsor high-profile restoration projects rather than infrastructure. The Campaign was thus conceived in two phases: the first would focus on infrastructure, including the upgrading of the water system, the installation of a sewer system, street paving, and the stabilization of building foundations; this phase was paid for largely by the Yemeni government. A second phase, to be launched after infrastructure was in place, would restore and rehabilitate key buildings in the city; funds for these projects would be sought from donor nations and agencies.

The approach of the Campaign—to treat historic centers as living entities, and to focus on infrastructure before restoration—set a precedent in Yemen: it established the role of conservation within a larger framework of development assistance.[49] But the campaign raised crucial issues that stem from different, and sometimes conflicting, approaches to conservation. The kind of conservation implemented in Sanaa—which deals not only with monuments, but with so-called minor architecture, urban elements, and even

48 R. Lewcock, mission report, April 1982, item 77. Lewcock and Bouchenaki recall that Dr. 'Abd al-Karim al-'Iryani, Qadi 'Ali Abu Rijal, and Qadi Isma'il al-Akwa' were key supporters.

49 These efforts have been sponsored by individual donor nations and by such agencies as the UNDP, the World Bank, and the Social Fund for Development. See M. Yoshida, "Conservation-Based Development of Historic Cities," pp. 143–70.

infrastructure—required different standards and methods, as well as new institutional, legal, and funding mechanisms. These issues were also at the heart of an on-going discussion within UNESCO, where they had been raised by the campaigns for Venice and Fez.

An Institutional Structure

In December 1984, a Board of Trustees chaired by the Prime Minister was created by presidential decree. It was composed of heads or representatives of various ministries—demonstrating that the safeguarding of the old city was a collective effort that cut across jurisdictions.[50] The Campaign to Save Old Sanaa was formally launched in the same month, two days before the Shibam Campaign. The appeal of UNESCO's General Director, recalls Bouchenaki, was timed to coincide with a meeting of Ministers of Culture of the Islamic World. In his speech, M'bow invoked the ancient history and unique architecture of the city:

> Sanaa, the prestigious capital of Arabia Felix [is] one of the oldest cities in the world. Legend holds, in fact, that it was founded by Ham [sic], son of Noah … Sanaa was a center from which the Islamic faith spread outward. Nevertheless, it continued to be a city of a variety of religious communities ….
>
> During that period Sanaa played an important role in the propagation of the Koran's message. A number of buildings that are still standing were erected during the lifetime of the Prophet Muhammad: one example is the Great Mosque, a jewel in the heart of the old city which was embellished and enriched under the Umayyad and Abbasid caliphs and the later Yemenite governors. In addition, Sanaa features pleasing vistas of caravansarais, *hammams*, *souks*, palaces, and dwellings surrounded by rich gardens, all harmoniously laid out with subtle ingenuity within the walls surrounding the city …
>
> But Sanaa is also is also a twentieth century capital, where the most up-to-date achievements of a contemporary metropolis stand side by side with traditional sanctuaries, *madrasas*, markets, and dwellings. It is the juxtaposition of these two worlds that gives the city its originality …

The character of the city is now threatened, said M'bow, by the migration of urban activities outside the ancient city. The authorities of Yemen have "infused Sanaa with new vitality, but are also determined to safeguard the historic, artistic, and architectural treasures of their venerable capital …." An emergency program had been designed not only to restore monuments and dwellings, but to upgrade infrastructure in the historic center. M'bow appealed for international solidarity in safeguarding "what can now be regarded as the heritage of all mankind." The plan of action designed by UNESCO and the YAR

50 Presidential decree #76, December, 1984. The Prime Minister at the time was 'Abd al-'Aziz 'Abdul Ghani. The Board consisted of Hussein al-Hobaishi, Legal Advisor to the President; the Mayor of Sanaa, then 'Abdul Wahid Jubari; Qadi Isma'il al-Akwa', Director of Antiquities; Qadi 'Ali Abu Rijal, then Vice President of the Presidential Council; and a number of prominent citizens including Sheikh Sinan Abu Luhum. For a list of members in 1987, see R. Lewcock, mission report no. 4, 1987, paragraph 15.

> is designed to safeguard the unique character of Sanaa, not only by preserving its historic centre but also by introducing those facilities it needs in order to adapt to the requirements of modern life … The financial and technical problems that must be overcome before these objectives can be attained are so massive and so complex that they call for a worldwide effort of solidarity.[51]

The formal launching of the campaign had been anxiously awaited in Sanaa, where work had already begun. The international campaign was intended to mobilize resources and technical assistance from other state parties to the World Heritage Convention; it would also make possible direct assistance from UNESCO in organization and coordination of the work.[52] After M'bow's announcement, the Campaign office was formalized as the Technical Office for the Preservation of the Old City of Sanaa (TOPOCS), responsible to the ministerial committee. The office moved from the Ministry of Municipalities to the Department of Antiquities; then, in the following year, it moved to Dar al-Jadid, one of the imam's palaces in the old city. The building had been made available to Antiquities for use as Campaign headquarters and had been restored with government funds.[53]

In addition to planning for the first projects, the staff prepared and submitted a nomination form for the World Heritage list at the end of 1985. Sanaa was added to the list in the following year, 1986, on the basis of its "exceptional value … and its extreme vulnerability." The criteria for the listing were iv, v, and vi, interpreted as follows:

> Criterion IV: Within its partially preserved wall [the old city of Sanaa] offers an outstanding example of a homogeneous architectural ensemble whose design and detail translate an organization of space characteristic of the early centuries of Islam which has been respected over time.
>
> Criterion V: The houses of Sanaa, which have become vulnerable as a result of contemporary social changes, are an outstanding example of a unique, traditional human settlement. Countless studies on the houses of Sanaa by urban specialists, architects and historians should not be a pretext for their destruction, even partial. The beauty of the urban landscape of Sanaa, whose overall appearance should remain intact, attests that they should be preserved integrally.
>
> Criterion VI: Sanaa is directly and tangibly associated with the history of the spread of Islam in the early years of the Hegira.

---

51 Amadou-Mahtar M'Bow, Sanaa, December 19, 1984.

52 In late 1983, Yemeni government officials were concerned that the campaign should be launched as soon as possible, so that UNESCO could provide oversight and coordination for work already in progress—that is, sewer installation and water system upgrading; the latter was being carried out on an apparently ad hoc basis by local cooperatives and citizens. The officials also hoped that when the campaign was launched, an expatriate Technical Director would be appointed to assist the Yemeni campaign director (R. Lewcock, mission report no. 2, 1984, p. 5). The expatriate post did not materialize as a full-time post, rather there was a UNESCO Coordinator (Lewcock) who undertook periodic missions.

53 The acquisition of a house in the old city for Campaign headquarters had been proposed in February 1982 by Werner Ligenau, German attaché with the Department of Antiquities. Dar al-Jadid was made available for the use of the Campaign in 1983; a budget was provided by the government for its restoration in 1985 (Lewcock, mission report no. 2, 1984, pp. 5, 23; mission report no. 4, 1987, para. 11; Yoshida, "Conservation-Based Development," p. 150).

The World Heritage Committee recommended that a "buffer zone" be established around the old city. It noted that municipal regulations for the safeguarding of the site had been drafted and adopted.[54] The latter point is incorrect: although guidelines were drafted and have been reissued at various points, they were not legally binding.

In 1986, a presidential decree made the technical office an autonomous entity, responsible directly to the Prime Minister. Renamed the Executive Office for the Preservation of the Old City of Sanaa (EOPOCS), it was placed under the direction of Dr. 'Abd al-Rahman al-Haddad, who had been the YAR's representative at UNESCO. Al-Ibbi and 'Oshaish, who had been running the Campaign office, served as Chief Architect and Technical Director respectively. The government allocated more than U.S. $1 million in 1986 to set up a budget for administration and pilot projects.[55] UNESCO organized a working group to follow and advise the campaign; the committee of foreign experts was headed by Lewcock, and representatives of the French, Italian, and German archeological missions in Sanaa were sometimes asked to assist.

The idea of making the Campaign office an autonomous body outside the ministerial system was proposed by UNESCO consultants. Elsewhere in the region similar efforts had been hampered when various ministries sought to control them; the only solution, recalls Lewcock, was the creation of an executive authority directly responsible to the Prime Minister.[56] It was expected that projects would be financed largely by foreign countries and international agencies; in a country so heavily dependent on foreign aid, projects like this were a means of survival. The General Organization of Antiquities and Libraries (GOAL) and Department of Municipalities were competing for control of the Campaign, and it seems that all parties were unhappy with the idea of an independent and powerful office. GOAL, in fact, was the official client for the Quaroni Bonifica's studies. But according to the late Qadi Isma'il al-Akwa', who was then director of GOAL, it was clear at the time that the scope of the Campaign was beyond his agency's capabilities.[57] The existence of a Board of Trustees composed of representatives from various ministries and authorities, as well as several influential citizens, was intended to promote conservation as an activity that cuts across jurisdictions and interests. But EOPOCS was perceived as infringing on the jurisdictions of other bodies—especially the Ministry of Municipalities, which had been home to the Campaign office. EOPOCS was also seen as infringing on the prerogatives of Antiquities, the ministries of Endowments (*Awqaf*), Public Works, and the newly created Capital Secretariat (*Amanat al-'Asima*, the Mayor's Office).[58]

---

54 UNESCO-WHC Documentation Unit Technical Evaluation, Yemen, no. C 385.

55 R. Lewcock, mission report no. 4, 1987, para. 11.

56 A similar conclusion was reached by the Quaroni Bonifica team and representatives of the Department of Municipalities and Housing and the Department of Antiquites, at a May 1984 meeting: "It was agreed that a project of such magnitude as the safeguarding of old Sanaa can only be successful if supervised by a council headed by no less a person than the prime minister himself" (F. Piepenberg, "Sanaa al-Qadeema: Challenges of Modernization" in A. Saqqaf, ed., *The Middle Eastern City*, p. 107). The competition between the ministries has been described to me by R. Lewcock, M. Bouchenaki, and others.

57 Al-Akwa' founded GOAL in 1969 and directed it until 1990 (R. Burrowes, *Historical Dictionary of Yemen*, pp. 25–6). The successor agency to GOAL is the General Organization for Antiquities, Monuments and Museums (GOAMM).

58 The Ministry of Public Works had been responsible for planning activities in Sanaa until the creation of the Ministry of Municipalities in 1979 (the two ministries merged for several years, then were separated in the early 1980s). In the 1990s, the Ministry of Construction, Housing, and Urban Planning was also working in the old city ("Preservation of Historic Cities in Yemen: Fact-Finding Mission Report," Association of Netherlands Municipality for the Ministry of Construction, Housing, and Urban Planning, April 1997, p. 11).

From the beginning there was redundancy of responsibilities and legal overlap, for example, between the conservation office and GOAL, which was responsible for the restoration of monuments. But the Ministry of Endowments claimed jurisdiction over mosques and other buildings in the historic core. GOAL's authority over monuments and other sites of historic value was based on the 1972 Antiquities Law, but securing that right in practice was a struggle.[59] There must have been concern early on that the work of these agencies was deficient. In 1987, Lewcock urged that work on monuments be approved by the Campaign office:

> It is recommended that the Governor of Sanaa issue a decree that al mosques and public buildings in the old city are national monuments and cannot be altered or repaired without the agreement of the Trustees of the National and International Campaign, represented by their Executive Office.[60]

In the end, EOPOCS was able to maintain control over projects involving the old city. But as its capacity and efficiency declined after 1990, other agencies became involved in the old city, especially the Mayor's Office. While GOPHCY was demoted to an agency within the Ministry of Culture in 1990, the Mayor's Office retained the special status that EOPOCS had in the 1980s, as an autonomous body responsible directly to the Prime Minister. Competition between GOPHCY and the Mayor's Office was acute in the early 1990s, sometimes reaching the point of open conflict, and hindering the implementation of projects.[61] The prerogative of granting building permits was especially contested—presumably because it involved lucrative kickbacks.

Conserving a Living Historic City: Strategy and Means

When veterans of the Campaign talk about the early years, one is left with the impression of a small, idealistic group of young individuals committed to saving a place they considered their patrimony. They remember the period—from the initiation of studies for the Campaign in 1983 to around 1988—as a kind of golden age. There was a high level of dynamism and cooperation, and commitment to a common goal. EOPOCS had no employees: rather all were secunded by ministries and offices represented on the Board of Trustees, and received excellent salaries. The architects, engineers, and other professionals at EOPOCS saw their mission as improving the conditions of the old city and at the same time, as inculcating new values—that is, encouraging the residents of Sanaa to appreciate their everyday environment as "heritage." EOPOCS staff faced many obstacles—practical, political, and social—for which, by their own admission, they were unprepared: none had practical training in the care of old buildings and cities, nor did they have a conceptual framework from which to proceed. Yet what they lacked in experience, they made up for in enthusiasm and energy. They came to understand conservation—in the unique form it was applied in Sanaa—through experience on the ground and through interaction with UNESCO consultants, especially Lewcock, whom they regard as a patron and mentor. In the mid-2000s, Campaign veterans still referred to Lewcock as the "godfather" of the old city.

59 This according to the late Selma al-Radi (personal communication).
60 Report no. 4 to UNESCO, 1987, p. 42.
61 See F. Mermier, "Sanaa, Métaphore de l'Etat Yemenite," p. 49.

As noted earlier, the strategy for the international Campaign developed by UNESCO consultants and EOPOCS staff aimed to achieve two interrelated goals: the improvement of living conditions so that residents would remain there; and the restoration and safeguarding of monuments. This strategy of "integrated conservation" was seen as especially suited to poor countries: because it brings together conservation and development; and because it reframes conservation in economic terms, especially as a means to promote tourist revenues.[62] Various tactics were developed to meet these goals, which were outlined in the initial and revised plans of action (1984 and 1987), and in the first five-year plan (1987). Conservation was to be applied at the scale of the urban fabric (houses, open spaces, streets, and gardens) and at the scale of individual buildings and monuments (mosques, caravansarays, the city walls and remaining gate, bathhouses, water fountains, and other utilitarian structures). The latter projects would be sponsored largely by grants and technical assistance from foreign donors and international agencies. In addition, between 30 and 100 private houses would be restored as demonstration projects. Several exemplary houses were to be restored with donor funds, but the repair and conservation of others would be financed through a loan fund.[63] The loan fund was part of a conservation strategy for the wider fabric, which would be upgraded largely with government funding and protected through a national conservation law, based on a master plan. The plan would outline building regulations—for new construction and alterations to existing buildings—and would also seek to preserve open spaces and urban gardens. The gardens were seen as fundamental elements of the old city and important to its environmental quality; they were severely compromised by neglect and intrusion by new construction.[64] The Campaign envisioned a variety of public and recreational uses for the gardens; gardeners, who inherited cultivation rights, would be organized as a corporation "under the authority of a new government body responsible for the upkeep of parks, gardens, and public open spaces."[65]

The improvement of living conditions would be achieved in three ways: installation and/or upgrading of infrastructure (street paving, water, sewer, as well as the burying of electrical and telephone lines); upgrading social and cultural services (especially sanitation, health clinics, and schools); and support and revitalization of the market and the general economy of the old city.[66] Traffic planning was also considered. While most felt that it should be limited in some way, some provision had to be made for vehicles in order to satisfy

62 In 1988 a UNESCO report defined "integrated conservation" as the integration of the objectives and methods of conservation and urban planning ("Project Findings and Recommendations," UNESCO-UNDP Restricted Technical Report, UNDP/YEM/88/006). See the paper by M. Welbank, one of the authors of the Shankland Cox study for Sanaa, outlining what was then perceived as the conflicting goals of development and conservation ("Conservation and Development," *Development and Urban Metamorphosis*, 1983, pp. 7–16).

63 A. 'Oshaish, interview. This issue will be taken up in Chapter 6.

64 In an edict of 1982 the Minister of Municipalities outlawed all new building and substantial alterations in the walled city. Building on the *maqshamas* was seen as particularly problematic, since they are such an "essential ... feature of the [old city's] character. This encroachment onto open land has already begun, as has the construction of concrete framed buildings, completely alien in character and scale."

65 B. Lane, *Pilot Restoration Projects*, pp. 52–4. A decade later, the GHK study proposed that gardens be reorganized as family cooperatives. But the authors recognized that the economic viability of the gardens was marginal; the gardens could be used for leisure, but "the social and cultural sensitivities of the local communities must be carefully considered" ("Cultural Heritage Report," vol. 2, pp. 27–9).

66 There were four schools in the old city in 1987, and two new ones were proposed ("Campaign Strategy, 1988–1997, p. 80).

residents and serve the market.[67] The Wadi Sa'ila—an important urban amenity—should remain accessible to cars, especially service traffic. But it should not become an alternate transit route, as earlier consultants had proposed: this would create development pressures along its banks, and definitively sever the two parts of the old city. (Upgrading of the Wadi became even more urgent as roads were paved, since the surfaces increased run-off and flooding; this affected housing foundations on the along the banks.)[68] Finally, the Campaign aimed to establish ongoing institutional support for EOPOCS and to mobilize national awareness and support. The latter aim would be met by establishing a division of Public Relations within EOPOCS in 1986. Other departments were Technical Cooperation and Studies; Projects; Planning; and Inspections.

It should be noted that UNESCO's role in the campaigns for Sanaa and other sites was that of a coordinating rather than a financing body. (This distinction was not initially clear to local policymakers, who hoped UNESCO would provide funds.) UNESCO assisted in the coordination of the Campaign and donor financing; its direct contributions were limited to short-term technical assistance, training, and equipment. Training was restricted to short-term missions and seminars, and equipment provided was minimal.[69] The success of the Campaign, then, depended on securing contributions from donors. The announcement of an international campaign was intended to throw the moral force of UNESCO and the World Heritage Committee behind certain sites that were deemed, in M'bow's words, the "common heritage of mankind"; this moral appeal would in turn attract international assistance.

The "state party" to the Convention—in this case, the government of North Yemen—was expected to demonstrate its commitment to conservation. Despite official sanction and support for a conservation office, government funding for projects was slow to materialize; it did not seem to be a high priority of policymakers, who were struggling with more pressing problems. In 1986 the German Ambassador assured EOPOCS staff that if they could convince their government to start the first project, donors would follow.[70] And in fact, as soon as the government committed funds for the first "demonstration project," several embassies promised funds for others: the restoration of several *samsara*s, a portion of street paving, and the reconstruction of the health clinic at Suq al-Baqar.

Despite consultants' emphasis on street improvement, it was not a foregone conclusion that paving and infrastructure upgrading should be the first project. But after considering several alternatives, it was decided that a streets project would have the greatest impact—producing immediate results and gaining the sympathy of residents for conservation.[71] The pilot area was al-Qasimi Street from the Sa'ila to al-Abhar Square: it

67 Various solutions to traffic management were proposed by UNESCO missions, restricting the volume and/or hours of traffic in certain areas of the old city. Numerous studies would be conducted over the years; a traffic plan was finally implemented in the spring of 2004.

68 Yoshida, "Conservation-Based Development," p. 160; M.B. Lane, *Pilot Restoration Projects*, p. 30. The progress of paving may explain why, as Yoshida notes, the improvement of the Wadi Sa'ila became a priority after 1988.

69 See "Project Findings and Recommendations," UNDP, 1996.

70 The ambassador at the time was Dr. Heinrich Reiners; he initiated the German contribution to the Campaign, the restoration of Samsarat al-Mansura (M. Petzet, "The Restoration of Samsarat al-Mansura as an Example of International Cooperation to Safeguard the Old City of Sanaa" in M. Petzet and W. Koenigs, eds, *Sanaa: the Restoration of Samsarat al-Mansurah*, Munich: Bayerisches Landesamt fur Denkmalplege, 1995, p. 9).

71 Among other candidates for demonstration projects were the restoration of part of the city wall and the restoration of an important building. But the former was sensitive in terms of conservation standards, and very visible; the latter would have limited impact.

was a visible area and a tourist route, and lined with important buildings as well as several urban gardens. A second area, to be implemented in the following year, would link the monumental city gate, Bab al-Yaman, to al-Abhar via the Great Mosque; an additional stretch was added to link Dar al-Jadid, the conservation office, to al-Qasimi Street. The project involved the installation of sewer lines, which had not yet been installed in this area and the upgrading of water pipes and connections.[72] At the same time, streets were regraded and manholes installed. Careful thought went into the selection of paving stones: a black stone with a round profile, commonly found in other highland towns, was chosen to discourage motorcycle traffic.[73] Paving was important not only to improve street quality, but also as an integral aspect of drainage; although storm drains were not installed, it was hoped that the sloped profiles of the streets would carry water to the Sa'ila.[74]

Consultants had repeatedly cautioned that the road work had to be carried out very carefully: the streets were narrow and the sewer pipes were wide; moreover, the foundations of the houses were very shallow, and the soil was already water-laden. Because of the risks, EOPOCS staff could not find local contractors willing to do the work. They finally found a contractor outside the city, who only agreed to the job at a very high price.[75] They themselves were on site around the clock, working as technical advisors, surveyors, and guards with the contractor and carefully monitoring excavations. Despite this care and attention, and efforts to explain the work to residents, the latter were suspicious and even hostile. It was only after there was visible progress, around 1989, that they started to accept the process.[76]

The Campaign planned to pave and upgrade infrastructure at only the main streets; connections to utilities and sewers would be made area by area. By the late 1990s, approximately 85 percent of the streets and squares of the old city had been paved, largely funded by the government (figures 3.2, 3.3).[77] The remaining areas have since been completed. Despite the success of the paving and infrastructure effort, it has not been

72 The water system was upgraded by installing a higher grade of pipe, which however still fell short of the specifications recommended by consultants. The recommended galvanized iron pipes were apparently not affordable; the project installed pipes in such a way that they could be easily removed and replaced later (Lewcock, report no. 4, 1987). The sewer system was projected to be complete by 1988.

73 R. Lewcock, interview. The choice of paving has been the subject of much contention. Most agreed that the rounded stones did not discourage motorcycle traffic, but simply increased vibrations which affected near-by buildings (for example, 'Adil al-Hamadi's technical report dated 5/19/88 on file in GOPHCY's Planning Department). In the mid-2000s the stones were being replaced by large, flat stones, textured using electric tools to simulate hand carving.

74 P. Helmore, "Water Supply, Storm Water, and Drainage," ICOMOS mission report, 1982, p. 68; former EOPOCS official, interview.

75 The contractor was described by a former official as a tribesman and "adventurer." The vibration of earth-moving equipment impacted foundations, producing cracks in the walls; sewer lines were laid at a depth of 40–50 centimeters, and in some cases foundations were only 60 cm. deep (former EOPOCS official, interview).

76 Amal al-Shaiba, interview. She used to visit the site of the first project at al-Qasimi Street with her husband, al-Ibbi. Residents would accuse them of destroying the streets; they were even fired upon when the news spread that concrete and steel were no longer allowed in the old city. Al-Ibbi used to talk with residents and try to reason with them.

77 GOPHCY promotional brochure, color, trifold format, Arabic; produced between 1997 and 2002.

without problems—especially due to the lack of coordination between work in different areas of the city.[78]

"People were suspicious of the Campaign," recalls Muhammad al-Saʿad, an urban geographer at the University of Sanaa, who was born and raised in the old city and lived there in 2004. "They thought it would change the city against their wishes." Some officials describe this suspicion as a "lack of awareness" or, less generously, as "ignorance." Although one former EOPOCS official says that he and his colleagues met with community leaders—*sheikh*s and *'aqil*s—in the lead-up to the Campaign—it seems that a concerted effort to involve the community never took place.[79] As coordinator of the Campaigns in both Sanaa and Shibam-Hadhramaut, Lewcock was struck by the contrast between the culture of the two cities. In the latter, community meetings were encouraged from the beginning, an expected part of the process—perhaps due to the prevailing socialist system. In Sanaa, however, decision-makers frowned on participatory meetings. Some architects at the conservation office—themselves from Ta'izz and Aden—continue to be struck by the different political cultures of Upper and Lower Yemen. They recall that EOPOCS leadership reacted negatively to their attempts to consult the community—an attitude which they feel hurt the conservation effort. This relates to the wider issue of the changing demography of the city, which will be discussed in the following chapter: while the built fabric represents an urbane and cultured past, few of the old families remain, having been replaced in great part by rural migrants. For local elites, and for many in the traditional middle class (community leaders, merchants, artisans, and others), the new arrivals have little community spirit, and compromise an old "urban identity."[80]

Funding Problems

During the first demonstration project, studies were underway for subsequent projects, including the restoration of several *samsara*s. Around this time, Lewcock began to withdraw from the Sanaa Campaign because of other professional responsibilities, especially his appointment as Director of the Aga Khan Program in Islamic Architecture at M.I.T. But he continued to attend working group meetings, and to undertake occasional missions for UNESCO. He had worked hard to persuade government officials to focus on infrastructure and municipal services. He called for a two-year delay on the conservation of buildings, so that all available resources could be directed toward infrastructure. "Restoration would really come too early at this stage," wrote Lewcock in June 1986, in his diary. "National campaign should state that it is not with restoration that one would normally begin such campaigns."

78 "Preservation of Historic Cities in Yemen," April 1997, p. 5. This Dutch report, apparently based on interviews with GOPHCY officials, also cites other problems, including engineers' and builders' lack of knowledge about underground conditions; and substandard work that resulted in damage to foundations, leaks in the sewage system, and insufficient grading for drainage. Some areas that were paved in the 1980s and '90s had to later be excavated and upgraded to new specifications. According to several architects I interviewed, lack of coordination between areas and phases of work and changes to drawings that had to be made on site have meant that the new systems remain in the memories of contractors; the absence of documentation is problematic for management and trouble-shooting.

79 In the mid-2000s Stadnicki noted that this was still the case; see "Vers un aménagement participative à Sanaa?"

80 R. Stadnicki, "Vers un aménagement participative à Sanaa?", p. 4.

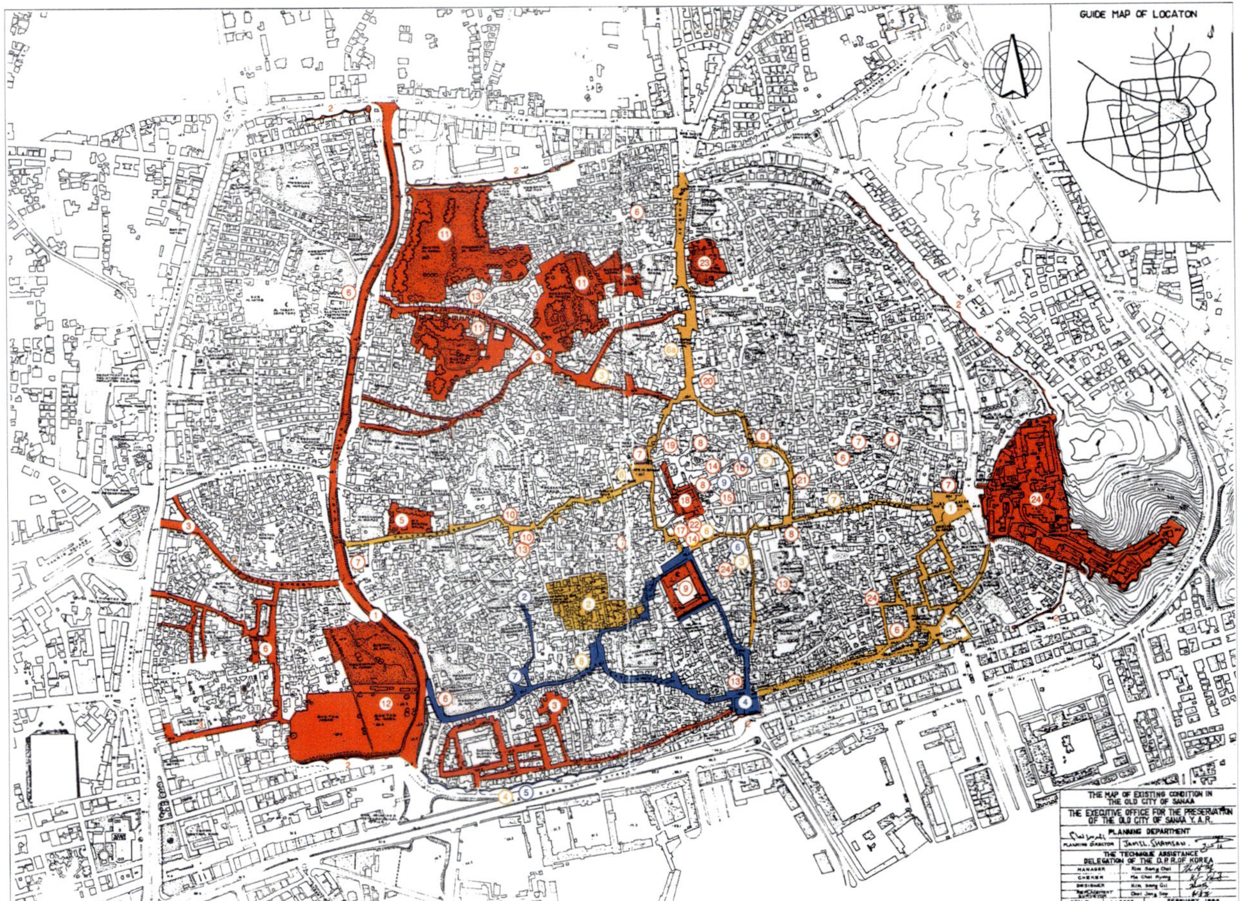

Figure 3.2 **Phases of street paving in the old city**

Source: M.B. Lane, *Pilot Restoration Projects for the International Campaign to Safeguard the Old City of San'a* © Aga Khan Award for Architecture/Courtesy of Architect.

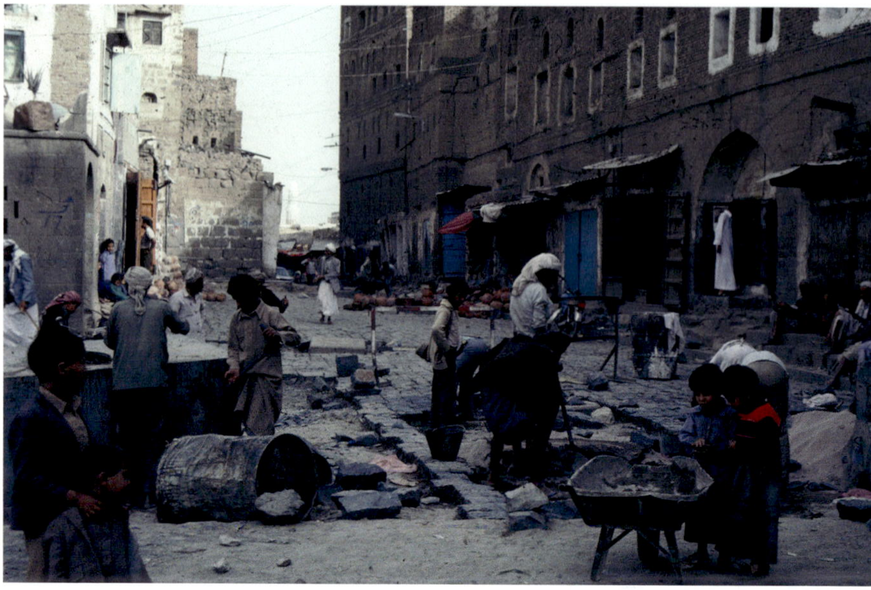

Figure 3.3 Paving and infrastructure work in the old city in the mid-1980s
Source: Photograph by Ronald Lewcock.

His approach was not accepted by most donor nations and policymakers, who were eager for show-piece projects. A certain momentum had been achieved: by late 1986 the conservation office had grown to 40 employees with additional expansion envisioned. In that year the office formally requested financial and technical assistance from a number of countries, and assistance was volunteered by several others.[81] Although support was sought from the oil-rich Arab states, funding came only from countries outside the region.[82] The YAR budget for conservation increased each year, but it did not keep pace with government contributions envisioned by UNESCO. In 1987 the budget for the vast scope of conservation activities was estimated at U.S. $201 million; about 40 percent would be contributed by the YAR; 10 percent would be covered by grants from donor nations, and the remainder by international loans.[83] But in an interview with a reporter for the Arab News Service,

81 For an overview of key initiatives and donors see appendix, and M. Yoshida, "A Concept on 'Conservation-Based Development' for Historic Cities," 1994.

82 Early on several Gulf countries pledged to support the Campaign, but by 1987 support had not materialized (D. Matar, "U.N. Plan to Save Old Sanaa Threatened," appended to Lewcock report no. 4, 1987; no date or citation). At a fund-raising meeting at UNESCO Headquarters in June 1989 participants wrote: "In order to ensure the active participation of arab (sic) countries in the financing of the Campaign which has, until now, been supported exclusively by countries outside the region, it is of paramount importance to set up a high-level mission … to a number of selected countries in Arab World, such as Kuwait, Libya, Qatar, Saudi Arabia and the United Arab Emirates" (on file at UNDP/ Sanaa). To my knowledge there was no support from within the region until the late 1990s, when some funding for conservation projects was provided to the SFD by the Arab Monetary Fund.

83 Lewcock, mission report, 1987, paragraphs 54–84.

EOPOCS officials estimated the budget to be higher, in the range of U.S. $1 billion. The article reported that the Campaign was threatened by lack of funds and the ongoing exodus of residents from the city. Little progress had been made since the Campaign had been launched in 1984, and so far there had been only symbolic financial contributions, mainly from Italy.[84]

As we saw in Chapter 1, conservation of historic areas in other cities in the region had been carried out by agencies like the UNDP or the UNOTC, in collaboration with UNESCO. Because Sanaa was framed as an International Safeguarding Campaign it was dependent on special contributions, which quickly proved to be insufficient. On the advice of both Bouchenaki and Lewcock, government officials sought funding from the UNDP, which had expressed interest in the Campaign. An "umbrella project" to provide support for the Campaign was proposed in July 1986, financed by the UNDP and executed by UNESCO. In May 1988 the UNDP approved a small project to promote the Campaign and provide institutional and technical support to EOPOCS; a second project, in cooperation with the ILO, would provide technical assistance and equipment for the National Handicraft Center.[85] The projects sponsored short-term consultancies, which were deemed to be more cost-effective than long-term missions. Although the use of United Nations volunteers and associate experts was proposed as a means to achieve continuity, only one associate expert was sent on a long-term mission, financed by the government of Italy.[86]

Around this time, the strategy of the international safeguarding campaign was reappraised at UNESCO: there was concern that there were too many campaigns for the organization to handle.[87] A team was sent to Sanaa to advise EOPOCS on new policies. A new action plan for the years 1988–1997, based on previous studies and reports, was developed by EOPOCS in conjunction with UNESCO; this plan continued to guide

84 Officials told Matar that Gulf countries had not yet acted on pledges to support the project.

85 "Draft Terminal Report/Restricted, UNDP/YEM/006," p. 1; and "Umbrella Project—Preservation of the Old City of Sanaa," UNDP, revised 27/4/87 (on file at UNDP/Sanaa). The two projects were identified as UNDP/YEM/88/006 (for promotion activities and institutional support); and UNDP/YEM/87/008, in cooperation with UNESCO and the ILO (for handicrafts; Abdo Seif, Senior Program Office, UNDP/Sanaa, interview).

86 Some 24 expert missions to Sanaa were carried out between 1988 and 1990. The missions included technical assistance for craft development at the newly restored Samsarat al-Nahhas (to be discussed in Chapter 6); as well as architectural studies for the central suq and Samsarat al-Baw'ani (by G. Barbato, 1988–1989) and for Samsarat Muhammad bin Hasan. The project also paid for the production of a promotional brochure (by B. Lane); for the production of a film; and for an exhibition of projects at UNESCO Headquarters (later installed in Sanaa, during the Regional Seminar on the Safeguarding of the Old Cities in the Arab World, 12/89; "Draft Terminal Report/Restricted, UNDP/YEM/006," p. 6). For the 34 UNESCO and UNDP missions carried out between 1985 and 1991, see "Preservation of the Old City of Sanaa: Project Findings and Recommendations," Annex A, 1996.

The UNDP project's concern with "integrated conservation" led to an international symposium to address various relevant issues which had not received sufficient attention: the need for socio-economic studies of the old city and research for a legal framework, especially regarding property rights. The symposium was held in 1991 and its recommendations were to be translated into a new action plan, but as of 1997, none had been formulated ("Preservation of Historic Cities in Yemen," April 1997, p. 9). This was partly due to the events of the intervening years, especially the restructuring of the conservation office, to be discussed below.

87 See above, Chapter 1. The situation must have been compounded by the financial crisis at UNESCO that resulted from the withdrawal of the U.S. in 1984 (D. Matar, "U.N. Plan to Save Old Sanaa Threatened").

conservation activities in the old city in the early 2000s.[88] It included cost estimates for some 20 priority projects, to be presented to national, bilateral, and international donors. The project proposals were incorporated into a detailed fund-raising brochure, which outlined work accomplished to date and projects in search of sponsors.[89]

By 1990, on the eve of unification, approximately 44,700 square meters of paving and infrastructure upgrading had been completed or were in process. Following the pilot project for al-Qasimi Street, the areas paved were al-Abhar Square to the Great Mosque and the conservation office; the area of Bab al-Yaman; the Great Mosque to Midan al-Luqayh (financed by the Netherlands); and the quarter of al-Nusayr (financed in part by Germany).[90] Several studies for garbage collection in the old city had also been undertaken.[91] The Ibn al-ʻAmir Elementary School—a new, concrete-frame structure with brick facade that incorporated local details—was substantially complete in 1990. Construction drawings for Suq al-Baqar clinic, including a building not in the earlier German project, were drawn by the North Korean team, but the clinic was not built until the mid-1990s after much debate and redesign. With the exception of two *samsara*s and a portion of the city wall, most architectural restoration work was in the schematic phase. The restoration of Samsarat al-Nahhas was substantially complete at the end of 1989, and the restoration of Samsarat al-Mansura was in progress. Funding for the craft training program to be housed in Samsarat al-Nahhas was provided by joint UNDP-ILO funding, which also funded a survey of the current state of crafts throughout Yemen. The Italian project for a model neighborhood was inaugurated with much fanfare in 1989, attended by hosts of Italian dignitaries and celebrities.[92] Several studies funded under the UNDP project were complete or in process;[93] the Swiss had begun studies for the restoration of Bayt al-ʻAnbasa and Bayt al-ʻAmri. The French conducted studies for the "restoration" of the southern wall, although much of it had been destroyed. By 1991 the wall had been substantially rebuilt, against the protests of some Yemeni and foreign professionals. They pointed out that the earthen (*zabur*) technique

---

88 A. ʻOshaish, "Toward a Strategy for the Preservation of the Urban Cultural Heritage of Yemen."

89 "Draft Terminal Report/Restricted, UNDP/YEM/006," p. 4; "Campaign Strategy and Action Plan, 1988–1997," prepared by EOPOCS and the assistance of the Cultural Heritage Division of UNESCO, August 1988. B. Lane was sent to develop project proposals with EOPOCS in 1988–1989. The brochure, *Pilot Restoration Projects*, was designed by B. Lane and M. Bouchenaki and paid for by the UNDP-UNESCO project YEM/88/006. Only some of the projects outlined in this brochure have been realized.

90 List compiled by GOPHCY, no date, on file in the Projects Department. According to Lewcock's notes from a 1986 UNESCO working meeting, the YAR signed a contract with China for paving from Bab al-Yaman to the Great Mosque (handwritten notes, dated June 10, 1986). But according to a list of paving projects at GOPHCY's Projects Department, this work was financed by the YAR. The list mentions some foreign funding for paving of the area from the Great Mosque to Maydan al-Luqayh, and the quarter of al-Nusayr. These projects were financed by Netherlands and Germany respectively; see M. Yoshida, "A Concept on 'Conservation-Based Development' for Historic Cities," p. 25, fn. 11.

91 W. Huisman, P. Verbuck, W. Vrins, "San'a Old City Garbage Collection and Disposal: an Evaluation Study," February 1989; this built upon earlier studies, including one by Peace Corps volunteers.

92 These *samsara*s and the Italian project will be discussed in Chapter 6. The Embassy of the Netherlands also rented and restored a private house in the old city, Bayt Sunaydar, intended for use as a commercial center for Dutch products and as a guest house.

93 For Samsarat al-Bawʻani, by G. Barbato, and Samsarat Muhammad bin Hasan (commonly called Bayt al-Mal), by Barry Lane and local architects, both conducted in 1988–1989.

used in reconstruction, by masons from the northern town of Sa'ada, was different than that historically used in Sanaa.[94] More important, remains of the ancient stone foundations were demolished during restoration.

Two Views of Conservation

Because of the dual focus of the Campaign—urban upgrading and building restoration—it is remembered in different ways. Residents experienced conservation largely in terms of the street paving, water, and sewer works, which transformed life in the old city. For them, conservation was not simply an aesthetic exercise, but a process that improved their lives and validated their environment (Figure 3.4). Yet many were disillusioned about the promises that accompanied conservation and World Heritage designation: in particular, the absence of funds and programs to conserve, or even maintain, the thousands of houses that make up the city fabric. Some professionals lamented that a conservation plan, which would have provided the basis for a law and certain institutional mechanisms, was never carried out. From 1985 Lewcock's reports mention a forthcoming conservation plan and the need for regulatory measures; and this is echoed in the revised Campaign Strategy for 1988–1997. According to Lewcock, a conservation plan was deliberately delayed in the beginning, because UNESCO agreed that emphasis should remain on infrastructure. This was in part to avoid mistakes of the Fez campaign, where many studies were conducted but there was little work on the ground. Regulatory measures were proposed, Lewcock recalls, but they were blocked by the Cabinet because of opposition from the Mayor's Office. Two decades later, an inventory of the Old City and other historic areas was finally initiated; it was completed in 2008. The inventory was to be followed by a conservation plan, which would deal with environmental conditions, socio-economic factors, issues of ownership, building regulations, and so forth. But the plan has been delayed, as a result of recent political turmoil.

Others identify the Campaign primarily with the restoration and reuse projects which, they feel, were driven largely by the interests of donor nations and agencies rather than local needs. "UNESCO's strategy was to compile a list of projects to sell to donors," said an Italian architect involved in the Campaign. "The donors left, disappointed, because there were no institutions to carry the work forward." The problem was two-fold: on the one-hand, there was a conceptual problem on the part of donor nations; on the other, local capacity was lacking. Campaign funds would have been better spent, he believes, in developing a conservation plan and a technical office that would provide meaningful assistance to owners. Franck Mermier, an anthropologist who has written extensively on the *suq* and who conducted studies for the Campaign, described the situation in the mid-1990s:

> The minimal coordination that takes place between the donor nations and GOPHCY is not inscribed in a comprehensive plan based on urbanistic, social, and economic criteria …

94 In the *zabur* technique from Sa'ada, clay is mixed with chopped straw, which gives the surface a shiny, textured finish. In Sanaa, the walls had been constructed of clay with a fine aggregate of pebbles—a technique still visible in extent fragments of the defensive walls, and other walls throughout the city. In addition, there was concern that the new mud casement would be structurally incompatible with the old remains over which it was laid. See T. Marchand, "Walling Old Sanaa: Reevaluating the resurrection of the city walls," *Terra 2000: 8th International Conference on the study and conservation of earthen architecture*, London: James and James, 2000, pp. 46–9. For a photograph of the walls before reconstruction, see P. Bonnenfant, *Sanaa*, plate 1.

> Only an external vision, marked by an often reductive aestheticism, seems to prevail notably among the foreign experts, usually architects, involved in these projects. The perception of the old city is sustained by the nostalgia for a golden age, a lost harmony, coupled with the specter of an urban archipelago isolated from the rest of the city. This ideology of heritage, that conceives the *intramuros* area as an open-air museum and its inhabitants as the last representatives of an endangered species, has colored the elaboration of most of the projects undertaken until now.[95]

The *samsaras*, for example, were to be restored and used for purposes that were clearly oriented toward elite or tourist audiences; a UNDP project aimed at reviving crafts appears to have been designed without consulting local practitioners, despite the recommendations of a Yemeni consultant.

Lewcock saw building restoration as a necessary part of the Campaign, but it had to come after infrastructure work. The shift to monuments restoration came too early, he says, perhaps because of pressure from embassies and ICOMOS. "EOPOCS should have told donors, 'We'll decide how to use the funds.'" If there was an "external vision" that guided donor interest and funding, it was not challenged by EOPOCS administration. Some who worked with the Campaign agree that aspects of programming were driven by foreign ideas that had little to do with the needs of the local community. "We didn't have experience, and we were trying to filter ideas from so many studies by foreign consultants," recalled a Yemeni architect. "Our approach did not account sufficiently for local, community functions that couldn't be integrated."

But in a sense the new functions, oriented to an elite and foreign audience, were attractive to local professionals. "We wanted to make the old city more presentable," says an architect who worked on Samsarat al-Mansura, which would become an artists' workshop and gallery. Different views of culture were at work here. While serving as the YAR's Permanent Delegate to UNESCO, 'Abd al-Rahman al-Haddad (later director of EOPOCS) wrote a treatise on national cultural policy, one of several such studies commissioned by UNESCO in those years. In the book, he describes the cultural revival envisioned by the revolution—which included both an elite Arab culture long suppressed by the imams, and a popular culture that expressed the national spirit.[96] The notion of a modern, elite culture—linked to Nasserism and pan-Arabism—informed one approach to the old city. As in Europe, the preserved old town—an historical relic—would become a symbol of a modern national identity, and a stage for events that express that identity. But in Sanaa there were two *kinds* of culture: an elite, Arab, forward-looking culture; and a popular or traditional culture. The latter would be difficult to integrate into the modern society envisioned by the revolution; but it played an important, even necessary, role in expressing the "national spirit."

For some, the most problematic aspect of donor aid was the tendency to entrust projects to foreign expertise. Few projects provided for the training and strengthening a local cadre of professionals, despite repeated recommendations from Lewcock. Nor was there much effort to involve the local building trades; there were calls for a "revival" of

95 F. Mermier, "Sanaa, métaphore de l'état yemenite," in Grandguillaume, ed. *Sanaa Hors les Murs*, p. 50 (translation mine). Yoshida echoes these sentiments: "Each donor country participates in preservation activities in its own capacity and works independently without attempting to cooperate closely with other countries."

96 *Cultural Policy in the Yemen Arab Republic*, Paris: UNESCO, 1982.

apprenticeship, even as the Campaign relied heavily on the expertise of master builders or *ustas*. Moreover, top positions at EOPOCS were often filled by political appointees who did not necessarily have expertise in relevant fields. These officials tended to leave design and implementation of projects to donor countries' architects and engineers. Most of the Yemenis in the conservation office were recent graduates of architectural and engineering programs, yet they were appointed as directors of departments. They supervised the work of foreigners, both resident and non-resident, but most did not do a great deal of design work themselves. Thus, with a few exceptions, they had little opportunity to acquire practical expertise.[97] Yet donor countries and international agencies were not always to blame for the dearth of investment in human capital: the Yemeni government did not make training a requirement of donor aid.

The role and personality of the director, al-Haddad, contributed to the problems of the conservation office, but opinions on this issue are sharply divided. A scholar with international connections (he served as a security officer in France for the YAR, and then as the government's representative to UNESCO), al-Haddad was a political appointee. He is generally remembered as an energetic and uncorrupt administrator, and a gifted politician who knew how to get things done. "Al-Haddad understood how things work in Yemen," recalled one professional—that is, through the influence of personality. Yet over time, he tended to concentrate power in his own hands—to the chagrin of the young men who had founded the office.[98] An internal struggle developed and by the early 1990s, had turned into open conflict: several staff members resigned, while others were transferred to other cities. According to one high-level official, al-Haddad "blocked the development of the [conservation] office." The continued efforts of the international community, he said, were the only thing that made up for the deficiency of the office and kept the Campaign going. According to another view, the dissidents in the office acted in a proprietary fashion and were not willing to play as a team.

The immediate issues at stake had to do with the organization of the office, particularly the decision-making process and the delegation of authority; but at a deeper level, they had to do with the mission and philosophy of the office—or lack thereof. The conservation office had *de facto* authority for the old city, but it was not backed up by law: unlike Fez, the old town of Sanaa lacked a distinct status within the municipality. Some local professionals and foreign consultants felt that EOPOCS should focus on documentation and technical assistance, and leave enforcement to other agencies. But the formation of a technical staff, with the required training, documentary base, and equipment, does not seem to have been

97 On these points see Yoshida, "Conservation-Based Development," pp. 157–9. The plethora of drawings by the North Korean team—which staffed EOPOCS' Planning Department—suggests that they produced construction documents for virtually all of the streets and infrastructure, and for many other projects including several sponsored by other countries. They also produced a detailed map of the old city (1989) which was updated and rectified on site and through aerial photography for the recent survey of the old city. During the Campaign, work by local professionals seems to have been limited to surveys of existing conditions, on-site supervision, and in some cases, the redesign of projects to conform to existing context (like Suq al-Baqar clinic). There were a few exceptions. At Samsarat al-Nahhas, Norwegian conservation specialists were brought in only after the project was well underway; and circa 1988 the Netherlands began to provide two or three scholarships per year to the Institute of Housing Studies (IHS) in Rotterdam for intensive graduate courses and accelerated one-year Master's degrees, with concentrations in conservation and urban management. This program however, required excellent English, so many talented candidates were disqualified.

98 According to one foreign consultant, the fact that the office was being run by energetic young men in their twenties was "a thorn in the side of the establishment."

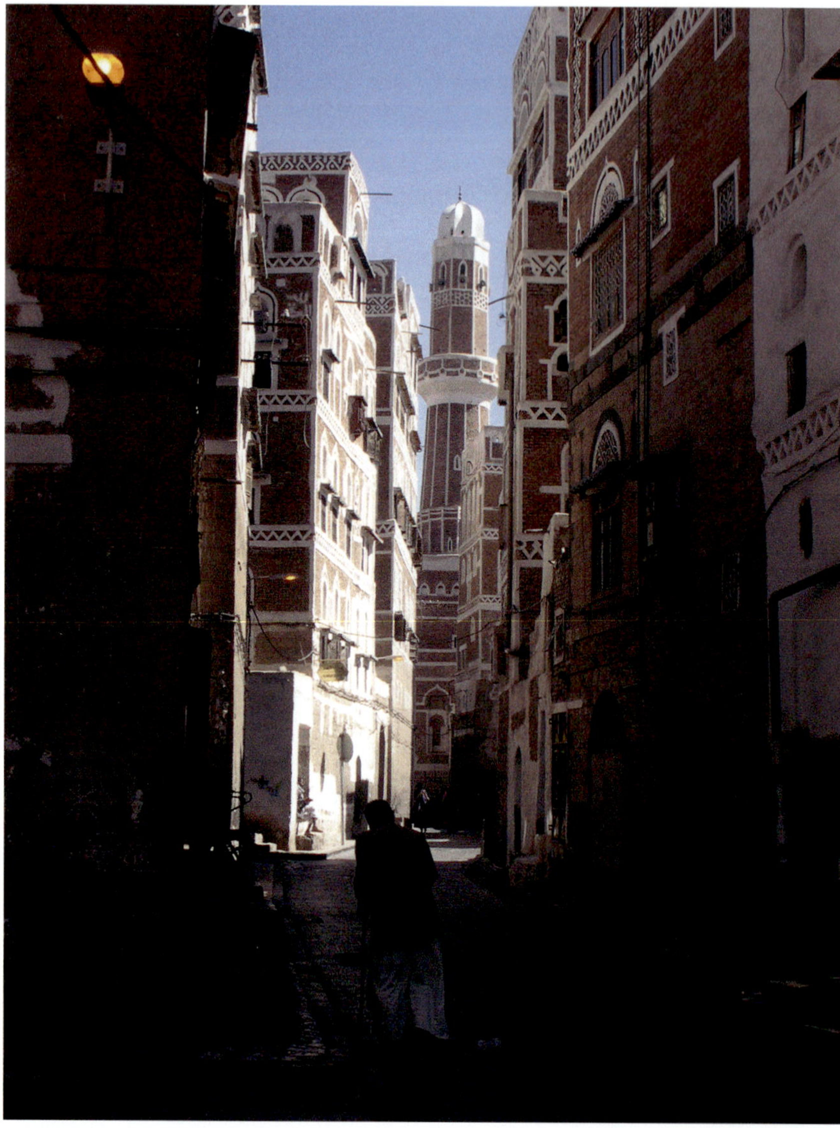

Figure 3.4 Al-Qasimi Street in the quarter of al-Abhar, the first area paved under the UNESCO Campaign

spelled out as a priority of the Campaign strategy and was not made a project in its own right. Although funds were raised for a number of projects outlined in the UNDP-sponsored brochure, not all were implemented. Informants say that the main problem was insufficient capacity at the conservation office. In 1999, 15 years after the inauguration of the Campaign, consultants noted that government organizations involved in conservation still had no clear function, and lacked coordination and transparency.[99]

There was also an unclear view of what conservation was, or should be, in the local context. Many of the projects involved problems of design and implementation, provoking controversy in the community and often within EOPOCS itself. At EOPOCS, the rebuilding of the city wall and the design of Suq al-Baqar clinic were especially contentious. Much of the controversy revolved around the question of appropriate methods of restoration and/or conservation. The choice of buildings and programs was set by donors, following a list developed by UNESCO and EOPOCS; and the technical approach was usually set by the experts they sent. Most Yemeni professionals in the technical office were aware of their limited background in conservation. While they questioned certain decisions, they did not have the power to impact them. When they were in the position of decision-making, they felt the lack of a conceptual framework; they knew that such a framework was well developed in Europe, and that it would have relevance to the Yemeni context. But foreign consultants came for short missions, and often lacked knowledge of the local context. Some were not interested in dialogue. But in many cases contact between foreigners and local professionals was restricted by the administration. "There was a separation of local and foreign experience," recalls a senior architect. As a result, the local staff worked largely on their own, without clear direction. One architect recalled:

> Eighty per cent of what we did in the technical office was a mistake. Engineers and architects did not have experience or training in conservation; they saw it as a cosmetic exercise—you can build in any materials as long as it looks the same. All our discussions in the technical office were at this level, cosmetic. We ignored basic questions like, why do we conserve this house? What kind of precedent does it set? What is the impact on public space, on people's attitudes?

A National Conservation Office

The situation was radically altered by events in 1990–1991. EOPOCS was restructured as a national office, part of the general reorganization of government that followed unification of the two Yemens on May 22, 1990. The office reflected wider political tensions, particularly the fear of southern socialists who had been integrated into the unity government. At the same time the Islamic party Islah gained in strength, supported by the regime as a counterbalance to the left.[100] "Things changed after unification," recalls a senior staff member. "There was more freedom and everyone took off their masks. Now you knew who was a socialist, who was a fundamentalist."[101]

---

99 H. Eckert and A. al-Maqrami, "Cultural Heritage Protection Project: the Social Assessment Report for the Historic Cities of Old Sanaa, Shibam-Hadramawt, and Zabid," 1999, p. 5.

100 The former governments of north and south remained in place in the form of parties: the General Popular Congress (the President's party); and the Yemen Socialist Party (YSP), led by southerners. For post-unification politics see S. Philips, *Yemen's Democracy Experiment*, pp. 47–62; and Dresch, *Modern Yemen*, pp. 186–91.

101 A socialist from Lower Yemen who had been with the conservation office since 1987 was fired by al-Haddad, although apparently not for political reasons. His case became a *cause celèbre* in the press.

The new General Organization for the Preservation of Historic Cities of Yemen (GOPHCY) was now responsible for the conservation of all "historic cities"—although its focus has been on the World Heritage Cities, Sanaa, Shibam, and Zabid (listed in 1993). While many feel that all historic towns and villages—which are distinctive examples of local and regional heritage—should be considered conservation zones, the capacity of GOPHCY was not up to this challenge. No longer responsible directly to the Prime Minister, GOPHCY was now a specialized agency within the Ministry of Culture and Tourism—in theory, with administrative and financial independence.[102] In practice, however, the Ministry of Finance, which retains a representative in each government agency and ministry, exercises significant control and takes a substantial percentage of all projects. According to informants, this is one reason why most international organizations and donors started to deal directly with NGO's or with the Social Fund for Development, which operate outside the ministerial system.

Several months later, in August 1990, Saddam Hussein invaded Kuwait. The lone Arab country on the Security Council, Yemen abstained in a vote to condemn the Iraqi invasion, provoking the ire of Saudi Arabia and the Gulf States. Saudi Arabia revoked the special status of Yemenis; other Gulf nations soon followed suit, and an estimated 800,000 Yemeni migrants were expelled. Their sudden return to Yemen led to the crash of an artificially inflated economy: they required housing and services, and flooded the labor market. The value of the Yemeni riyal began a sharp decline, resulting in a dramatic decrease in income and prices.[103] At the same time, most donor nations and agencies cut off aid in protest against Yemen's position on the war. The new institution of GOPHCY had been deprived of its special authority, and had an expanded mandate and dramatically reduced resources; it was now also deprived of the donor funding that had helped overcome opposition to conservation.[104] The Campaign had been implemented by executive decree; there was still no legal basis for its authority.

For its part, GOPHCY's leadership undertook its own, internal restructuring: several senior technical staff who did not support official policies were sent, or asked to go, to far-off posts to manage branch offices. Meanwhile salaries of the conservation office began to fall, like those of other government offices. The absence of donor funds, scarce resources throughout the government, and political troubles—the short-lived civil war of 1994—meant that little money was directed toward conservation. A few projects went forward: the restoration of Samsarat al-Mansura (Germany); the upgrading of Mu'adh quarter (Italy), including the restoration of Bayt Sari' and the construction of a restoration laboratory (later reconceived as the Center for Architectural and Technical Studies, CATS); and the restoration of Bayt al-'Anbasa (Switzerland). Several of these projects were embroiled in controversy. The clinic at Suq al-Baqar was completed in 1995. Other projects that had been projected did not get off the ground. The North Korean team, which had drafted many of the projects and specifications, left circa 1994.

The main activities carried out by GOPHCY in Sanaa were paving and infrastructure. Much of the city had been paved by 1997, although there continued to be problems with

102 A veteran of the Campaign recalls that al-Haddad lobbied to retain the special relationship with the Prime Minister's Office, but failed.

103 On these events see Dresch, see *Modern Yemen*, pp. 186–7 and pp. 207–8.

104 F. Mermier, "Sana, métaphore de l'état yemenite," in G. Grandguillaume, et al., eds, *Sanaa Hors les Murs*, p. 52.

burst pipes and collapsing houses.[105] A number of projects were undertaken in Shibam and Zabid, although a project for a brick factory which would have addressed a critical material shortage in Zabid was abandoned. Al-Haddad remained Director of GOPHCY, assisted by a deputy (*wakil*). He offered his resignation in 1993, but did not resign until 1995; he died later that year. Some staff lobbied for the appointment of one of the dedicated founders of the office as al-Haddad's replacement. But the position was given to a political appointee. GOPHCY was again restructured in what was essentially an internal coup. The primary change affected the department of Technical Cooperation and Studies, which had conducted studies and provided counterparts for donors. The functions of this office were now to be divided between the new Center for Architectural and Technical Studies (CATS), housed in a building designed by the Italian team, and the Planning Department. The senior architect at the Technical Cooperation Office was appointed director of International Technical Cooperation, restricting his role to public relations. He refused this offer and resigned.

As the efficacy of the conservation office declined other players emerged to fill the gap, notably the Mayor's office and, more recently, the Social Fund for Development, an extra-ministerial organization underwritten by the World Bank and funded by a variety of donors. In 1995, the IMF approved a program of "structural adjustment" for Yemen and funds slowly started to trickle in from various sources.[106] Shortly thereafter the World Bank began prepare a major cultural heritage initiative for the World Heritage Cities of Sanaa, Shibam, and Zabid, with funds contributed by Japan. A British firm conducted a preliminary study, assisted by a local consulting firm and several professionals who had worked with GOPHCY.[107] After submission of the report in August 1999 the World Bank aborted the initiative, primarily because it felt that institutional capacity was lacking. However, a decision was made to reroute the funds through the Social Fund for Development. The SFD was deemed to have sufficient capacity and autonomy, as well as the ability to hire and pay experts. With the restorations of several urban gardens, which began in 1998, cultural heritage became a distinct program sector at the SFD; most public conservation projects are sponsored directly or indirectly by the agency. The SFD works closely with both GOPHCY and GOAMM: the two offices jointly initiate projects and some GOPHCY employees work

105 Requests for compensation for damaged or collapsed houses continued to be submitted to the conservation office through the mid-1990s especially in the area of Bab al-Shu'ub, where the phase one system had not yet been upgraded (on file with GOPHCY Project Department).

106 Credit and grants were extended by the IMF, the World Bank, the European Union, and the Arab Monetary Fund (Dresch, *Modern Yemen*, pp. 207–8).

107 The study by Gilmore Hankey Kirke Ltd. was conducted from 1997–1999 for the World Heritage Cities of Sanaa, Shibam, and Zabid, with funds contributed by Japan (Gilmore Hankey Kirke Ltd., in association with PanYemen Consulting Services, "Cultural Heritage Protection Project, Project Preparation, Final Report," 6 volumes, August 1999).

In the early 1970s cultural tourism was pursued as a development strategy by the World Bank at various sites in southern Turkey, Senegal, and elsewhere. Later in the decade the Bank reassessed the strategy and closed its tourism department. Some critics saw tourism projects as neocolonialist, with side effects that corrupted local morality (like increased prostitution). In some places, where trained personnel, foodstuffs, and other services were inadequate they had to be imported, leading to low returns for local industry and labor markets. The World Bank renewed its interest in cultural heritage in the mid-1990s largely because of the interest of president James Wolfensohn (who served 1995–2000; G. Brizzi, interview). On World Bank activities in this sector during the 1990s see M. Cernea, "At the Cutting Edge: Cultural Patrimony Protection through Development Projects," in the proceedings of a World Bank-sponsored conference on cultural heritage (I. Serageldin, et al., eds, *Historic Cities and Sacred Sites: Cultural Roots and Urban Futures*, Washington, DC: World Bank, 2001, pp. 67–88).

as consultants to the SFD. Professionals and others interested in conservation were generally pleased that the SFD had infused new life into the conservation effort, with the help of funds from the World Bank, other donor agencies, and foreign nations. But some noted that the creation of such agencies—based on World Bank operational standards—exacerbate the very problems that the Bank identified, by encouraging an internal drain of resources and skills away from government.[108] Having little power, GOPHCY appeals for assistance to UNESCO and other parties which traditionally work through government counterparts.

"Weak institutional capacity" is still cited in virtually every consultant's report—not only in conservation, but in planning and other sectors. Yet the shape of foreign aid, and the failure of government to impose conditions on that aid, contribute to the situation. Only rarely are deeper structural issues addressed. "'Structural adjustment' is primarily an adjustment of balance sheets," wrote Paul Dresch in 2000. "To address the more substantial structures would, in Yemen's case, mean dismantling the state itself, which is not sought by international agencies or by other governments."[109] EOPOCS was a transitional entity with no legal basis other than the executive decree that created it. The emergence of a profession with a shared ideology, knowledge base, and critical process is inseparable from the institution that embodies and reproduces its structures.[110] The failure to create such an institution is a source of deep frustration for many professionals. They tend to fault not the

108 The SFD was established in 1997 (law no. 10) to undertake community-based projects. It was transformed into a major implementing agency through the infusion of World Bank Funds. It is administered by a Board of Trustees, headed by the Prime Minister and including representatives of various ministries. According to a program officer, the SFD was conceived as a "social safety net" to mitigate the effects of structural adjustment, but as of the mid-2000s its work did not come close to meeting needs.

The main interest of the SFD for the World Bank is that it enjoys autonomy from the government bureaucracy, and thus can be held to international procedures and standards. Donor funds are routed through the Ministry of International Cooperation, which interfaces directly with the SFD—bypassing the Ministry of Finance. The Bank helped to create another, similar agency in Yemen, the Public Works Unit (*Wahdat al-Insha' al-'Amma*) which is responsible to the Prime Minister and the Ministry of Public Works. Such semi-autonomous agencies, responsible directly to the Prime Minister—as was EOPOCS itself—are similar to the Central Planning Office, which was established in the 1970s (it was still active in the late 1980s, when the German government channeled commodity aid money through it for the restoration of Samasarat al-Mansura; see below, Chapter 6). Burrowes observed that such dynamic entities are created periodically by the system in order to bypass the cumbersome bureaucracy, but over time they are absorbed by the system (personal conversation). The World Bank helped create similar agencies elsewhere in the region.

Even though the contributions of donor states and agencies cannot be traced to individual projects, in the mid-2000s most preferred to channel their funds through the SFD. EU funding was still limited; the Arab Fund, based in Kuwait, financed many of the urban garden restorations. Some donors sponsored longer-term documentation and planning projects. Two of these have were financed under the World Bank's $5,000,000 Cultural Heritage Trust from Italy, which required the use of Italian firms: the inventory directed by D. Pini; and a cultural heritage inventory and archiving project for the Hadhramaut, conducted by an Italian firm called MEMAR (2001–2002). From 2000–2004 the Netherlands financed an Urban Cultural Heritage Program (UCHP) through the Public Works Unit and the Ministry of Public Works; its main achievement was the drafting of a conservation plan for Zabid. The UCHP was dissolved in 2004 with a reduced foreign aid program from the Netherlands.

109 *Modern Yemen*, p. 208. See also I. Glosemeyer, "The Development of State Institutions", in R. Leveau, et al., eds, *Le Yemen Contemporain*, Paris: Editions Karthala, 1999, pp. 79–100.

110 H. Radoine, personal conversation. See also Magali S. Larson, *The Rise of Professionalism*, Berkeley, 1977.

foreign missions, whose values and ideas often coincided with their own, but their own government. For some these feelings were tied to the general disillusionment that followed the revolution:

> We were young guys, looking for real work. We were all *muhandisin* (architects or engineers) … there were no more class divisions. No one was better than the other. Heritage [*turath*] was the property of all—a symbol of our unity, not of our differences.

The problems of the conservation office—which impeded the design and implementation of projects and the enforcement of regulations— slowed the emergence of a unified discourse. The shape of conservation has been influenced, negotiated, and challenged by various actors and groups—like the "democracy experiment" itself.[111] In the long term, this may lead to a different and stronger kind of conservation, with roots in social practice and a diversified institutional base. Some of the ingredients of this process will be examined in the following chapters.

111 "[T]he relative freedom to debate has remained one of the most conspicuous aspects of the Yemeni political experiment" (S. Philips, *Yemen's Democracy Experiment*, p. 49).

Chapter 4
The Old City as Jewel, as Stigma

In the mid-1980s a remarkable project was undertaken: a new dam was built at Ma'rib, the site of the ancient Sabean dam and temple complexes, and the heart of a poor tribal area far to the east of Sanaa where state control has been weak. "Despite two thousand years that separated the two dams and their very different technical conceptions," writes Jean Lambert,

> the modern dam was unanimously presented by the media and the *vox populi* as the 'reconstruction' of the ancient dam: it represented the resurrection of ancient times (*ayam zaman*) and the return to a Golden Age of the kingdom of Saba, of which the dam was the greatest achievement.

Executed with foreign funding and expertise, the project was framed in terms of ancient and modern myths, perceptively described by Jean Lambert. Unification talks were going on at the time, and 'Ali 'Abdullah Saleh portrayed himself as the reunifier of the ancient Yemeni state. The "rebuilding" of the dam developed the idea of Yemen as birthplace of Arab civilization, establishing its place in Arab-Islamic Qur'anic and biblical genealogy—and hence in Western civilization. At the same time, the project was touted as a modern miracle that would help achieve self-sufficiency in food production: a sign of contemporary brilliance that matched that of the ancestors. In political rhetoric and artistic interpretations, the rebuilding of the dam was seen to prefigure a "restoration" of the nation, particularly the anticipated reunification of the North and South. The fall from grace—the destruction of the dam—and the subsequent decline of the Yemeni people occurred because they forgot God and the example of the ancestors. In the commemorative plaques at the new dam, as well as a contemporary play produced in Aden, this notion of fall was assimilated to disunity and civil war: in contrast to traditional accounts, which were essentially pessimistic, the modern renditions saw salvation in the reestablishment of national unity.[1]

Concurrent with the dam project (1984–1986), the cities of Sanaa and Shibam-Hadramaut were nominated by North and South Yemen as World Heritage sites: each was to assume its place as symbol of ancient Yemeni civilization. These projects also referred to a national "restoration":

> To be successful, the [UNESCO] Campaign will need to develop, within the people of the old city of Sanaa and throughout the Yemen Arab Republic, a sense of participation in the restoration of a national spirit as symbolized by Sanaa al-Qadima.[2]

---

1 J. Lambert, "Saba et le barrage de Mareb: récit fondateur et temps imaginé," *Peuples Mediterranéeans*, no. 56–7, July–Dec. 1991, pp. 141–59, 168 The dam was financed by a $90 million gift from Sheikh Zayd, the late ruler of the UAE, who traced his ancestry to Yemen.

2 "Campaign Strategy and Action Plan:,"1988–1997, August 1988, co-authored by UNESCO and EOPOCS.

The idea of an ancient Golden Age, then, was a cornerstone of national heritage in Yemen as it had been for many other nations. But there was a new twist in the era of "world heritage": the identity of a nation was no longer unitary; rather, it embraced a regional pluralism, identified with various urban, vernacular, and natural sites. This new discourse of diversity within unity had political overtones, and was also related to the growing movement of cultural tourism. Yemen was tremendously rich in these kinds of sites, all of which could now be embraced as representative of the nation. Sanaa, the site of an ancient urban civilization, had as its counterpart the city of Ma'rib, capital of a desert and tribal civilization that had controlled the fabled incense trade. Both places could be appropriated by all as common heritage and representative of a shared genius; they were often presented together as emblematic of the nation, in both official and popular discourse.[3] Heritage would not erase tribal and regional identities, but rather would bring them into the national fold.

If, as Benedict Anderson asserts, nationalism can be seen as a process of forgetting, heritage transforms old oppositions into "cultural diversity." In this, heritage worked in parallel to the political promises of the regime, which emphasized pluralism and, since unification, the decentralization of power.[4] In Yemen as elsewhere, heritage has become an important part of state discourse. But as Michael Herzfeld reminds us, the formation of collective identity is not a "top-down" process: rather, it is constructed through a process of cultural engagement.[5] This is especially true of Yemen's heritage project since, as we saw in the last chapter, various factors have impeded the emergence of a unified discourse. Conceptions of heritage and the past are particularly rich and varied: in government circles and outside them, official representations mix with and are colored by living memories and by traditional practices and values that are still a vital part of the present.

In this chapter, I will discuss some of the ways in which heritage is represented by residents with various relationships to the old city. I will then look at several places where definitions of heritage are contested: in the tower house, which has in many ways become emblematic of the old city; and in the *suq* and urban gardens. In these places social and economic changes—many of which can be linked to the UNESCO Campaign—have led residents and other interested parties, such as the Ministry of Endowments, to contest GOPHCY's judgments regarding historic value. The chapter will conclude with an early attempt by conservation professionals to define the "historic monument" in the Yemeni context.

Heritage and Origin (*asl*)

In the myth of the Ma'rib dam and its modern renditions, failure to follow the example of the ancestors is implicitly linked to the influence of foreign ways: the diversification of cultures and a sullying of cultural origin.[6] Origin (*asl*) is a key concept in Yemen: in both tribal and non-tribal society, one's origin is determined by lineage. This origin is a source of honor (*sharaf*), the social representation of the male individual, which also

3 For example, in the twin commemorative stamps issued by the government in the 1980s (Steve Caton, personal conversation). See also the oral account related by J. Lambert, "Saba et le barrage de Mareb," pp. 146–7.

4 Hence Saleh's well-known statement: "The state is part of the tribes, and our Yemeni people is a collection of tribes" (cited in Dresh, *Tribes, Government, and History in Yemen*, p. 7).

5 M. Herzfeld, *Cultural Intimacy: Social Poetics in the Nation State*, Chapter 1.

6 J. Lambert, "Saba et le barrage de Mareb," p. 150.

includes one's land and one's house; in public rhetoric, ancestors pass their honor on to their descendants.[7] Where tribal affiliations are weaker or forgotten—in urban areas, and in western and southern parts of the country—consciousness of kin group and lineage are still strong. One identifies oneself as part of a *bayt*—a family line and literally, house—which is also the source of one's honor.[8] In her study of the village of Wadi Dhahr, near Sanaa, Martha Mundy observes:

> In the speech of older women the vision of social order takes the form of a tripartite division: men of religion, men of the sword and the plough, and men of service. Images of rank are cast in terms of origins. Houses (*bayt*, pl. *buyut*) possess origins; men are known and ranked as members of houses.

While the religious elite trace their origin to founders of the faith, both townsmen (*'arab*) and countryfolk (*qabili*) trace their origin to a place "where [the] patronymic house is known for its relation to the land, its honorable occupations and marriage alliances."[9] Most Yemenis continue to identify themselves in terms of the place of their family's origin, even if they have not lived there for several generations or more.

There is no word in Arabic for "ethnic": in Yemen, the terms that are used to approximate it have to do with origin and descent.[10] It has been suggested that in some parts of the Middle East, the tribe and the family are models for the imagined community of the nation.[11] This echoes an earlier process that may have occurred in towns and cities: as heads of important tribes and families assumed leadership, the community was modeled, consciously or unconsciously, on the patriarchal family. In most highland towns in Yemen communal identity tended to be linked to the surrounding tribes, which trace their origin to a common and ancient ancestor. By contrast, argues Mermier, in Sanaa a distinctive urban ideology or *citadinité* emerged, distinct from the city's tribal context.[12] This was perhaps a means to overcome the tribal loyalties of early settlers.[13] Like its tribal counterpart, this urban ideology also posited common ancestors: Sam, who links Sanaa to a series of pre-Islamic and biblical Prophets, and his descendant Qahtan, the common ancestor of South Arabians.

Saba al-Suleihi, an architect who worked with the UNESCO Campaign, sees the facades of Sanaa as the public representation of a collective and authentic identity:

---

7 P. Dresch, *Tribes, Government, and History in Yemen*, pp. 39, 43–54.

8 P. Bonnenfant, "La Maison Microcosme," in Bonnenfant, ed., *Sanaa: l'architecture domestique et société*, Paris: CNRS Editions, 1995, pp. 50–56; and G. Vom Bruck, G., *Islam, Memory, and Morality in Yemen*, New York: Palgrave Macmillan, 2005, p. 65. See also Vom Bruck, "A House Turned Inside Out: Inhabiting Space in a Yemeni City," *Journal of Material Culture*, vol. 2, no. 2, July 1997, p. 144.

9 M. Mundy, *Domestic Government: Kinship, Community, and Polity in North Yemen*, New York: I.B. Tauris, 1995, pp. 39–40. For the link between the patronymic group and the physical house in North Yemen, see, pp. 93–7.

10 Day, "Power-sharing and Hegemony," pp. 94–5. The closest terms, says Day, are *'irq* and *salala*; he is probably referring to *'araqa*, which H. Wehr defines as "deep-rootedness; ancient ancestral line"; and *asala*, "nobility of descent, purity of origin" (*Dictionary of Modern Written Arabic*, J. Cowan, ed., Ithaca: Spoken Language Services, 1976).

11 A. Shryock, *Nationalism and the Genealogical Imagination: Oral History and Textual Authority in Tribal Jordan*, Berkeley: University of California Press, p. 318.

12 F. Mermier, "De l'Usage d'un Concept: la Citadinité a Sanaa," *Peuples Méditerranéeans*, no. 46, Jan.–March., 1989, p. 31, 34–7.

13 S. al-Suleihi, "The Lyrical Facades of Sana'a," unpublished Master's thesis, Department of Architecture. M.I.T., 1992, p. 17.

> Authenticity is established by this direct connection with the legendary past and by the belief that [Yemenis] carry the pure South Arabian blood in their veins. The fact that legend … connected Sanaa with the venerated ancestor of the country, with emphasis on place rather than race, suggests a physical dimension of nobility; it is the very spot where Shem [Sam] chose to build his lofty Ghumdan, the romantic ideal for all houses in the city …. In this sense, Sanaa becomes an architectural manifestation of the genealogy of Yemeni tribes gaining, therefore, distinction [in the merging] of legend and political power, as the seat of the central state.[14]

Like the family, the city (and nation) derive their authenticity from origin and lineage, which include the concept of purity. That which is authentic and original (*asli*) precedes the introduction of Western ways and norms, and hence encompasses the notion of the "folk."[15] The origin of the city is the source of its honor and also of its divine protection. Architecture, and particularly the house, is a key component of urban culture, by virtue of its refinement and its origin. As such, al-Suleihi writes, the Sanaani house is an "incomparable piece of rhetoric."[16]

Transferred to the larger associations of city and nation, heritage or *turath* (from *w-r-th*, "to inherit") evokes a complex of social and emotional ties, including the achievements of the ancestral line. *'Ayb*, or public disgrace, damages *sharaf*; hence the condition of the old city was seen as a disgrace of collective honor. As such, heritage assimilates the concept of *sharaf* to present concerns.[17] The mythical founding of Sanaa and the lineage of its founder are constantly evoked in conservation literature. This literature brings together two myths of authenticity: UNESCO's notion of the city as representative of a past culture, a city of "the Islamic ages" that is "untouched by time"; and a local myth of the city as a place of shared descent, values, and customs. In the former, emphasis is placed on material integrity as the criterion of authenticity. But significantly, both myths attribute special status to the house, *bayt*, a word that also means family and lineage. While the dam at Ma'rib would be "restored," the old city of Sanaa would be "re-antiquated" or rather, retained in its antiquated state.

In many parts of the world "re-antiquating" is a "shocking destruction of the familiar … not continuity but radical change."[18] But in Sanaa, residents were not separated from the "antique style" by decades or centuries of change. In a sense, conservation validated the familiar and the indigenous. It built on *citadinité* by validating Sanaani urban culture which, as al-Suleihi suggests, may find its clearest expressions in architecture—in particular, the composition of elaborate facades. Yet, as we shall see, for many in Yemen preservation (*hifadh*) does not necessarily mean absence of change.

14 S. al-Suleihi, "The Lyrical Facades of Sana'a," pp. 16–17.

15 E. Davis and N. Gavrielides, "Statecraft, Historical Memory, and Popular Culture in Iraq and Kuwait," in Davis and Gavrielides, eds, *Statecraft in the Middle East: Oil, Historical Memory, and Popular Culture*, Miami: Florida International University Press, 1991, p. 143. The authors describe this phenomenon in Kuwait where, in the absence of ancient sites, folk heritage (*al-turath al-sha'bi*) has been encouraged by the state.

16 S. al-Suleihi, "The Lyrical Facades of Sanaa," p. 26. Al-Suleihi argues that much as poetry served as the vehicle for tribal identity and ideology, architecture plays a parallel role in Sanaa.

17 On this point see P. Dresch, *Tribes, Government, and History*, p. 40.

18 M. Herzfeld, *A Place in History*, p. 14.

The Old City as Symbol

As in other countries, the historic core of the capital city has become emblematic of the larger metropolis, and of the state itself. The timeless, unchanging nature of old Sanaa is constantly reaffirmed, despite the fact that its social composition has changed dramatically, and the tower houses have been divided into apartments. The old elite and the traditional middle class—shop owners, artisans, religious authorities, and others—left the old city in the 1970s and '80s, and were replaced by families of rural origin. The religious composition has also changed: until 1962 the city was overwhelmingly Zaydi; it is now predominantly Sunni, of the Shafi'i school.[19] Members of the old middle class who have retained shops and positions in the old city complain of a rural invasion—and the consequent disappearance of the old Sanaani identity, which they claim to represent. "We've been complaining for five years about the invasion of rural people," said a jeweler in the Suq al-Milh in 2004. "They move into our houses, they steal our jobs, and they dirty our streets."[20] Yet, as sociologist 'Abdo 'Ali 'Othman noted, the old elite and the traditional middle class have become more accepting of their traditional "others," if only because they have no choice.[21]

The elevation of old Sanaa as a symbol of the nation transcends sensitive historical issues: the role of the city as the dynastic center of the later Zaydi imams; the tensions between Upper Yemen and Lower Yemen; and the traditional antagonism between the city and its tribal hinterland.[22] The concentration of resources and interests in the capital city means that various groups identify with it. Tribesmen, who were at least in theory once unwelcome in Sanaa, have become important citizens and patrons. Many fought on the side of the Republicans in the revolution. They were rewarded for their support in official and unofficial ways; as they have settled and invested in the capital, they have appropriated the city in symbolic terms. Two prominent tribal leaders were given houses in the heart of old Sanaa as compensation for their own family houses that were destroyed in other towns. One is Sheikh Sinan Abu Luhum, an important revolutionary figure of the Bakil tribal confederation, and one of the most powerful tribal politicians in the YAR during the 1960s and '70s. He took a special interest in the city, founding the Committee for the Improvement of the Capital (*lajnat tahsin al-'asima*) in 1975. He was granted a distinguished house on al-Abhar square; several decades later he was persuaded to cede the house to the conservation office. Renovated in the late 1990s, it served as the Women's Handicraft Center for several years and later became the Museum of Sanaani Popular Life. The other is the late Sheikh 'Abdullah ibn Husayn al-Ahmar, former head of the Hashid confederation, speaker of Parliament, and leader of Islah (the Islamic opposition party). Extensions to the rear of his

---

19 F. Mermier, "The *Suq*s of Sanaa: Changing Functions and Symbolic Centrality," in M. Gharipour, ed., *The Bazaar in the Islamic City: Design, Culture, and History*, Cairo: American University in Cairo Press, 2012, p. 154; J. Lambert, "Consommation de masse et tradition a Sanaa: vers une culture urbaine," in G. Grandguillame, et al., eds, *Sanaa hors les murs*, pp. 116–17; and 'A.'A. 'Othman, interview.

20 Related by R. Stadnicki, "Vers un aménagement participative à Sanaa?", p. 52.

21 "The City and the Other," unpublished paper presented at a conference co-sponsored by the Ministry of Culture and Sanaa University, August 2004.

22 In 1994, shortly after unification, Mermier wrote that the "mosaic character of the city, which has not yet given rise to a true urban culture, finds itself confronted by the threat of a tribal environment that can impose its rules in the heart of the city, if not the state itself" ("Les Fondations Mythiques de Sanaa et d'Aden," *Revue du monde musulman et de la Méditerrannée*, vol. 67, special issue, "Le Yemen, présent et passé de l'unité," p. 37).

house, which is across from Sheikh Sinan's, reveal an attempt to appropriate the "Sanaani style." An architect described this addition as an example of the coarse, *nouveau riche* "tribal taste" that has invaded the city. "In Sanaani," he said, "it would be described as *jadhib*, 'catching the eye'"—and therefore unrefined.[23]

The UNESCO Campaign seems to have renewed appreciation for the local architecture. Additions to tower houses, once clad in cut stone, are now more often clad in traditional *yajur* (fired brick); this evolving taste can often be seen in the same building (Figure 1.3). The tower house has become emblematic of the Sanaani style—as witnessed by the ubiquitous plaster models that are popular among both Yemenis and foreigners (Figure 4.1). Some officials and professionals attribute this renewed interest and pride to conservation. But it builds on an old discourse. "Sanaanis are very aware of their architecture," said a middle-aged scholar, who was born and raised in the city. "They will look at something you do and tell you it's not right." As I toured the old city alone, or with an architect or inspector, residents would sometimes offer their opinion: "This stone is cut incorrectly," said one. "It should be cut by hand, not with a machine." A young worker at an old house in Bir al-'Azab, that was being restored by the Ministry of Tourism, showed me around the site with obvious pride—pointing out various disfiguring (*mushawwih*) elements that had been removed, and would be restored to their original state. He had worked in construction, but this was his first experience in old-style work; he is not from a family of builders. Yet he had a fairly sophisticated understanding of local architecture. His family's house in Bir al-'Azab is of concrete block; they planned to add a story in "*yajur*, in the old style; we'll clad the block below in stone." The new brickwork that can be seen throughout the city is not beautiful, he noted: "There's too much ornamentation (*nuqush*)." It should be restricted to the decorative brick courses (*hizam*, pl. *ahzima*), he said, that divide floors from one another.

The focus on Sanaa has had important consequences, which were perhaps not foreseen by the writers of the World Heritage Convention. The privileging of certain sites—especially the three World Heritage Cities of Sanaa, Shibam, and Zabid—means that resources and studies are concentrated there. Hundreds of other important sites throughout the country, and even the old neighborhoods that surround the old city of Sanaa itself, have been neglected. When conservation projects are sponsored for other sites, they may be carried out in the absence of local studies. As such, they may be restored according to an idealized "Sanaani style" with which architects are familiar. This is one of several ways in which the heritage enterprise may, in effect, homogenize diversity. "Heritage is a fashion," says one architect. "It's more dangerous to our historic cities than modernism, because it combines resources with ignorance."

Until recently, the Sanaani style was defined by master builders, or *usta*s. But aspects of their practice have suffered due to new actors, technologies, and forms; nor has a strong architectural profession developed to fill the gap. Under the imams, officials and builders experimented with new institutional typologies, using local materials and building on Ottoman precedents. These experiments were largely abandoned in favor of what some see as a kind of "postmodernism" derived from residential architecture. Features like multi-lite windows (*qamariyya*s) and decorative brick courses have been extracted from their domestic context and applied to commercial and institutional buildings—despite precedents for these types in the city. Suq al-Baqar clinic is a prime example: initially designed by the North Korean team in a modernist idiom, it was redesigned by architects at EOPOCS in

23 On Sheikh Sinan Abu Luhum and Sheikh 'Abdullah ibn Husayn al-Ahmar, see R. Burrowes, *An Historical Dictionary of Yemen*, pp. 6–7, 22–4.

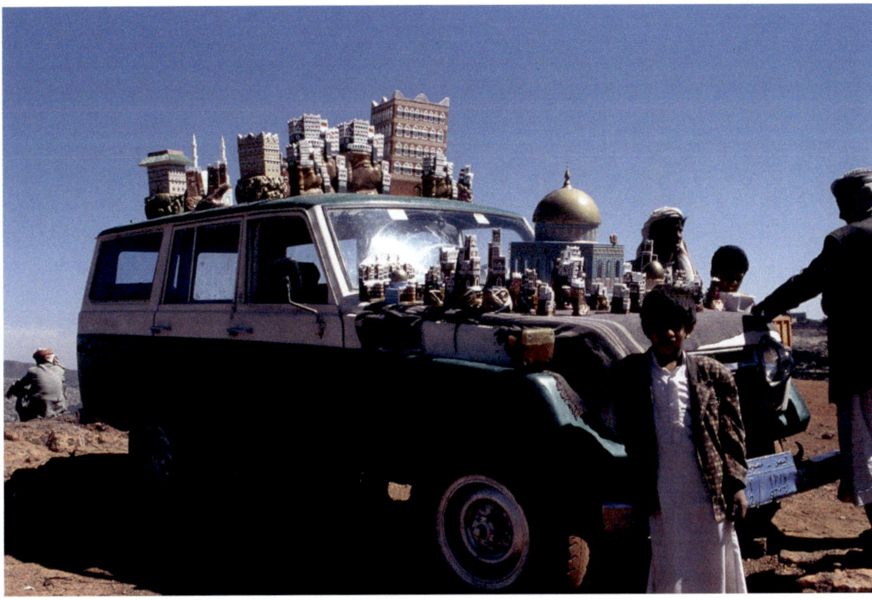

Figure 4.1 Plaster models of tower houses on sale at Wadi Dhahr, a scenic location near Sanaa

an attempt to be contextual (Figure 4.2).[24] One suspects that the domestic facade is used in this way because it is extroverted, in contrast to the sober and introverted mosques and caravansarays—and more immediately identifiable as "native." The Sanaani style is used for new buildings not only in the capital, but throughout the country. Foreign consultants and some local professionals, particularly those familiar with the historical style, see this work as "kitsch." But in Yemen neither clients, nor their contractors and builders, have been schooled in architectural modernism—which, like conservation, holds that contemporary building should be distinct from that of the past.

Jewel or Stigma?

The thing that immediately strikes a visitor accustomed to the gentrified historic cities of Europe and the U.S. is that the old city of Sanaa seems surprisingly "normal." There is an obvious mix of activities, from small shops and workshops, to the pervasive fabric wholesalers, to tourist shops, and shops where electronics and cell phones are sold. There is also an obvious mix of social groups, from the working poor to a middle class that

24 On Suq al-Baqar clinic see Yoshida, "Conservation-Based Development." Precedents for institutional buildings include the Ministry of Health on Tahrir Square. Built under Imam Ahmad, it was inspired by Ottoman models: builders translated decorative elements into multi-colored stone.

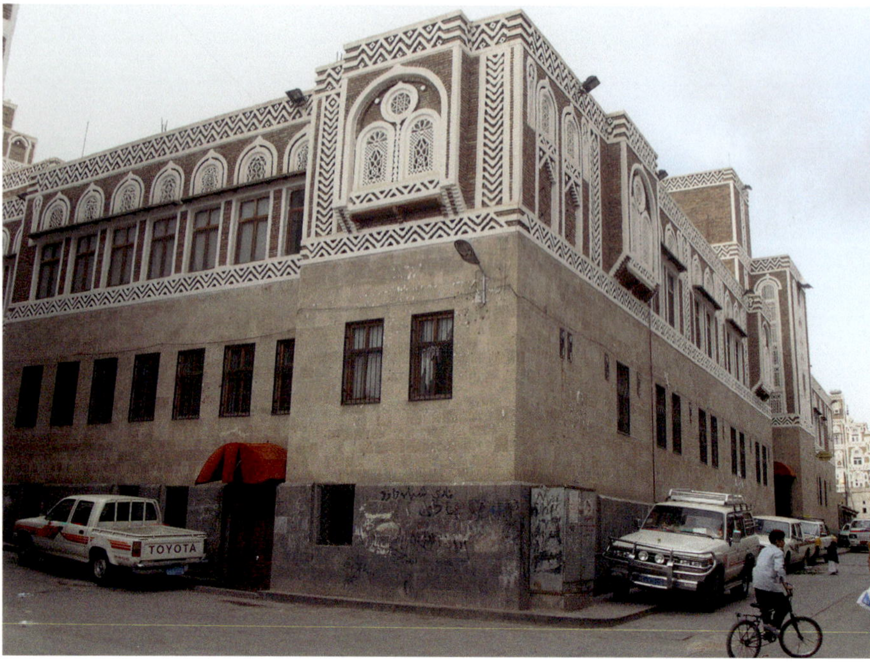

Figure 4.2 The clinic at Suq al-Baqar, designed by local architects during the UNESCO Campaign

includes traditional, and changing, social statuses.[25] If one group is underrepresented it is the political-intellectual elite, who have chosen to live in the new districts. Despite the many problems and limitations of the conservation effort, one is struck by this very old urban fabric, an architectural "jewel"—a phrase that I often heard from officials and residents—whose physical decline has been slowed or arrested. For residents, the generally well-preserved houses and restored gardens, the clean streets equipped with modern infrastructure, are a source of pride. They have seen their "historic" city—which is still their everyday city—validated by the international community and by the state.

Consciousness of an "old Sanaa" developed gradually, as new districts were built in a radically different idiom. As late as the 1970s the old city *was* Sanaa, and the new districts were outside the city (*kharij Sanaa*).[26] But this original Sanaa was not restricted to the central core—it included the old suburbs that were still partially surrounded by walls. There is thus a kind of disconnect between the official designation of "old Sanaa" and personal and collective memory. The heritage of Sanaa is more varied, and much of it is of very recent origin—especially in old inner suburbs like Bir al-'Azab and al-Qa'a,

25 During my field work I visited a number of families who appeared to be extremely poor. An interesting case is a row of one-room, concrete block units, built in the open space between two grand old houses in the quarter of al-Nahrayn. Occupied by rural migrants from the countryside—the woman I spoke with lived in the single room with her husband and three children—the complex is known as a *samsara* in the neighborhood, evoking the transient character of the inhabitants.

26 According to elderly residents, cited by Daolatli in his mission report of 1981.

which were the training ground for the current generation of master builders, their fathers, and grandfathers. "UNESCO's Sanaa is not the *ustas*' Sanaa," observed one architect. The limitation of historic designation to the central core has contributed to a unitary reading of Sanaa's history, architecture, and people.

The isolation of the historic core has been reinforced by the continuing focus on the city walls. Portions of the wall and the sole surviving gate, Bab al-Yaman, were restored or rebuilt under the UNESCO Campaign. In 2005 a project was proposed to rebuild the stretch of wall adjacent to Bab al-Yaman, which would have meant removing buildings and shops and compensating the owners. The walls, as we have seen, are a very old symbol of Sanaa: their gates embellished by talismans, they were seen as a symbol of divine protection.[27] The focus on the walls is motivated by their iconic image, but also by their value as a defensive structure—no longer against raiding tribes, but against modern encroachments. In the words of one architect, the walls are essential to preserve the city as an historic monument, safeguarding it against "contamination of the undisciplined architecture of the [new] urban sprawl."[28] The stark contrast between the refined architecture of the old city and the prosaic development that surrounds it makes it difficult not to sympathize with this statement. Yet as Trevor Marchand notes, the walls not only reinforce the edges of the old city as a physical space, but as a space of "traditional" identity.

In the parlance of UNESCO, old Sanaa is representative of a traditional urban settlement, "untouched" by the passage of time. The rhetoric has been adopted by local residents, as it confirms their image of the city and of themselves. The mass exodus of the well-to-do and much of the middle class in the 1970s resulted in dramatic upheavals in demography and social relations. Though few of the original families remain, residents of the old city told me that traditions are still alive. The ideal appears to perpetuate practice. The old city is known as a more "traditional" place, where women would not think of going without the face veil and where foreigners should dress conservatively.[29] Social relations in the old city—which, according to the elderly, are nothing like they were in the old days—are still close enough that they are stifling for some young professionals used to modern notions of privacy. An artist moved to al-Zubayri Street from the old city because it was difficult to work there. "People were always stopping by, wanting to draw you into social relations," he said. "They don't respect your privacy."[30] At the same time, the compactness of the fabric, the pleasant open spaces and gardens, and the restriction of vehicular traffic contribute to a sociability and sense of calm that many find lacking in the new districts. A middle-aged man of tribal origin who lives in a distant suburb told me that he came to the old city to walk, and to visit with friends at the teahouse at Qubbat al-Mahdi. "I go because of the sense of calm (*raha*) I feel there," he said.

The idea of the "old city" seems to validate traditional social norms in the face of rapid change. Residents sometimes use the language of conservation to defend conservative religious and social values—especially when they feel that their city is being co-opted by

27 F. Mermier, "Les Fondations Mythiques de Sanaa et Aden," p. 134.

28 Lecturer at Sanaa University quoted by T. Marchand, "Walling Old Sana'a," p. 49.

29 In 1995 Lambert reported that young women who wanted to be rid of the veil sought to move to the new districts ("Consommation de masse," in G. Grandguillaume, ed., *Sanaa hors les Murs*, p. 129).

30 Elderly residents relate stories of the old days, when different branches of the family often inhabited contiguous houses. One woman in her eighties recalls that women used to move freely across rooftops, and knock on party walls when they needed something from their neighbors. A widow, she still lives in her husband's house in the old city, but no longer knows her neighbors. "The old way of life is no more," she said.

elite interests. This intersects in unexpected ways with the agendas of international actors like UNESCO. A case in point is the Bustan al-'Amri, a garden attached to the *waqf* or endowment of the mosque of Qubbat al-Mahdi.[31]

During the UNESCO Campaign, a public garden and "ecomuseum" were proposed on a site that included the Bustan al-'Amri and the adjacent gardens of al-Sultan and al-Ingad. The *marna*' (well enclosure) of Qubbat al-Mahdi would be restored and operated by a camel; the water channels would be extended to irrigate the public park. This project was envisioned as part of a larger plan to rehabilitate and beautify the banks of the Wadi Sa'ila.[32] According to one source, the proposal was appealing to some residents who lobbied authorities to implement it, but never succeeded.[33] In late 2003, in preparation for Sanaa's year as Cultural Capital of the Arab World (2004), the Ministry of Culture built an open-air theater in the garden. The idea was to hold cultural events along the wadi banks, using the old city as the *mise en scène*. But initially, the events did not involve the residents themselves—and in at least one case, actually excluded them. At a classical music concert in March 2004 performed by a German orchestra, admission seems to have been restricted to Yemeni officials and the foreign community. The spectacularly lit house facades along the Sa'ila formed the backdrop for the event, but the shadows of young Yemeni men seated along the wall created a disturbing, pseudo-colonial scene, with the natives excluded from their own territory (Figure 4.3).

Residents of the old city and others protested against construction of the theater—arguing that the use of *waqf* property for entertainment went against the original intention of the patron (*waqif*). Moreover, the garden was attached to a mosque: entertainment and music were not appropriate for the site. The controversy was covered in the press, and provoked a debate in Parliament.[34] On one of my rounds with inspectors, we were approached by a young man who demanded to know why GOPHCY was not discussing the matter. "Any modernization (*istihdath*) in the old city of Sanaa is forbidden," he said. "It says so on the Internet" (i.e., on UNESCO's website).[35] According to a GOPHCY official, residents found UNESCO's proposal for a public garden appealing, and more compatible with the *waqif*'s intentions. In their effort to preserve these intentions—and, implicitly, to resist the coopting

31 *Qubba*, which means "dome," refers to the tomb of the patron, a separate structure in the complex. The mosque itself is often referred to by the name of the tomb. On the mosque, which was built by the Imam al-Mahdi li-Din Allah al-'Abbas in 1750–1751, see R. Lewcock, et al., "The Smaller Mosques of Sanaa," pp. 383–4.

32 In the Campaign brochure, the three gardens are referred to collectively as Bustan al-Sultan; see M.B. Lane, *Pilot Restoration Projects*, pp. 54–5. According to the brochure, Bustan al-Sultan had been abandoned for many years: "[i]n view of its prime position, the site is under heavy pressure for commercial development." At the time Bustan al-'Amri was still cultivated. In the intervening years, much of the western portion of Bustan al-Sultan was built up, and open spaces were used as a *qat suq*.

33 Stadnicki, "Vers un aménagement participative à Sanaa?, pp. 57–8. In 2001 surveyors found this garden, as well as Bustan al-'Amri, under cultivation and in good condition. Bustan al-Ingad, the southwestern portion of the site, had not been cultivated for some years due to lack of water. In the late 1990s the municipality created an asphalted *suq* here (M. Barcelo, ed., *Les jardins de la vielle ville de Sanaa*, pp. 70–74).

34 *Al-Sahwa* (paper of *Islah*, the Islamist opposition party), February 12, 2004.

35 The inspector told me that GOPHCY opposed conversion of the garden into an open-air theater. The Ministry of Endowments (*Awqaf*) initially supported it, he said, but changed its position in the face of strong opposition from residents. In May 2004 the Social Fund for Development was considering a proposal to demolish the theater and restore the garden to its former state, but the proposal must have been dropped.

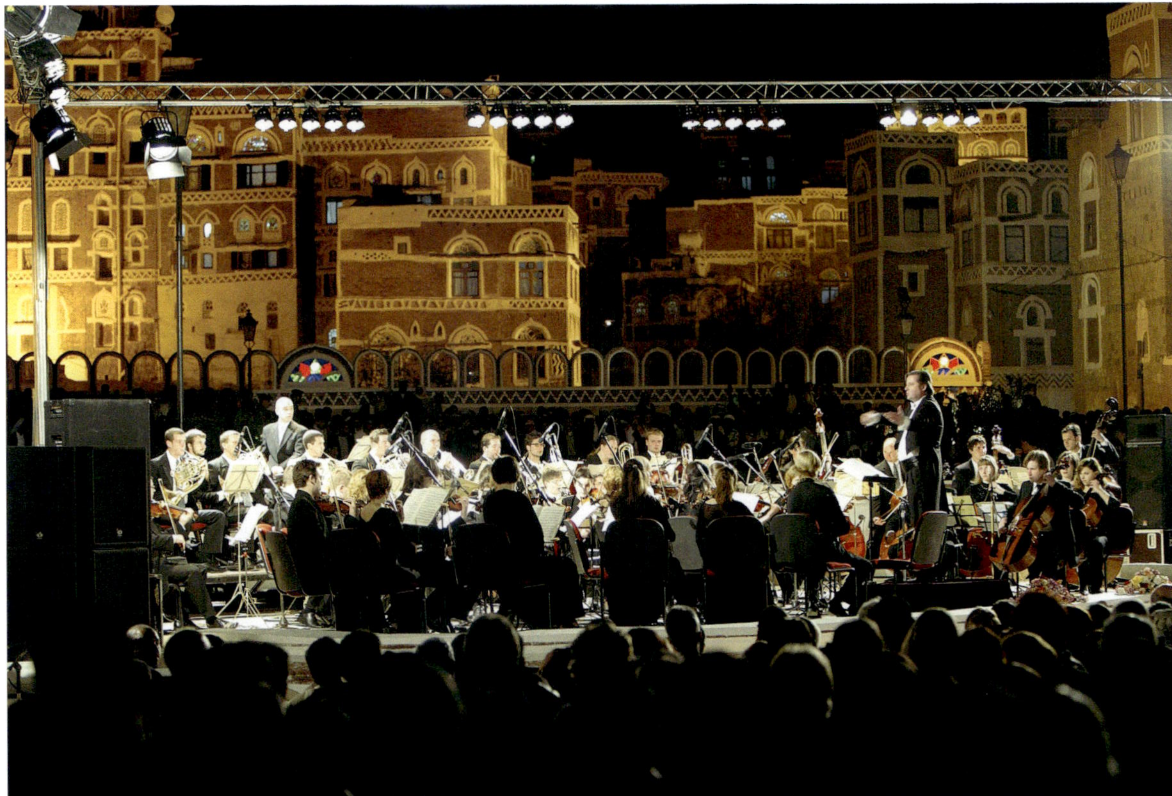

Figure 4.3 **Facades along the Wadi Sa'ila served as a dramatic backdrop for a concert by a German orchestra in the amphitheater at Bustan al-'Amri, during Sanaa's year as Cultural Capital of the Arab World (2004). Attendance was by invitation, excluding most residents of the old city who watched from the banks of the wadi**

Source: Photograph © 2014 Marcos Abbs.

of their city by an elite and alien program—residents invoked the authority of UNESCO alongside Islamic law. During the month of Ramadan that year, in what may have been an effort to appease residents, the theater was used for performances of religious music (*tawshih*) by groups from different parts of Yemen.

As we saw in Chapter 1, conservation is heir to an older tradition in the social sciences and the humanities that equates social and physical form. This has been taken farther in recent years, with the adoption of the new category of "intangible heritage"—a category that includes not only music, poetry, and crafts, but also social practices and customs.[36] It is unclear if certain cultural practices seen by some as incompatible with modernity—for example, the seclusion of women—are to be acknowledged as part of Yemen's intangible heritage. The phrase "customs and traditions" (*'adat wa taqalid*) appears to have taken on new currency in the old city: it is cited with both reverence and ambivalence, sometimes by the same person.[37] This ambivalence extends to the old city itself. In the popular imagination, the old city is portrayed as a bastion of Yemeni cultural identity, and also as backward and conservative: its inhabitants are blamed for "keeping the country in the dark ages of the Zaydi imams."[38]

As we have seen, the ambivalence toward pre-revolutionary Yemen was shaped in relation to the foreign and the modern. Modernizing Egyptians and Yemenis returning from abroad compared Sanaa unfavorably with foreign cities. Through the eyes of foreigners, the old districts came to be valued as evidence of a local genius that contrasted favorably with the faceless architecture of modernity. A new pride in the local and the taditional was assimilated by the modernizing elite; yet they continued to feel embarrassment at their country's late and incomplete modernization, of which the old city was a potent reminder. As Jean Lambert observed in the mid-1980s:

> In the often depressing conditions [of the present] it is hardly surprising that an image is being formed of a Golden Age of the old days of the city of Sanaa ....[T]his mythic memory ... has as its corollary a great ambivalence toward the real city. One oscillates between a contempt for a backward, but recent, past, whose sequels are still too perceptible, and the valorization of a more distant and idealized past ... This ambivalence is perhaps also inherent in the state of abandon of the old city, in this transitional period in which UNESCO has begun to restore it: the gaze of the foreigner is thus a determining factor in the content of one's self-image ....[39]

As a form of institutionalized and collective nostalgia, heritage draws on the nostalgia of individuals and families; here, it is alternately expressed as pride and shame. The shame is both societal and personal, since many educated Yemenis were themselves raised in

36 For a critique of intangible heritage see Chapter 1.

37 One young woman—who belongs to what 'Abdo 'Ali 'Othman calls the "traditional urban middle class"—explained and reaffirmed the requirement that a woman obey her husband, which is equivalent to obeying God. Later in the conversation, she expressed sadness that her mother would not let her become a hairdresser and cosmetologist, areas in which she felt gifted. Cutting hair, as a hairdresser or barber, is a profession associated with the *muzayyin* class, not suitable for a woman of her stock. "These *'adat wa taqalid*," she said. "I wish they did not exist." On the *muzayyin* see R. Burrowes, *Historical Dictionary of Yemen*, pp. 360–61.

38 Related by Marchand "Walling Old Sanaa," p. 49. On this theme see also Herzfeld, *The Body Impolitic*, pp. 5, 30.

39 "Consommation de masse et tradition à Sanaa," p. 98.

these "backward" conditions. Many seem to feel a perceptual and even temporal divide between the new areas of Sanaa and the old city—which perhaps corresponds to an inner divide, between one's sense of self and one's past. This may explain the often shocking comments I heard from several officials and intellectuals during my fieldwork. On the one hand, they affirmed the need to preserve the old city as a "living museum." "Even the people," one official told me, "are antiquities (*athar*)." At the same time, they disparaged the backwardness and ignorance of the people who lived there, blaming them for the decay of the old city. As residents and inspectors bitterly noted, however, the biggest "violators" are those with access to power. These privileged individuals incorporate memories of the old city in their villas, which they describe as "mixes" of the traditional and the modern. Yet they rarely visit the old city, seeing it as a world apart.[40] This attitude contrasts markedly with that of old city residents, who seem to move with remarkable ease between the old districts and the new. Many of their relatives have moved to new areas because the family homes could no longer accommodate them.

Despite the pride felt by many residents of the old city, there is for many a sense that the opportunities brought by modernity are happening elsewhere. The attachment to the ancestral home may be diminishing among some of the younger generations. While praising conservation and the old architecture, one young man said to me matter of factly: "those with means, leave the old city." As the inhabitants of the old city become poorer, Lambert noted, their resentment is increasingly aimed at those with means who have made their homes outside the old city, "those villas that officials (*al-mas'ulin*, literally 'responsible ones') build for themselves in the new quarters."[41] Yet others—educated single men or, less often, families—have returned to the old city or are thinking of returning. They cited the urban life and sociability they feel is missing in the new districts; a few younger individuals cite the qualities and "spirit" of the old houses. But their number remains small: living in the old city continues to carry a social stigma, especially for the educated. A professor at Sanaa University who grew up in the old city returned to the family home to live there with his wife and children. But his colleagues looked down upon this, as incompatible with his status. There is also an intermediate type of residence: a man who lives in the new part of town may return to the family home to socialize, or to get away from his wife and the restrictions of family life. The *mafraj* of the old houses, which often has a panoramic view of the old city, is considered the preeminent place to chew *qat*.[42]

Official support for conservation seems to follow the flow of donor interest and funding. Enthusiasm waned when foreign aid stopped in the early 1990s. It revived to some extent later in the decade with the interest of the World Bank and other donors, and the creation of a Cultural Heritage Unit in the Social Fund for Development. In the mid-2000s, government support seemed half-hearted and sporadic, spurred only by the

40 Since moving to the new districts, many families returned on Fridays to visit relatives living in the old family house. Now it is generally middle class families who make these weekly visits. The better off a family, the more likely it is that all members would have relocated to the new districts.

41 Cited by Lambert, "Consommation de masse et tradition a Sanaa," p. 93.

42 There is a culture of *qat* chewing, which includes particular postures and protocol, and recommends certain aesthetic and environmental conditions. For example, Sanaanis believe that while chewing one's body is more susceptible to drafts, hence windows cannot be opened (even with the copious cigarette smoking that accompanies *qat*). The use of the old family house for male activities has also occurred in the old city of Jeddah, Saudi Arabia: while families have relocated to modern villas, men return to the old houses to socialize, and sometimes set up offices there (Hasan Uddin Khan, personal conversation).

promise of funding or threats by UNESCO to remove Sanaa and other sites from the World Heritage List.[43] The situation grew worse after the recent uprising: donor interest in cultural heritage projects diminished in light of more urgent priorities. (At the time of this writing, the restoration of the Great Mosque was the only project underway in Sanaa.) Yet, as the director of the Cultural Heritage Unit noted, restoration projects are labor intensive and provide critical employment, especially in remote areas.[44] The inconsistency of support was lamented by professionals, some of whom had been committed to the old city for decades. Funds could always be found for international conferences, they noted, but rarely to help property owners with maintenance and restoration. "Conservation is just for public relations (*lil 'ilaqat al-'amma*)," said a veteran of the Campaign—that is, it's just to please outsiders. "We need to move discussions of conservation from the board rooms of officials and foreign experts, to the streets of the old city."

The Sanaani House: Attitudes and Adaptation

The family home is a fundamental institution in the South Arabian peninsula, representing the unity and permanence of a lineage. The term *bayt* is used to signify the house and the family that resides in it. It also refers to a form of ownership—the family endowment—that aims to maintain the indivision of property, and thus reinforces the tie between the family line and the house.[45] Many residents of the old city of Sanaa appear to have a deep emotional attachment to their houses, particularly when the house has been held by a family for generations. In other cases, the house may have been acquired only a generation or two ago—but may still be seen as evidence of the continuity of the family line. Conservation has touched these emotions, and also begun to transform them. Thanks to media coverage of the conservation effort, residents have become increasingly aware of their houses, as well as aspects of their everyday life, as heritage (*turath*). "Now everyone is aware of *turath*," we heard earlier from a builder's wife. "Before, it was something they just did."

As noted earlier, the term *turath* derives from the verb *w-r-th*, "to inherit." *Turath* associates the idea of *wirth* or family inheritance with the collective inheritance of society and, as such, follows and essentially translates the French term *patrimoine*. Many residents appear to have internalized this association, which makes the conservation of the city and its houses incumbent not only upon residents, but upon the government. Yet in some cases, *turath* conflicts with deeply held values of private property as enshrined in customary law—and some owners have challenged GOPHCY's legal authority on that basis (Figure 4.4).

43 In 2001, a special presidential commission was formed to address various threats to Sanaa's historic fabric cited by UNESCO (the High Committee for the Preservation of the Old City). The most recent UNESCO warning was in May–June 2013, as reported in the Wall Street Journal (http://blogs.wsj.com/middleeast/2013/05/28/battered-by-revolution-old-sanaa-could-lose-its-world-heritage-designation/tab/print/) and *Al-Masdar Online* http://almasdaronline.com/article/47061 (the latter mentions both Sanaa and Zabid; both articles accessed 8/13/13). In Yemen and other countries affected by the Arab Spring, government oversight has been sporadic or non-existent, leading to an increase in "violations."

44 Nabil al-Makaleh, Director of the Cultural Heritage Unit, SFD, personal communication, July 2014.

45 P. Bonnenfant, "Maisons. Voisins, et Exterieur," in P. Bonnenfant, ed., *Sanaa: Architecture Domestique et Société*, pp. 65–6.

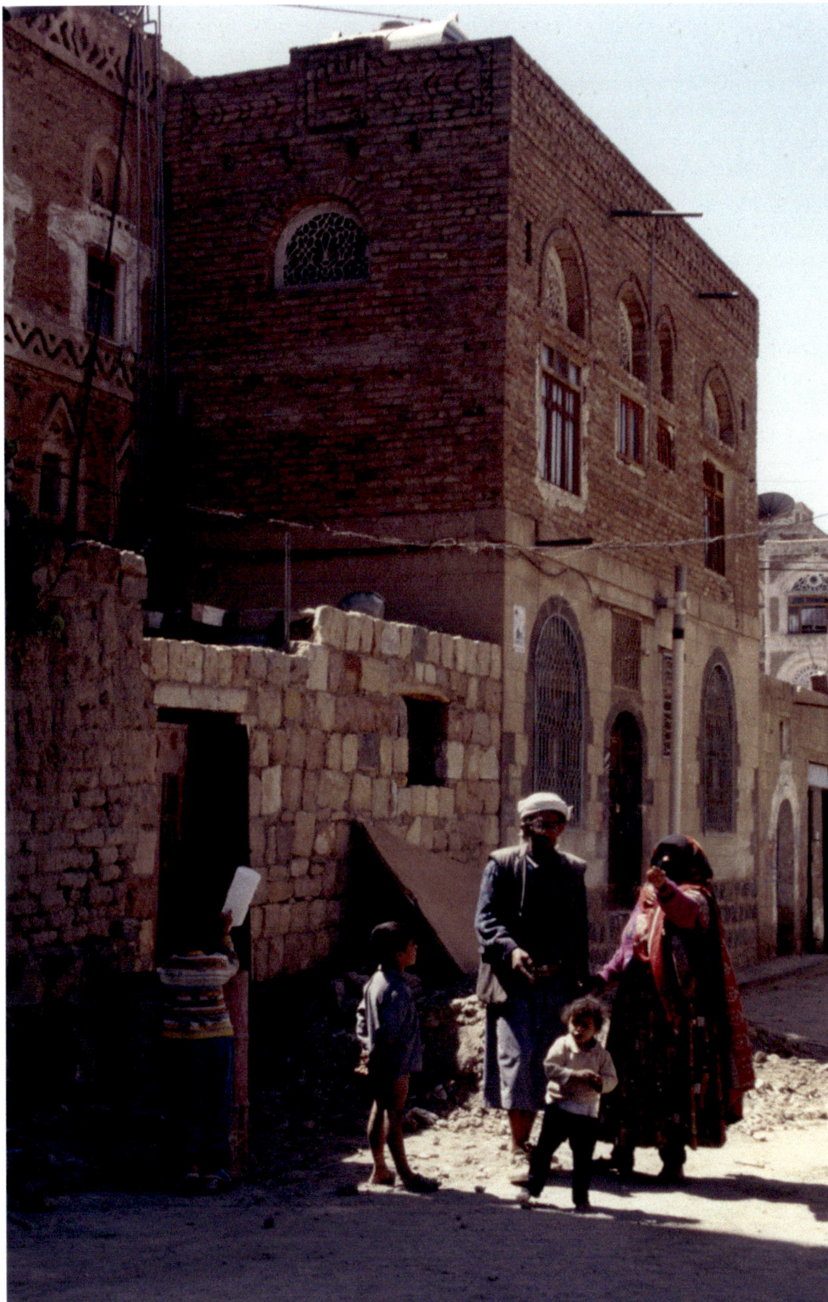

Figure 4.4 A woman yelling at an inspector, who has told her to take down a non-conforming wall

Here again, both sentiments are often expressed by the same individual. "The buildings of the old city are from our forefathers," said the assistant to the *'aqil* of the quarter of Bab al-Yaman. "Of course we must preserve them. They show the genius (*abkariya*) of our forefathers." He assured me that he had not changed his house at all, except to install a modern bathroom. But when I asked him about a proposed prohibition against vehicles in the old city—a prohibition of which he approved—he almost inadvertently said: "They want to make the old city a museum (*mathaf*), and this is difficult." When questioned about the term "museum," he went on to explain that Europeans and Arabs from other countries come to Sanaa to see something different, something that could be seen nowhere else. If the government wants to make the old city a museum, he said, it would have to buy the houses from their owners—but the owners would never agree to sell. To illustrate his point, he recounted the story of Imam Yahya's attempt to expand the Great Mosque. In order to do this, the Imam needed to buy five or six houses that stood in the way of the expansion—but the owners refused to sell. Even an autocratic ruler—with the aim of expanding the most important mosque in Sanaa—could not impinge on the rights of owners.

The multi-story houses of Sanaa—especially the larger ones, which have been taken as the prototype by scholars and architects—were often inhabited by several generations of a family, including the families of married sons (Figure 4.5).[46] Additional floors would be added to the house as sons married, or new structures would be built in the courtyard or on adjacent land. The tendency to expand vertically means that, in many cases, lower floors are often very old while upper floors are often less than a century old. The facade, in a sense, describes the historical development of taste—with larger windows at the top reflecting modern preferences. But because builders who add to and renovate houses tend to work in a style similar to that of their predecessors, it is difficult to date houses (or floors) by style.[47] Through the 1970s, before the UNESCO Campaign, new houses and additions were sometimes built entirely of traditional materials. Now, the typical addition is concrete block with brick veneer, and uniform window openings decorated with plaster (*nura*).

A number of scholars have explored the social and symbolic significance of the house, including the use and distribution of space, and the way it is supported by decorative programs.[48] Based on Bourdieu's analysis of the Kabyle house, Paul Bonnenfant identifies a series of oppositions in the Sanaani house—high and low, light and dark, right and left—which he relates to wider cultural themes. Privileged spaces, he suggests, are associated with height, light, and the mind, and hence with male society; these contrast with the low and dark spaces,

---

46 For a detailed discussion of the tower house and its construction, see R. Lewcock and R.B. Serjeant, "The Houses of San'a," in *San'a*, pp. 436–500; and P. Bonnenfant, ed., *Sanaa*. Most studies have focused on large or moderately large tower houses—houses that are five stories or taller—as the standard. Yet many small houses exist throughout the city, their size perhaps determined by family income and status. They are not only shorter, but their plan is smaller; the vestibule, or *hijra*, found at each level of larger houses is often absent—and perhaps was not needed if the house was to be inhabited by a single branch of the family (see below). Different branches of a family often inhabit contiguous buildings, which may have been built on what was once common property. (An elderly resident of al-Fulayhi told me that she had refused to buy a house across the street because it was not contiguous with the family's property.) To my knowledge, there has so far been no study of house sizes by period and area, which would provide important information on the city's history.

47 R. Lewcock and R.B. Serjeant, "The Houses of San'a," in *San'a*, p. 487.

48 See especially P. Bonnenfant, "La Symbolique de la Demeure" and "Maison Microcosme" in P. Bonnenfant, ed., *Sana'a: Architecture Domestique et Société*, pp. 534–8 and 554–9; G. Vom Bruck, "A House Turned Inside Out" ; and Naim-Sanbar, "L'Habitat Traditionnel a Sanaa: Sémantique de la Maison," *Journal Asiatique*, nos 1–2, 1987, pp. 79–113.

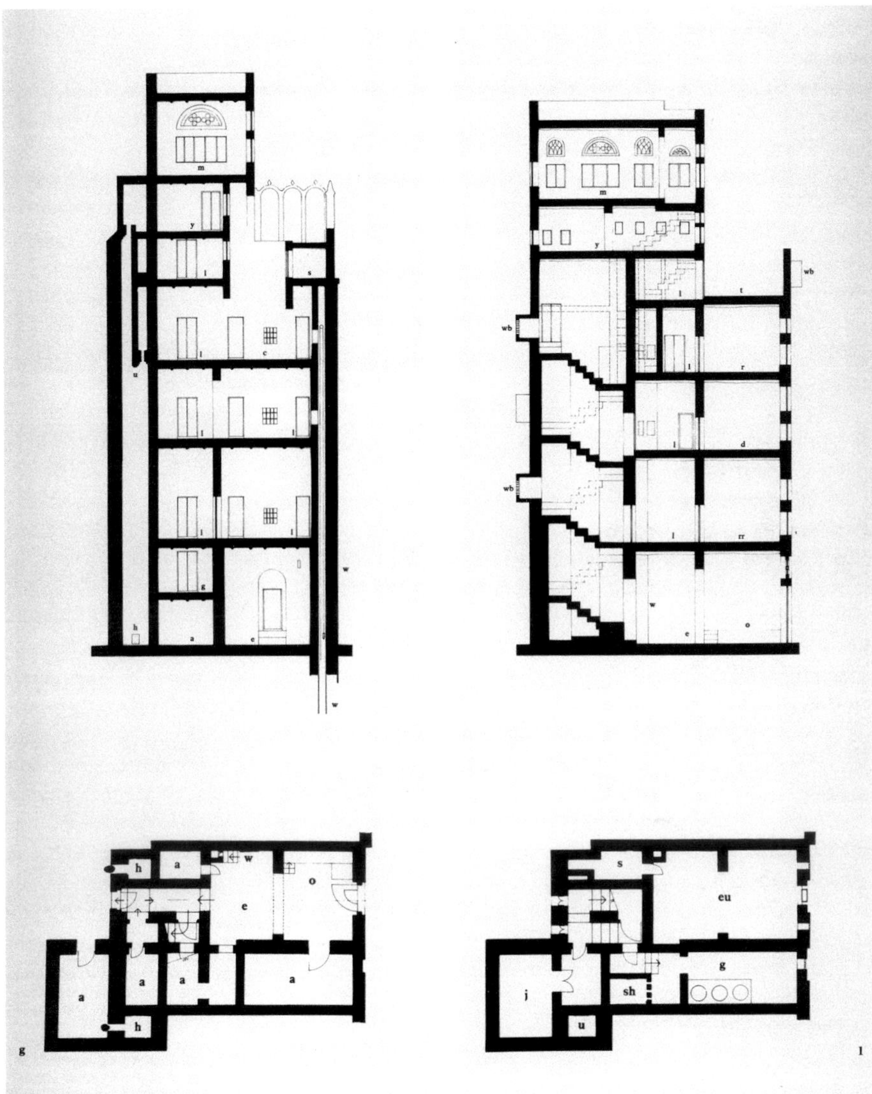

Figure 4.5 Plans and sections of a house

Source: San'a: An Arabian Islamic City. Reprinted with permission of Ronald Lewcock.

which are the realm of women and the bodily functions. Others argue that local perceptions of value and hierarchy are more nuanced and fluid, influenced by environmental, socio-cultural, and functional considerations. Gabriele Vom Bruck, for example, argues that Bourdieu's analysis is not applicable to the Sanaani house. Based on her study of houses of the old elite, she argues that domestic space is not gendered. Rather, the use and meaning of space depends on who occupies it at a given moment; and on his or her gender, age, and status within kinship and social hierarchies. As such, "[s]pace comes into being through practice; cultural meanings thus invoked are principally stable and contextual."[49] Since women are not allowed to interact with unrelated men (non-*mahram*), the quality of "public" or "private" is determined not by the space itself, but by the presence of a woman's body; it follows her body, inside the house and outside it. Nor is the "inside" of the house associated with women—since men use the house as a work and a gathering place. Naim-Sanbar studies evolving meanings and perceptions of the house through the terms that residents use to designate space. She points out that there is no simple equation between privilege and height. The kitchen—*dayma* or *makan al-mar'a*, "the place of women"—is on an upper floor (albeit for functional reasons).[50] The room at the top of the house is valued above all as a "place from which one looks" (*mafraj*): it represents not an interiorized conception of mind, but rather one that is activated through contemplation of the outside world. While the *mafraj* is usually reserved for male gatherings, women may also arrange to use it (Figure 4.6). The emphasis on view—along with verticality, extroversion, and shared gardens—challenge the Orientalist stereotype of the inward-looking "Islamic house."

Builders note the close relationship between the plan dimensions of a house (width and depth) and its height. The relationship is both aesthetic and structural: the height is related to the depth of foundations, but also to the width of the *qutba* (literally "pole" or "axis"), the column around which the central stair core is constructed.[51] In the case of tall houses, the walls of the house may be several meters deep at the base. The mass of the house absorbs sun during the day and releases it at night, keeping the interior at a relatively constant temperature. In fact, the value of a house is determined by the orientation of its primary living spaces. According to a custom that is often cited by residents and builders, a south-facing house (*'adani*, the direction of Aden), is said to be a complete house (*bayt kamil*); a house that faces west (*gharbi*) is called half a house (*nuss bayt*); one that faces east is called a quarter house (*rub' bayt*); and one that faces north (*qibli*, the direction of Mecca) is not a house (*ghayr bayt*). Services are usually placed along the north wall—for example, toilets, which are connected to a shaft that terminates at ground level. The favored south and west facades—those that receive maximum sun exposure—are reserved for living areas. In the past, it was most important that the *mafraj* face south; but now living rooms and bedrooms are considered more important, and thus should have a southern or western exposure. In Sanaa the privileged direction of Mecca (*qibli*) is outweighed by environmental factors that influence quality of life.

The ground floor of the house, often with a mezzanine, was traditionally used for the storage of foodstuffs from the countryside and sometimes to keep animals. It is here that one finds one or more wells (*bir*) from which water was drawn to upper floors, and the grist mill (*madhan*), where wheat was ground into flour. On the exterior the ground floor is expressed as a tall and forbidding stone plinth with minimal openings, that contrasts to the

49 "A House Turned Inside Out," p. 165.

50 The room was so identified by men interviewed by Naim-Sanbar. Lewcock and Serjeant write that women were not restricted to a particular part of the house, although the kitchen is largely their preserve, along with an adjacent terrace screened from outsiders' views by high walls. Naim-Sanbar says one floor of the house was usually reserved for guests and another for women, but I did not find evidence of this in my field work.

51 This discussion is indebted to Lewcock and Serjeant, "The Houses of San'a," in Serjeant and Lewcock, eds, *San'a*, p. 441; and to conversations with Nadia al-Kawkabani, architect, and Ahmad al-Rawdhi and 'Abdullah al-Rawdhi, builders.

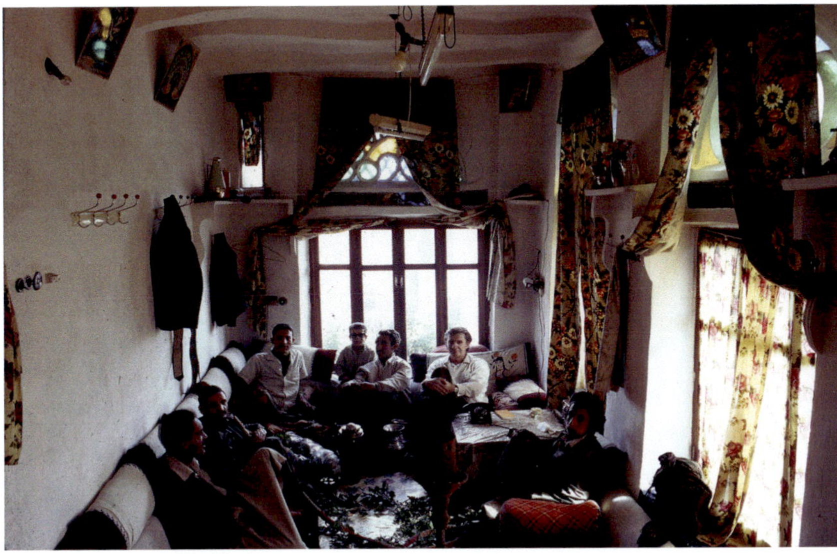

Figure 4.6 A qat chew in a mafraj in the old city, circa 1982

Source: Photograph by Ronald Lewcock.

more domestic floors above, in fired brick and plaster (Figure 4.7). A series of thresholds control access to various parts of the house, required to preserve the seclusion of women. From the central hall at ground level, one gains access to a stair that serves the upper floors; at each landing is a small room that separates the stair hall—where unrelated men may circulate—from the private realm of the family.[52] The term for this space, *hijra*, derives from the Arabic root *h-j-r*, to forbid or to deny access; in South Arabian languages *h-j-r* meant "to protect," in ways that included the magical.[53] The upper floors consist of a series of living rooms, including the *makan al-wasat* ("place in the middle"), usually on the second floor, which served as a common space for the family.

Although rooms were generally furnished in a similar matter—with continuous cushions around all walls—they are distinguished by their size, orientation, and ornamentation. According to residents, the most formal of the sitting rooms, the *diwan*, ideally runs the length of the house; it thus has light on three sides. At the very top of the house is the

---

52 It is still customary for unrelated men shout "Allah! Allah!" as they ascend the stairwell, to avert women above to their presence. The women would then retreat into private quarters, or don the veil—a kind of "threshold" that restricts perception of the woman's body.

53 N. Naim-Sanbar, "L'Habitat Traditionnel a Sanaa," pp. 87–91; and Vom Bruck, "A House Turned Inside Out," p. 144. *Hijra* also denotes a settlement that is neutral ground or refuge, a place where one seeks protection. Sanaa itself is believed to have been a *hijra* (R.B. Serjeant, "San'a, the Protected, *Hijrah*," in Lewcock and Serjeant, *San'a*, pp. 39–43). Similar words that indicate protection are attributed to the city and the house; for example, *hima*, which indicated protected grazing lands outside the walls (p. 40) also refers to the "protected" perimeter of the house. The idea of protection is echoed in decorative elements on the facade which probably derive from, and may still be used as, talismans (P. Bonnenfant, "La Symbolique de la Demeure," pp. 534–8).

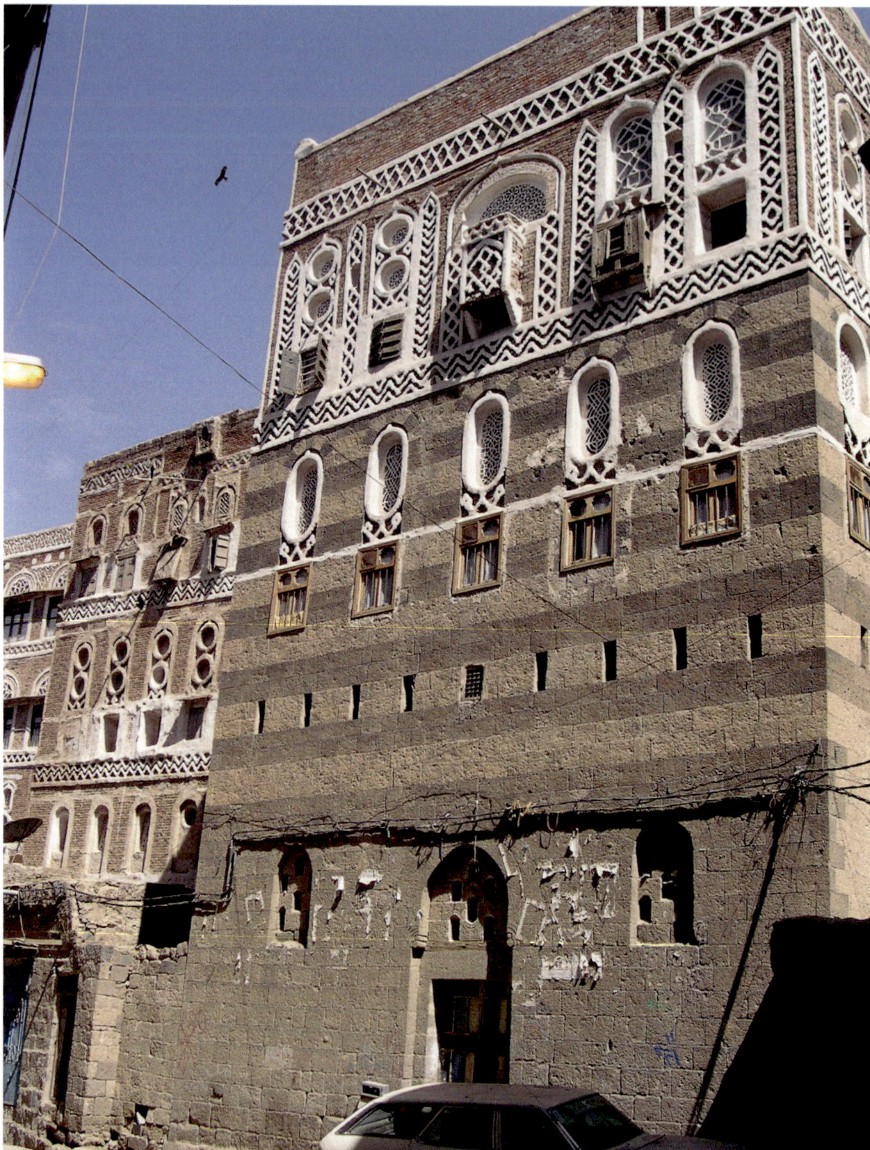

Figure 4.7 Bayt al-Mahdi, in the area of al-Zumar

mafraj which, like the *diwan*, often has refined plaster ornamentation. Its windows are markedly larger and more colorful to optimize views and aesthetic effects, particularly in the hours of waning light. (The hour before sunset—*al-sa'a al-sulaimaniyya*—is a time of contemplation rather than conversation; it corresponds to the sense of melancholy that sets in after several hours of chewing *qat*.) Within these rooms, the privileged position is "the head of the room" (*ra's al-makan*), that is, the short wall farthest from the door.[54] The ornamentation of these rooms contrasts with the plain surfaces of functional rooms, where food is stored and prepared and human waste eliminated. The kitchen (*dayma*) is usually just below the top floor, to facilitate ventilation; now that most kitchens have been modernized, the old kitchen is recalled as the place of the wood-fired oven (*tannur*) where bread was baked, often several times each day.[55] The *dayma* typically receives little light: it has few windows, and is often located on the northern wall.

In recent decades the interior organization of the house has been transformed in ways that reflect changes in the family and the wider society. Since at least the 1970s there has been a tendency for families to organize as nuclear units. In many cases, sons leave the family home when they marry—something that was unthinkable a generation ago—and rent or build in the new districts. When they remain in the old city, they usually live with their wife and children on a single floor of the house; if space and finances permit, they install a kitchen and build partition walls to create additional rooms. In larger houses the *diwan* and, according to Naim-Sanbar, even the *hijra* may be partitioned to create separate rooms, or to allow for a kitchen or modern bathroom. When several branches of the family continue to occupy a house, they may eat a common meal during the day, but in the evening each branch tends to stay on its floor; according to one resident, this tendency has been reinforced by television. When a man has more than one wife, each woman and her children may live on a separate floor of the house; but more often, the second wife lives in another house. The transformation of the tower house into an apartment house was noted by the Quaroni Bonifica team. Accordingly, they modernized and renovated a "model house," Bayt Sari', for use as several apartments (to be discussed in Chapter 6). Indeed, the apartment house model seems the only way the tower house can remain relevant. Yet multiple kitchens and bathrooms result in a four- or five-fold increase in water usage on site—exacerbating the problem of groundwater, which destabilizes foundations.

Division of the tower house into separate apartments contrasts with the old arrangement, where the entire family lived on all floors. Children often slept together, with grandparents. Women shared a single kitchen, and would often take turns cooking and cleaning.[56] Now, residents told me, sharing a kitchen is no longer acceptable because each woman wants her own kitchen. Moreover, by assigning each nuclear family to a floor, it is easier to control numerous and unruly children. The last point is significant, if one recalls that the birth rate—or rather, the number of children who survive infancy—has skyrocketed in recent decades. Changes in lifestyle and preferences are often cited as factors that impel residents to leave the old city. But according to the overwhelming majority of residents with whom I spoke, the main reason people leave the old city is expanding families and crowded houses (*izdiham al-buyut*). This was a point rarely noted by officials and conservation professionals.[57] As we have seen, Sanaa historically met the needs of an expanding

54 J. Lambert, "La musique dans la maison tour: harmonie et dissonance," in P. Bonnenfant, ed., *Sanaa: Architecture Domestique et Société*, p. 169.

55 N. Naim-Sanbar, "L'Habitat Traditionnel," p. 99. At the time of my fieldwork many women still baked bread, usually once daily, in a gas-fired oven.

56 During my field work I visited numerous houses that were reorganized, loosely or formally, into apartments; the old arrangement was related to me by elderly residents. I also saw a number of small houses owned or rented by poor and working class families that were partitioned into apartments, with up to eight children living on a single floor.

57 Only two officials noted this point, Qadi 'Ali Abu Rijal and Hussein al-'Amri. "Families of

population through densification. But now, planners say, change cannot be accommodated in the traditional way: density should not increase, in order to maintain a livable ratio of open to built space. New construction in open spaces is prohibited by the conservation office, and vertical additions have been severely restricted, because of structural concerns or because the added height will adversely affect the skyline.[58] But the question of how population pressure is to be accommodated—or not—has not so far figured into conservation planning; there continues to be an emphasis on aesthetic concerns, a tendency that World Heritage guidelines encourage.

Owners are usually proud to show a foreign researcher around the house; they immediately point out obsolete facilities—the well, the mill, and the wood-fired oven—as signs of *turath*, relics of historical interest. But some complain about the house's deficiencies, including small rooms, lack of light, the many steps and the separation of functions on different floors; many residents wish they had a private garden.[59] When women shared a kitchen, this was seen as a major inconvenience. Kitchens and bathrooms are in a sense the locus of modernity, and most have been partially or fully modernized. In the 1970s Carla Makhluf reported that many Yemeni women saw the introduction of modern kitchens and bathrooms as one of the most important signs of progress since the revolution.[60] Two women I interviewed noted that modern furnishings and finishes make it easier to clean the house: "This is why women in the old city are old at age forty," said one, "because the work is tiring, and makes us old."[61] Although these women say they prefer to sleep on the floor, many people prefer beds—which leads to a new kind of functional zoning within the house. But there are few beds in the old city: the matrimonial bed—which is now often specified in marriage contacts (again, by women!)—is not easily accommodated in the small room reserved for a couple's use. (I saw only one such bed in an upper room of Bayt Sari'; it must have been built or assembled in place.) Similarly one rarely sees tables, sofas, and chairs—while in new houses at least some rooms are designed for such pieces.[62]

In raising the seated individual above the floor to socialize, to eat, and to sleep, furniture affects an entire complex of cultural norms. In the old houses interior spaces—and therefore,

---

two or three children used to be the norm, the only ones to survive birth and illness," said al-'Amri. "Now families of ten children or more are common."

58 In the mid-2000s a UNESCO mission reported that there was a moratorium on all building pending the implementation of a conservation plan; this was confirmed by 'Abdullah Zayd 'Isa, Director of GOPHCY. But inspectors told me that when a resident applied for a permit for a vertical addition, it might be approved if it was consistent with the height of neighboring houses.

59 The desire for a garden may be a new one. Although many houses front on a *maqshama*, they do not necessarily have direct access to it; even if they do, access and use are not private. Residents may have represented their houses more positively to me as a foreigner, assuming that I was interested in *turath*. During her fieldwork with a German-led survey in the mid-1990s, al-Kawkabani found that even residents who repaired and maintained their houses did not like them.

60 *Changing Veils: Women and Modernisation in North Yemen*, London: Croom Helm, 1979; cited by Naim-Sanbar, "L'Habitat Traditionnel," p. 105.

61 When a female GOPHCY employee expressed admiration for a traditional bathroom that had still not been replaced, the owner of the house (a woman) said to her, "You like it because you don't have to use it every day!" While plumbing and appliances have been introduced in these spaces, modern finishes have generally not been applied. One woman described the modern bathroom she envisioned "covered with white tiles." This image from European or American interiors must be attractive: if bathrooms are not finished in *qaddad* (waterproof plaster) and continually maintained, they tend to retain moisture and smell.

62 New houses of the affluent and educated middle classes often provide for both lifestyles: they usually have living and dining rooms, and also designate one or more rooms for traditional floor seating.

the exterior facade—were designed according to the height of the individual, seated on a floor cushion. The width of rooms was determined by the bearing capacities of wooden beams, which in turn relates to the preferred social distance between interlocutors, who are seated on cushions facing each other. When furniture is desired, builders must raise the height of window sills and heads, and therefore the ceilings. This changes the proportions of the windows and of the facade.[63] The problem bears directly on the conservation of houses and guidelines for new additions. The traditional hierarchy of window openings—which become larger and more numerous on upper floors—is seen by historians, architects, and builders as a defining characteristic of Sanaani architecture.[64] But the distribution of window openings reflected the collective living arrangements of an extended family, with different uses on different floors. Now the tower house has become, in effect, an apartment building—with each floor dedicated to the *same* use, that is, the dwelling of a nuclear family unit. Can the old hierarchy of the facade be conserved if it no longer relates to interior functions? Should a family living on a lower floor have fewer, and smaller, windows in an attempt to preserve a facade in its historical form?

The loss of hierarchy seems to be a modern theme, and in Sanaa it can be seen everywhere—from facades with uniform windows to pop music which, as a builder noted, is driven by a constant rhythm rather than by melody and structure.[65] Everywhere modern conservation is practiced, difficulties arise where inside meets outside, or where content impacts form. These are usually resolved by giving priority to exterior facades, which define public space and are thus of primary aesthetic and historical interest. In addition, a distinction may be made between the main facade and side facades, with more stringent guidelines applied to the former. But the logic of this approach is not obvious to some professionals, who see "tradition" embodied as much, or more, in a building's interior. One veteran of the Campaign recalls that in a 1991 seminar led by the Quaroni Bonifica team, an Italian architect said that "anything can happen on the inside [of the house], as long as the outside remains the same." He and several colleagues objected. "If the house is a whole that reflects the life of a society," he asked, "why should different standards be applied to inside and outside?" These feelings were echoed by Ahmad al-Rawdhi and 'Abdullah al-Rawdhi, *usta*s who work specialize in traditional construction, and also build "new *turath.*" (Their father was Muhammad al-Rawdhi, a lead *usta* for the UNESCO Campaign, whom we will meet in the following chapters.) They counseled me on several occasions to spend time inside the houses if I wanted to experience the old city. "Tourists, too, should go into the houses if they want to know what Sanaa is," said Ahmad. "It's not just about the external appearance (*al-mathhar*)." On other occasions, he maintained that buildings should be faithfully kept "as they were" on both the inside and the outside.

---

63 For more detail on this point, see Chapter 5. In new houses, all rooms have higher window sills. Rooms furnished in the traditional manner—with floor cushions—have to adapt to the higher sills: the standard height of the back cushion (*wisada*) has therefore increased from 20 to 45 cm (N. al-Kawkabani, personal conversation).

64 At a 2004 training session at CATS—part of the UNESCO-led survey and inventory of the old city—the loss of hierarchy on facades was identified as a major threat to the historical character of the city. See also D. Pini, *Inventory of the Historic City of Sanaa*, pp. 53–4.

65 A builder who also plays *'oud* remarked: "Today young people (*shabbab*) are influenced by TV. They are interested only in rhythm, not in the melody and how it goes with the words." Although the builder did not draw the above parallel between loss of hierarchy in music and architecture, he agreed with me when I proposed it. For analogies between architecture and music, see Chapter 5.

"What you see of the old city is all for show (*décor*)," said one resident. "Inside houses have changed completely." But this is probably an overstatement, as the structure of the tower house makes it difficult to reconfigure interior space. I saw few houses that had changed significantly, except for modernized kitchens and bathrooms and the erection of interior partitions. Few models have been proposed for the adaptive reuse of houses. Where substantial change is desired, owners prefer to demolish and rebuild; clandestine demolitions have increased alongside property values.

Limits of Adaptation

Despite misgivings about the house interiors, many residents seem to be sympathetic to conservation. Most of those with whom I spoke were aware of regulations that prohibit changes to facades, and require use of traditional materials in any renovation or new construction. They generally approved of the regulations—often citing the need to maintain *turath*, "the work of our forefathers (*ajdadna*)."[66] Their approval of conservation—even as their houses become less suited to modern lifestyles—suggests that they may in fact be slowly dissociating outside from inside, with the facade embodying *turath*. Yet they seemed to have a more fluid interpretation of *turath* than the one described in official guidelines. Adding additional floors, or building annexes to house married sons, is after all traditional, intrinsic to *turath*. The owner of a well-preserved five-story house in al-Nahrayn hoped to build a one-story annex—of brick, "in the old style"—in the adjacent courtyard for his son when he married. "Many generations have been born in this house," he said. "I will never leave, I will die here … we old people will preserve these houses." Building in open spaces, however, is now prohibited in the old city and few exceptions are allowed.

Conservation professionals argue that the traditional mechanisms of adaptation no longer work. They speak of preserving the "character" of the city, but this often translates into practical planning concerns: for example, the need to maintain a balance between built and open space, which has decreased dramatically in recent decades—especially in the urban gardens, dozens of which have been restored by the Social Fund. Vertical additions are banned, at least in theory. They present structural problems that are recognized by both architects and builders: the shallow foundations of the tower houses have limited bearing capacity, especially now that the soil is water-laden. Upper stories are often built in concrete, a rigid cap on the masonry and wood structure below; with settlement and movement, walls may crack and ultimately fail. But for both local professionals and UNESCO, aesthetic and historical factors are at least as important as structural concerns: vertical additions adversely affect the skyline, an important aspect of the city's character. This is not, in fact, so different from traditional practice: a good builder, I have been told, assesses the surrounding context—literally, the "air" (*jaw*) around the house—and builds what is appropriate to it.

According to a study conducted in 1999, about two thirds of residents said they were able or willing to repair their houses; the figure correlated roughly with the number of houses that were reasonably well maintained.[67] Most residents with whom I spoke recognize the value of traditional materials and techniques and would like to continue to use them. Some were aware of the value of local architecture as a system—its strength, its environmental

66 These findings are echoed in Eckert and al-Maqrami, "Social Assessment Report."

67 Eckert and al-Maqrami, "Social Assessment Report," p. 25.

properties, and its artistic qualities. But according to planners, the cost of maintaining the massive tower houses is beyond the means or priorities of most families—especially if maintenance is to be carried out according to stringent conservation guidelines. Traditional materials like stone are in short supply, or are produced differently. Brick (*yajur*) and hand-hewn stone (*hajar muwaqqas*) are significantly more expensive than new materials like concrete block, which is produced locally. (Architects said that labor for traditional construction work is more expensive, but builders I spoke with disagreed.) Several builders saw the lack of suitable brick as the single largest obstacle in conservation. The old-style brick is thinner, they said; now made only by special order, it is more expensive. But even on special order, the thin brick does not perform like the old wood-fired brick, which was less porous. Builders therefore generally use new (or standard) brick, and the dimensions of this brick affect the proportions of the facade.[68] "GOPHCY doesn't care about what kind of brick we use," said one builder. "They are only interested in the size of openings—and this is their biggest mistake." Stone is in short supply, and lime is no longer produced in the old way.[69]

Owners often opt for hybrid solutions, applying decorative brickwork and stone cladding on a new constructive system (Figure 4.8). Walls are typically built of concrete block instead of stone and fired brick; where stone is used, it is often machine-cut (*manshur*) rather than hand-finished, although machine-finishing is technically prohibited in the old city. Ceilings and floors are made of milled woods of standard sections instead of what builders call *baladi* ("domestic," typically *tanab* for beams); floor surfaces are cement on plywood. Interiors are often finished in traditional plasterwork. The use of structural concrete, hidden from view or disguised by brick veneer, has been much debated among conservation professionals. Some see it as a natural evolution of building practice, while others feel it changes the character and qualities of the architecture.

Such hybrid solutions are usually approved by inspectors who, following local and international guidelines, are primarily concerned with the exterior of buildings: as one builder put it, "GOPHCY is only interested in exterior 'make-up' (*maquillage khariji*)." The "traditional" facade of the Suq al-Baqar clinic was applied to a concrete frame under the supervision of a general contractor (Figure 4.2). One of the *usta*s who worked on the facade told me that he did not enjoy the work. "I want to be independent," he said. "I want to know what I'm building on top of." The reduction of "tradition" to veneer—a process that occurred in the nineteenth century in Europe and the United States, as part of the industrialization of building—ultimately undermines the critical potential of the architecture as a constructive and environmental system, and the local knowledge on which it is based. By focusing on image rather than substance, conservation guidelines may contribute to the demise of the heritage they purport to preserve.

---

68 The old brick was 4 cm X 20 cm; new or standard brick is 7 cm X 16 cm. In 2005 the latter cost YR 13, as opposed to YR 17 for special order, old-style brick.

69 According to a *qaddad* master, lime (*nura*) is now fired with *salit*, an automotive oil. A mason told me there are three types of stone in Sanaa: *nuqumi* (from a quarry on Jabal Nuqum); *haddi* (from a quarry in the village of Hadda), and *habash*, black volcanic stone. The quarry on Jabal Nuqum was closed when a military installation was installed; the quarry at Hadda is also closed. "When we need *nuqumi* stone we use *bahluli* [from the area of Bani Bahlul]." On the road outside Sanaa, he said, you can pull over at various vendors' sites and buy a wide variety of stone.

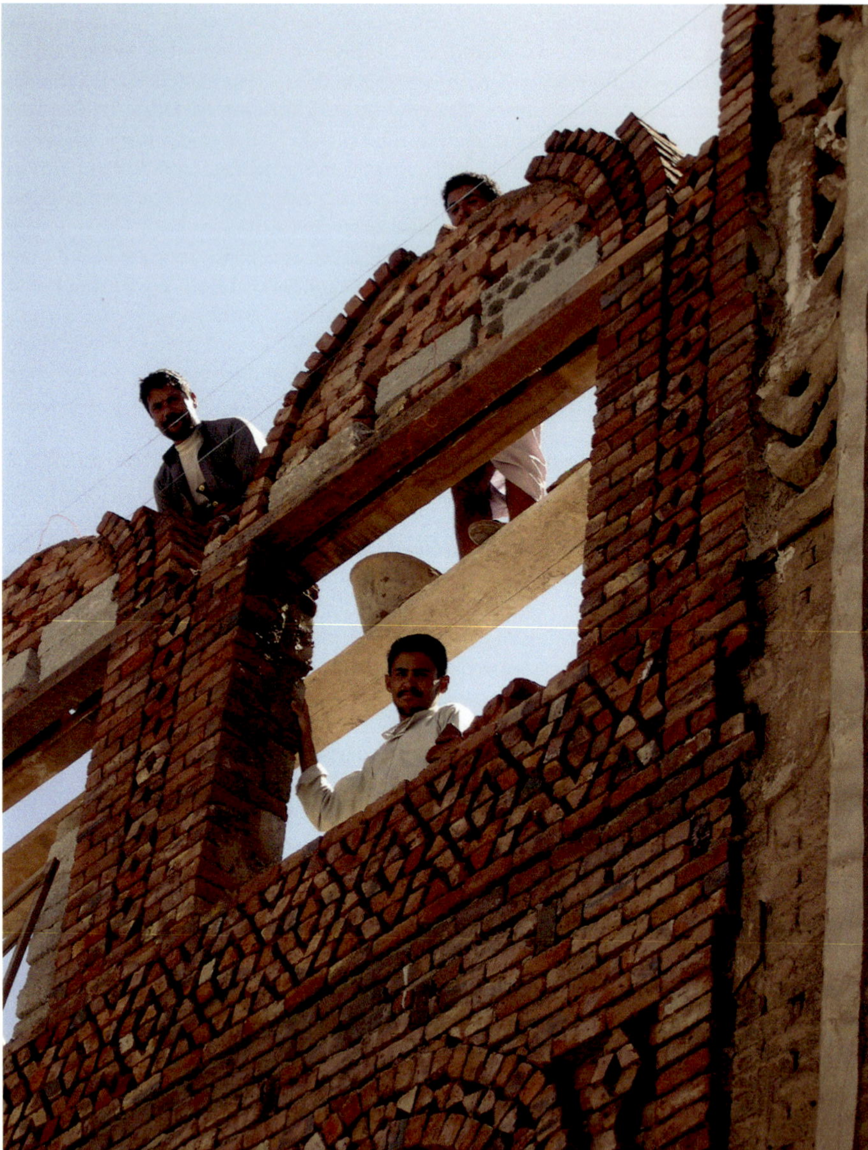

Figure 4.8 **In additions in the old city, traditional brickwork is usually applied as a veneer on top of concrete block**

There are other disincentives to maintenance, in particular absentee and shared ownership.[70] Inspectors told me that inheritance disputes were one of the main causes for the neglect of houses. Upon the death of the owner, the house is supposed to be divided among the owner's children. Since most heirs no longer reside in the house, division of property is often delayed. The relatives who remain in the house are often poor or elderly, and do not have the means to maintain or repair the building. In some cases owners rent their houses, as a whole or by floor, to unrelated families. In this case the incentive to maintain the house may be even less, since rents are low—in dramatic contrast to the high commercial rents in the old city.

Owners with whom I spoke were attached to the family houses, but were also awarc of their monetary value which had skyrocketed since the UNESCO Campaign. Largely as a result of street paving and infrastructure, the value of houses has appreciated dramatically. Because there has been little market for the houses themselves—the well-to-do, the educated, and the upwardly mobile middle class continue to prefer modern houses in the new districts—the old city has not yet gentrified. Few families have been able, or willing, to sell. But some individuals envision turning the family home into capital. A young woman in al-Fulayhi quarter told me that when a foreigner made a generous offer on their house in the 1990s, none of the family wanted to sell. But now, she said, they might sell if the price were right. On another occasion, we were talking with her elderly father, who said that he would never agree to sell the house. "It is impossible for a Sanaani to sell his house to a stranger," the daughter agreed.

Suqification

Given economic circumstances and high commercial rents, there has been pressure to put portions of the house to productive use, especially the ground floor. This level is no longer used to store foodstuffs from the countryside, so it is functionally obsolete. It opens onto the street, a convenient place to install a shop. But this requires changes in both form and use that contravene conservation guidelines. The facade of the ground floor and mezzanine—a tall, heavy stone base with minimal openings—is considered an important aspect of the aesthetic of Sanaani architecture and a determining characteristic of the streetscape. Sometimes the stately portal of the house is used as a shop entrance; in other cases, new openings are created. Both solutions are rejected by inspectors on aesthetic grounds. There are also structural considerations: new openings in the massive stone walls are often not designed to carry the tremendous load of the floors above. This often leads to differential settlement and cracking.

The use of the ground floor for shops is part of a process that has been called "suqification": the spread of commerce into areas that were largely, or exclusively, residential (Figure 4.9). This is a common phenomenon in conservation zones in many parts of the world, due to rising rents. According to the planners who coined the term, suqification violates traditional zoning in the city which had ensured familial privacy—that is, the privacy of women. When a shop is opened in the ground floor, they argue, the family living above is impelled

70 Ownership (*mulk*) is traditionally the preferred status of occupation; it should be full and complete ownership, rather than co- or shared ownership (Bonnenfant, "Maisons. Voisins, et Exterieur," in P. Bonnenfant, *Sanaa*, pp. 65–6). But this ideal is rare in practice.

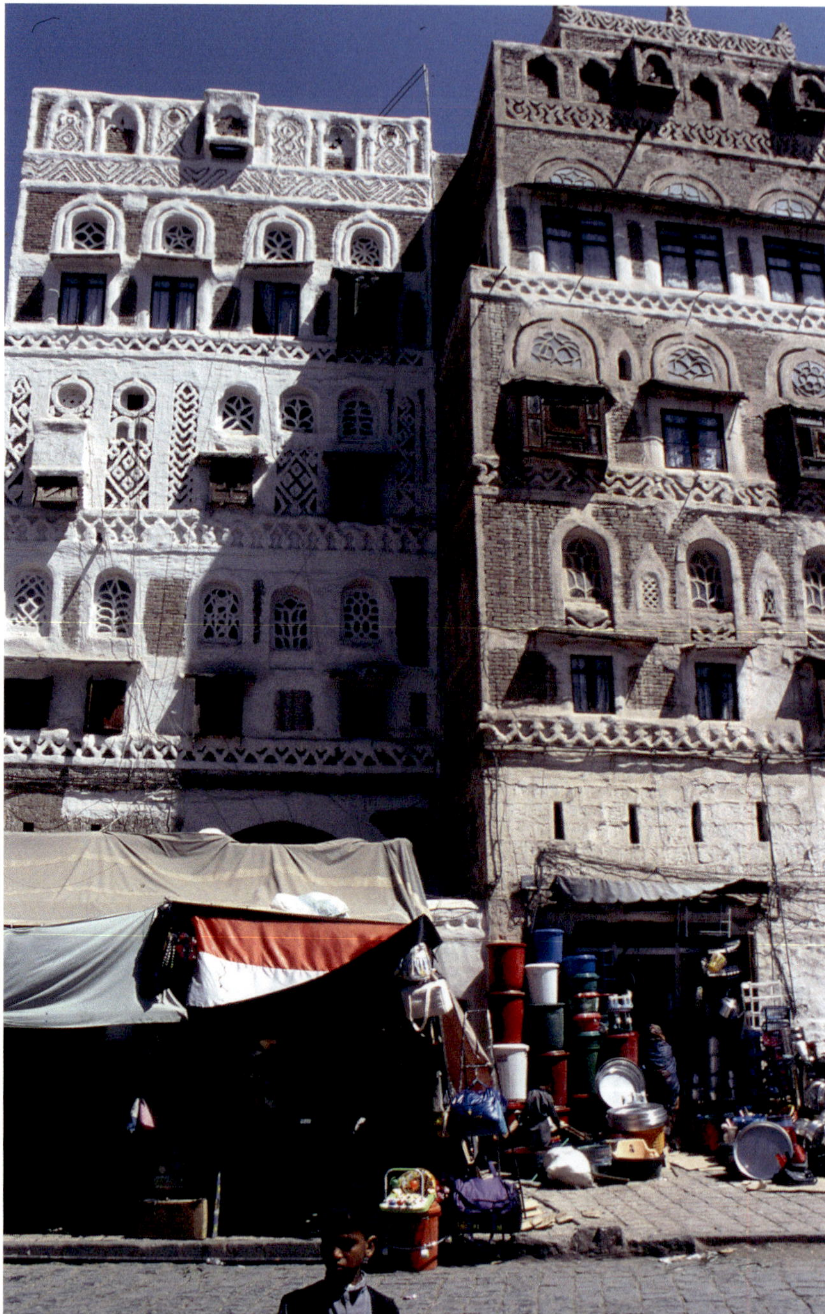

Figure 4.9 Suqification: the ground floor of these houses in al-Zumar have been transformed into shops

to leave the house. As a result, the house falls into disrepair and ultimately collapses.[71] But the theory is questionable. Groundwater is the most frequent cause of collapse. Ron Lewcock also doubts that traditional zoning was so rigid: certainly in the twentieth century, well before the advent of conservation, commerce existed in some residential areas. The area called al-Zumar, the northern end of the spine that runs from Bab al-Yaman through the market to Bab al-Shuʻub, probably always had some commerce, but was thoroughly suqified before the revolution. Shops were built in front of many houses, or installed in the ground floor. As of the mid-2000s, many of these houses were still owner-occupied. The notion of rigid functional zoning may derive from writing on "the Islamic city," which often assumed the rigid separation of residential districts based on familial privacy.[72]

Yet the negative perception persists. "No one would ever agree to live above a shop," an architect told me—a point that was echoed by several builders. But some residents I met did not agree. Often a homeowner installs a shop himself, I was told. He operates it or assigns it to the care of a young family member; in some cases, he rents it to a merchant unrelated to the family. In many cases the owner continues to occupy the house with his family. When asked if shops in residential areas compromise privacy, residents generally responded in the negative. "Women come and go freely," said one man in the quarter of al-Nahrayn. "We are like one family here." Some residents, including several women, said that shops make life more convenient: one cited the old adage that proximity to the *suq* is desirable.[73] These contradictory reports may perhaps be explained by the disjuncture between the image and the reality of the old city. Values and customs are changing, in part as a result of the market economy. Increasing numbers of homeowners are renting floors to unrelated families, for example, which would have been unthinkable a few decades ago. The old values and ways continue to be confirmed verbally, though they are often contradicted by practice.

As the immediate interface between the city and the family, the question of the ground floor captures the dilemma of conservation. Suqification has been highlighted in various UNESCO warnings, and GOPHCY has responded in various ways, for example, by attempting to enforce a moratorium on the opening of shops in houses. During my fieldwork, I witnessed acrimonious confrontations between homeowners and officials at GOPHCY headquarters, with the former pleading that they needed the income from a shop to provide for their families.[74] Their resentment was fueled by the large-scale violations visible throughout the city, carried out by those with influence (*wasta*). One architect pointed

71 This theory of *suqisation*—usually translated into English as "suqification"—was developed by Jacques Feiner and Hadi Eckert. Feiner elaborated the thesis in his dissertation, "La vielle ville de Sanaa: analyse morphologique comme fondement de la sauvegarde patrimoniale," Doctoral thesis no. 1652, Department of Architecture, Ecole Polytechnique Fédérale de Lausanne, 1997.

72 Eckert and al-Maqrami, for example, wrote: "… the strict physical separation between residential areas and the suq area [is] … a major distinctive feature of historic Islamic cities" ("Social Assessment Report," p. 29).

73 On this point see P. Bonnenfant, "Maisons, Voisins, et Extérieur," p. 25. Attitudes toward the *suq* must have varied according to social status and lineage. The *suq* was seen as an inappropriate place for members of the old elite, with the exception of certain highly regarded professions like book-binding. It was a place of rough behaviors; an interface between the urbane world of the city and the world of visiting tribesmen. See F. Mermier, "The *Suq*s of Sanaa," p. 151; and G. Vom Bruck, "The House Turned Inside Out," p. 155.

74 The potential income from a shop is enormous. In the mid-2000s in active commercial zones like as-Sukkari (the road just north of and parallel to the southern wall), key money could be hundreds, or even thousands, of U.S. dollars; monthly rent reached \$500–\$600. For comparison, the daily wage of an unskilled laborer in the same period rarely exceeded USD 5. To my knowledge there has been

out that the ban on shops sidestepped an important issue—the functional obsolescence of the ground floor:

> In all the conferences and symposia on the old city no one has ever mentioned this. They only talk about how the ground floor is essential to the historic character of the city. But it is impossible for us to leave it without a function … If the idea of conservation had been developed here, rather than abroad, it would be completely different.[75]

The key, many professionals feel, is to give people alternatives. "Instead of prohibiting openings in the ground floor," said a senior architect at GOPHCY, "perhaps we should help owners design the openings, so that they are aesthetically acceptable and structurally sound." Such an approach would add a prescriptive element to the largely proscriptive guidelines.

The authors of the UNESCO Campaign, both foreign and local, maintained that the value of the old city of Sanaa lies not its monuments, but in the some 5,000 houses that make up the fabric of the city. Yet no means has been found to conserve this fabric. During the Campaign, there were repeated calls for a small loan fund to assist homeowners, but a fund was not established.[76] Over the years there have been some attempts to assist homeowners, but beneficiaries and funding have been limited. Conservation professionals agree that it is a key issue. "If historic cities are the property of the collective, even the world, should individual homeowners be made to bear the cost of their maintenance?" observed 'Abdullah al-Dailami of the SFD.[77] "At least a city like Venice derives some benefit from its status as a world heritage city," said one resident. "They have lots of tourists, but what benefits have we seen?"[78]

The designation of *all* houses as historic raises the question of conservation standards and methods: it suggests that the conservation of historic cities should perhaps be reconceived as *maintenance*, which was traditionally carried out by residents and builders, rather than experts. "Residents had a spiritual tie to their houses," said a professional who now lives in the old city. "This tie translated into practices (*suluk*), like watering down the earth in front of one's house, and renewing plaster (*nura*)." This is perhaps why residents

---

no study of rents in the old city, which would have to be analyzed alongside ownership and other demographic factors.

75 The Quaroni Bonifica studies envisioned the reuse of the ground floor for living space, and the 1978 Shankland Cox report recommended that the ground floor be used for parking. According to Lewcock, both conservation professionals and local policymakers rejected the latter idea (mission report #2, 1982). Ahmad al-Ibbi reportedly studied the problem of the ground floor.

76 In 1989, for example, a working group urged "… that the government should as a matter of urgency implement their plans for a Campaign loan bank to support and provide incentives to private owners to restore their buildings in the old city in accordance with standards set up by the Campaign Executive Office. … Also some funds should be made available to Campaign office to allow it to intervene in reconstruction of houses in danger where, for some reason—disputed ownership or inheritance—residents are unable to undertake the work themselves …." (Third International Working Group, 24–26 January, 1989, GOPHCY Department of Projects and Planning, folder labeled *Dirasat 'Ama*).

77 Now Managing Director of the SFD, al-Dailami was Director of the Cultural Heritage Unit in the mid-2000s.

78 Contrary to popular perceptions, Venice is not a World Heritage city. Although it was the site of one of the first UNESCO International Safeguarding Campaigns, it has never been inscribed on the World Heritage list.

and builders have so easily embraced the idea of conservation, which seems to validate familiar practices.

Merchants and the Ministry of Endowments

It has been often noted that conservation—which aims at slowing or arresting material decay—is in fact a powerful agent of change. In the old city of Sanaa property values rose dramatically when streets were paved and infrastructure installed. Yet it would be a mistake to attribute the changes in the old city—particularly the booming economy of the *suq*—to conservation alone. Since the 1970s, consultants have been concerned about the centrality of the old city, located at the convergence of important urban and regional arteries. In the years following the revolution, a modern government and commercial district developed just to the west of the old city, in the area of Tahrir Square. Consultants feared that this would exert considerable development pressure on the historic core.[79] This pressure—along with demographic changes, the decline of traditional trades, and certain administrative reforms—have led to a complete transformation of the old *suq*.[80] However, it is also true that the centrality of the old city, combined with conservation, have kept the *suq* alive—albeit it in an altered form.

In the early years of the Campaign, virtually all foreign consultants saw the *suq* as the economic backbone of the old city and a vital part of the conservation effort, although they approached the problem differently. At the time the *suq* was in decline, a process that had in fact begun decades earlier.[81] It accelerated after the revolution, with increasing demand for consumer goods. Merchants and artisans were hurt by imports, competition from new shopping areas, and outmigration from the old city. Commerce was hurt by conditions in the *suq* itself—unpaved streets, lack of infrastructure, and traffic congestion.[82] Officials estimate that when the Campaign began, approximately one third of the shops in the market had closed, especially around Suq al-Nadhara. The Campaign turned this situation around. Thanks to infrastructure work and street paving, shops reopened—selling a combination of cheap imported consumer goods, and certain traditional products of high social and symbolic value.[83] Like the rest of the old city, the *suq* was not gentrified or oriented toward tourists, whose numbers remained low. In the late 1990s Eckert and al-Maqrami described the *suq* as a "shopping area for the poor." As commerce expanded

79 Eckert discusses the relationship between the historic core and the surrounding city in "The Chief Urban Functions of the Old City of Sana'a Today" (UNDP Technical Report, Paris, 1992). He looks at the old city in its wider urban context, and implications for traffic and development. The issue was also treated in the Berger Kampsax and Shankland reports of 1978 (see above, Chapter 3); in volume one of the Bonifica Quaroni studies (Proposal Phase, 1981); and in Ron Lewcock's reports for UNESCO. For more a more recent account see D. Pini, et al., *Inventory of the Historic City of Sanaa.*

80 For a recent assessment see F. Mermier, "The Suqs of Sanaa"; see also his earlier work, *Le Sheikh de la Nuit, Sanaa: organization des souks et société citadine*, Arles: Actes Suds, 1997.

81 Muhammad Jazim, one of the leaders of the crafts survey conducted with UNDP funding, dates the decline of the *suq* to the British presence at Aden. "The British saw Aden as a *suq*," he said. "They sold goods cheaply there. Local products could not compete … southern crafts died before the northern ones." Some informants cited the mass emigration of Jews in 1948 as the beginning of the decline of handicrafts, especially silverwork. For different approaches to the *suq*, see Chapter 6.

82 R. Lewcock, Report no. 3, 1986; Studio Quaroni S.r.l./Bonifica S.p.A, "Safeguarding of Sanaa Historical Center," Proposal Phase, 1983, volume IV.

83 F. Mermier, "The Suqs of Sanaa," pp. 157–9.

into surrounding neighborhoods, residents complained about traffic congestion, noise, and disturbance. "Old Sanaani people," they wrote, "blame newcomers to Sanaa responsible [*sic*] for dirty streets and waste accumulation everywhere," while others trace the problems to the lack of municipal services."[84]

The proximity of the *suq* to the modern center allowed for complementary functions: services like post offices, banks, government offices, hotels, and the like are located in the area of Tahrir, but not in the old city.[85] In the following decade, skyrocketing land values and commercial rents in the *suq* and adjacent center displaced certain market activities. By the mid-2000s, the central core had come to be dominated by fabric wholesalers—a development that must be related to the historic primacy of textiles.[86] The development was worrisome to the conservation office, because of the scale of these wholesale operations and the financial power behind them.

When a tower house is used for commercial purposes, the impact extends well beyond the facade. Limited interior spans, thick walls, and minimal openings do not easily accommodate commerce, especially the wholesale trades. In some cases, houses are altered significantly or demolished. This is the most worrisome aspect of suqification: the tendency for large merchants, notably fabric wholesalers, to transform houses into warehouses, with retail on the ground floor. In some cases, a merchant would acquire several houses (in one case, an entire block) and demolish them. In their place he would build a concrete building organized around a central corridor, similar to the small shopping arcades in the new districts. This type—commonly called a *samsara*, after a traditional commercial building type—is usually clad with a stone facade inspired by traditional motifs, including the ubiquitous *qamariyya*s. Demolitions are often clandestine: interior beams and floors are removed little by little, often at night, to destabilize the structure; it either collapses, or is condemned and slated for demolition. Fines clearly have little meaning in the current market, where expected income may lead a merchant to invest tens of thousands of dollars.[87] In one case, a magnificent old house on the square of al-Fulayhi was demolished in this way, and a hotel built in its place (Figure 4.10). In another case, the well enclosure (*marna'*) of al-Tawashi mosque was destroyed, and a commercial building built in its place. Such incidents have provoked drawn-out struggles between GOPHCY and merchants, and with the Ministry of Endowments (*Awqaf*) which holds title to many of the properties. While there have been sustained and even heroic attempts by conscientious employees at GOPHCY, others in the agency are suspected of collusion.[88]

84 Eckert and al-Maqrami, "Social Assessment Report," pp. 26–9. "Old Sanani people blame newcomers to Sanaa responsible [*sic*] for dirty streets and waste accumulation everywhere," while others trace the problems to the lack of municipal services (pp. 26–7).

85 F. Mermier, "The Suqs of Sanaa," pp. 156–7.

86 Of the textile *suq* Mermier writes: "… the traditional urban power center was in the textile suq. It was formerly the most prosperous *suq* in the market and included many representatives of the urban elite …." ("The Suqs of Sanaa," p. 152). On the displacement of market activities see Stadnicki and Touber, "Le Grand Sanaa," pp. 39–40.

87 In one case, a merchant reportedly paid USD 20,000 to buy and demolish a small shop that blocked access to a *samsara*.

88 Corruption is endemic at all levels of government; one professional quipped bitterly, "In Yemen we have no rules, except for bribes." Some GOPHCY employees note that the biggest violations are committed by those with the most influence—and it is assumed that these cases involve lucrative bribes. Those lower down in the hierarchy feel justified in extracting what they can, turning a blind eye to smaller scale violations that are less offensive. Bribes are often the only way underpaid employees can make ends meet. "It's very simple," explained an informant. "If GOPHCY does not support its

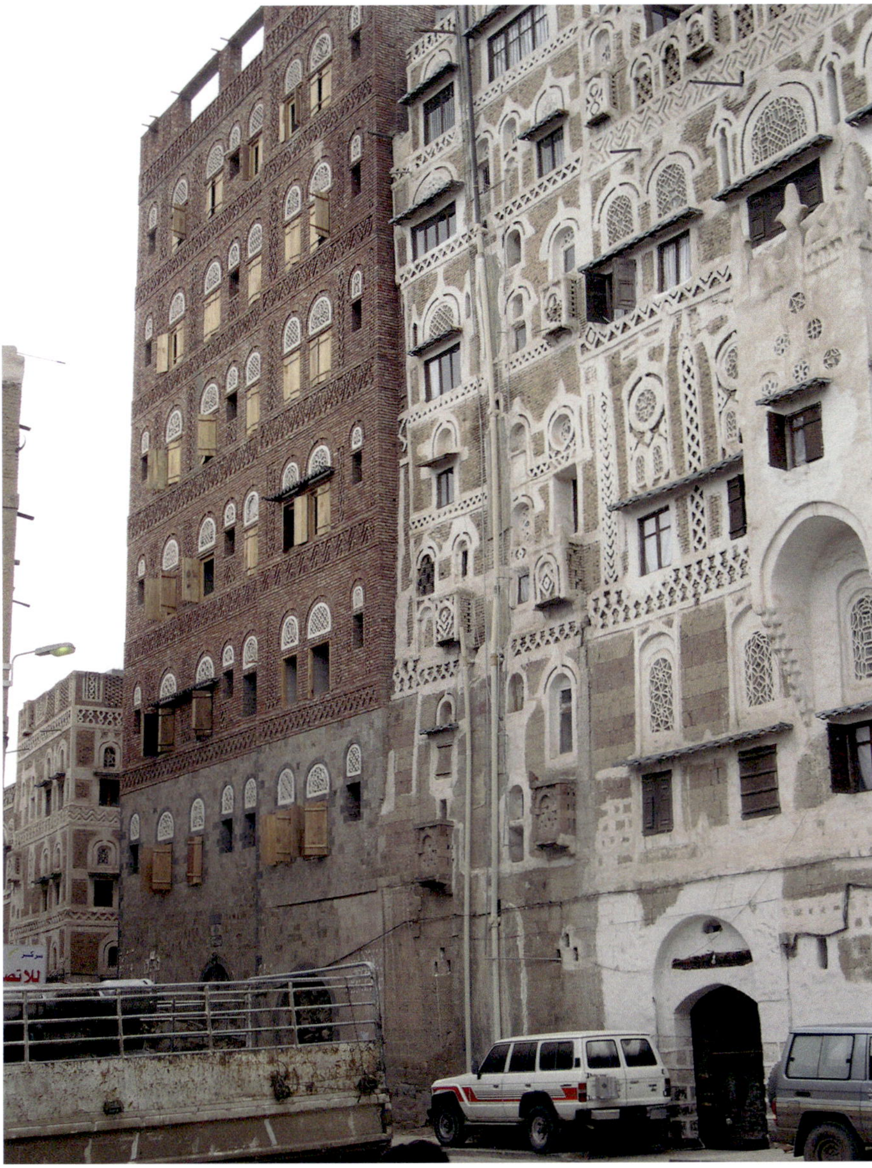

Figure 4.10 On the left, a new hotel built on the site of a demolished tower house in the quarter of al-Fulayhi

One family in particular was said to dominate the *suq*, transforming the northern sector into a continuous commercial zone. Several family members acquired properties in three phases, from the late 1980s through the early 1990s. In the first phase, one individual acquired and demolished an old *samsara*, known as al-Rubu'i, to the north of al-Shahidayn Mosque, replacing it with a concrete shopping arcade.[89] Soon after, two brothers bought land on the south side of al-Zumar square, where there were a series of shops in various states of ruin. New construction was started behind the shops, but stopped due to a dispute between the brothers. Two other family members began to acquire houses on the north side of a street that extends westward from al-Shahidayn square, just south of the new shopping arcade, with the intention of creating a textile *suq*. Working at night with large crews, they removed much of the interior structure of the houses. For a time, GOPHCY managed to halt the work. According to informants, officials subsequently offered to work with the family to transform the houses into a kind of *suq*, with a series of uniform, arched openings at ground level. But a GOPHCY official I interviewed maintained that the agency had not been involved.

The same family attempted to gain control of two urban gardens. The last remaining unbuilt areas of the old city, the gardens are subject to severe development pressures; they were already of concern during the Campaign.[90] Professionals call them the "lungs" of the old city: they provide greenery and open space within the dense fabric, allowing the city to "breathe" (Figure 4.11). These vegetable gardens (*maqshama*, pl. *maqashim*) and orchards (*bustan*, pl. *basatin*) are endowed properties, usually attached to a mosque. Sustained by recycled water and waste, they were an important source of fruit and vegetables within the walls before the revolution. In recent decades their area diminished, and by the late 1990s many of the gardens were partially or completely abandoned. A variety of reasons can be cited, but the main factor is probably the dramatic decrease in the water table.[91]

---

employees, how can they be expected to report violations? If the salaries were good, as they were in the early days, there would be no violations in the old city." One inspector told me that he earned YR 11,000 per month (at the time, equivalent to U.S. $56), while his rent was YR 12,000.

89 The demolition of Samsara al-Rubu'i is mentioned by F. Mermier, "Sanaa, métaphore de l'état yémenite," in G. Grandguillaume, ed., *Sanaaa Hors les Murs*, p. 49. According to one version of the story, told to me by a professor at Sanaa University, the family took advantage of the holiday at the end of the pilgrimage month to demolish the *samsara*. When GOPHCY learned of the incident it dispatched several armed guards to the site. But the family showed up with three cars—"guys with guns"—thus outnumbering the government side. This anecdote, noted the professor, illustrates the dilemma of GOPHCY and of government offices generally. According to a guard who has worked at GOPHCY for many years, this particular family can "call anyone—GOPHCY is terrified of them."

The following account of phase two demolitions is based on conversations with GOPHCY employees and several informants outside the agency; the merchants were out of the country and thus unavailable for comment. An energetic and well-trained inspector was reportedly transferred out of the agency for his opposition to this and other projects. I was unable to locate GOPHCY's files on the case, but I did locate plans for the new "suq" described below, signed by an architect at GOPHCY.

90 See for example M.B. Lane, *San'a: Pilot Restoration Projects*, pp. 52–5.

91 On the gardens see the important study edited by M. Barcelo, *Les jardins de la vielle ville de Sanaa*; and the GHK study, especially the diagram of causes and effects in the decline of the gardens ("Cultural Heritage Report," vol. 2, p. 26). In the 1970s Varanda reported that the water table was already dropping at a rate of 2.5 meters per year, drying up existing wells and prompting the drilling of new and deeper ones; the installation of electric pumps exacerbated the problem (Varanda, "Tradition and Change," p. 202). Other reasons for decline of the urban gardens include the cessation of night soil collection; the low social status of gardeners; and the availability of better-paid farming and other employment. Certainly the introduction of modern sewer and water systems played a role: in some

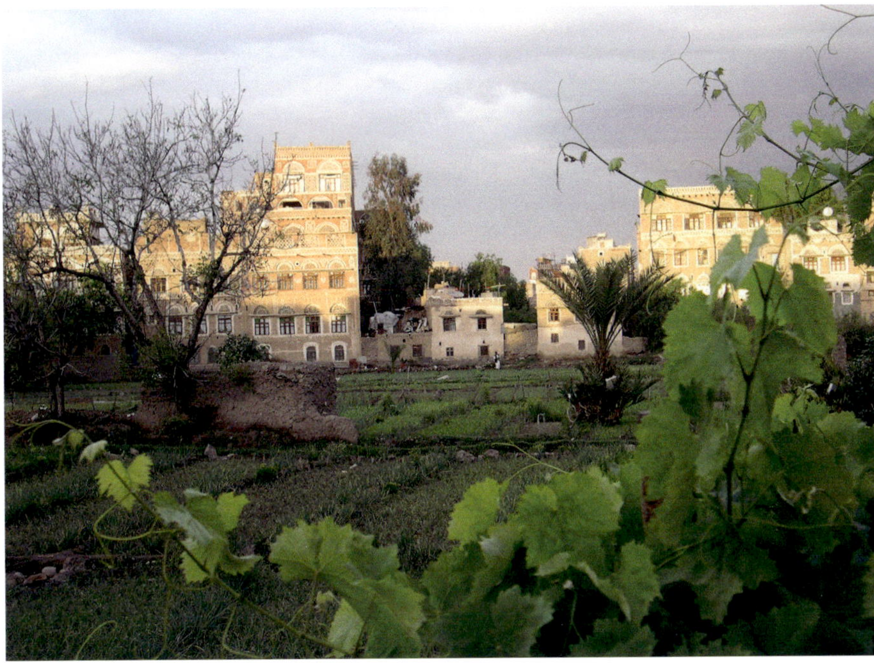

Figure 4.11 Maqshama al-Hurqan, one of the urban gardens restored by the Social Fund for Development

Abandonment of the gardens led to encroachment by surrounding households, already pronounced in the 1970s, and to the use of the gardens as garbage dumps. In their effort to consolidate their holdings around the northern edge of the *suq*, members of the merchant family acquired cultivation rights (*haqq al-yadd*) for the gardens of al-Khudayr and Bab al-Salam from the Ministry of Endowments. When they moved itinerant merchants onto the sites, GOPHCY officials feared that permanent structures would follow. They sent a memo to the Ministry instructing it to evict the itinerant merchants, and to strictly prohibit any construction in the gardens. The memo was later reinforced by a decree from the Prime Minister's Office.

In the late 1990s the Social Fund began its work in cultural heritage with a series of garden restorations, which had been outlined in the Campaign strategy a decade earlier. These projects reinforced the edges of the gardens with retaining walls, intended to prohibit encroachment by surrounding households or other parties. The water supply and sewage systems were rehabilitated where required, and the storage tanks or *birka*s (Arabic plural *birak*) were restored with *qaddad*, a traditional water-proof plaster. By mid-2006 36 gardens had been restored, and four more were in progress. Many professionals saw the garden restorations as the most successful in the conservation effort. But some acknowledged that the gardens had lost their *raison d'être*, both as productive farms within the walls and as

gardens gray and black water from the mosque were not separated, depriving the garden of irrigation water. This was addressed in some places by the SFD garden restorations (N. al-Makaleh, interview).

income-generating property for religious endowments. At the time of this writing, the fate of the gardens is again in question. The SFD initiative stopped several years ago, because of diminishing water supply to the mosques and lack of government oversight. The restored gardens have deteriorated and the SFD has been unable to rehabilitate new ones.[92]

One of the saddest incidents in the *suq*, according to professionals, is the "restoration" of Samsarat al-Baw'ani (Figure 4.12). According to popular tradition and the opinions of conservators, it is probably the oldest *samsara* in Sanaa; until the 1970s it was used as a *samsara* for oil. In a mission report of 1989, Lewcock urged that emergency funds be allocated to stabilize the building, which was in danger of collapsing, and to conduct preparatory studies for a careful restoration:

> Samsarat al-Baw'ani is the oldest samsara building in the suq and probably one of the oldest buildings in the city. Its conservation is a matter of urgency … With regards to the philosophy of conservation, such an important building demands the minimum of alteration and the maximum of strengthening and conservation. This may necessitate special techniques for the consolidation of the surface of the stonework without replacing more of it than is absolutely necessary for structural conservation.[93]

At the time, Giancarlo Barbato had recently documented the building with funding from UNESCO/UNDP. His studies were accompanied by a restoration and reuse proposal which envisioned the building as a "Centre for Cultural Studies." Like Lewcock, Barbato emphasized the historical importance of the building and the need to conduct a careful restoration and archeological investigation of the structure.[94]

According to informants, a GOPHCY official "sold" Barbato's drawings in 1995 to the influential merchant who controlled the *samsara*. The merchant initially agreed to follow the restoration plans, but then began to build according to his own wishes; he wanted to improve the space and enlarge the shops for his subtenants. Certain individuals at GOPHCY wanted to work with the merchant, inspired by recent thinking about private-public partnerships in conservation.[95] They modified the plan to suit his requirements—for example, by connecting the courts at ground level (Barbato's plan would have connected the courts only at the first floor level). According to the former project architect at GOPHCY, he and his colleagues insisted that the original structure be respected and that traditional materials be used throughout. In the end, the merchant ignored these plans and prescriptions. He completely transformed the interior space, erecting concrete block partitions and covering surfaces with cement. He reconfigured the plan and built new entrances like those of other

92 N. Al-Makaleh, personal communication.

93 R. Lewcock, mission report no. 5, 3/89, items 50–51. The local tradition that Samsarat al-Baw'ani is the oldest *samsara* in Sanaa is substantiated by architectural evidence: a number of unique features suggest that the building—especially the southern portion—probably dates from the mid-fourteenth century or earlier. See G. Barbato, "Restoration Project of the Samsarah al-Bawani: Centre for Cultural Studies" March 1989, item 6; and his unpublished paper, "The Methodological Approach for the Restoration of the Samsarah al-Bowani and Samsarah al-Majjah," Symposium on the Integrated Urban Policy for the Old City of Sanaa, Sheraton Hotel, Sanaa, 15–19 December, 1991.

94 G. Barbato, "Restoration Project of the Samsarah al-Bawani," section 10. The accompanying plans, sections, and elevations are reproduced at 1:50 in B. Lane, *Pilot Restoration Projects*, pp. 62–3. According to Lane, the *samsara* was to be used as a workshop for the revival of textile trades.

95 As presented, for example, at the Institute for Housing and Development Studies (IHS) in Rotterdam where numerous professionals have received training with scholarships from the Dutch government.

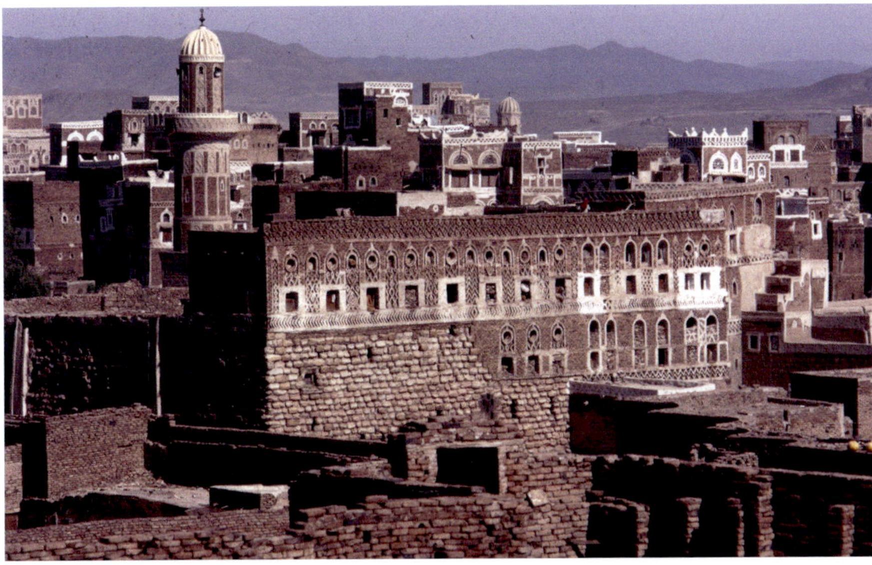

Figure 4.12 View of Samsara al-Baw'ani in the early 1980s

Source: Photograph by Ronald Lewcock.

shopping arcades. The Director of Planning and the project architect went to the Director and Deputy Director of GOPHCY, who were, however, unable to stop the work.[96]

The Ministry of Endowments, which controls the vast majority of historic properties, was a key player in this and similar episodes in the old city. As noted in Chapter 2, the institution of *waqf* was a means to provide services to the city: rents from income-generating properties like *samsaras* would be used to support religious and charitable structures. As in other countries, *waqf* properties in Yemen were reorganized under a ministry, which continues to be directed largely by a religious establishment—although new blood was introduced in a recently created "investment sector."[97] In the old system, an overseer (*nadhir*) managed the income of each complex; now rents from all *waqf* properties go to the Ministry and are redistributed according to its schedule of priorities. According to officials, the Ministry is primarily interested in maintaining mosques. As we saw earlier, there is institutional overlap: mosques are *waqf*, but they are also "monuments" under the jurisdiction of GOAMM; and GOPHCY oversees *all* work in the old city.

The role of ministries of endowments in historic cities, and their different understanding of history and utility, is a critical and understudied aspect of conservation in the Islamic

96 It is not clear if plans on file in the Planning Department are those drawn by GOPHCY architects, or if they are plans submitted by the merchant, whose name appears in GOPHCY's title block; there are no signatures of GOPHCY employees on the drawing. The ground floor plan of this set shows original columns and staircases demolished and relocated, new entrances on the east and north elevations, and a connection between the courts at ground level. According to a former GOPHCY official, the agency did propose a connection at ground level.

97 An inspector at the Ministry commented that the leadership is old guard (he motioned with his hands, indicating the headgear of the traditional religious elite); but the investment sector is staffed by "young guys (*shabbab*), like you and me … you can talk to them."

world. These ministries are heir to a distinctive institution, although historically it was not without its problems and abuses.[98] One scholar has suggested that the *waqf* system was an indigenous form of conservation, driven by concerns of social utility rather than history and aesthetics.[99] Indeed, the Ministry of Endowments in Yemen often invokes social utility—what Alois Riegl called "use value." It claims moral authority as the traditional caretaker of historic cities, a role that GOPHCY has usurped. But some professionals at GOPHCY say that its real, and only, interest is financial gain. The Ministry and GOPHCY have clashed over additions to mosques and other religious buildings, which are now, at least in theory, prohibited. A particularly high-profile dispute involved a proposal, backed by then-President Saleh, to expand the Great Mosque. The project would have widened the street from Bab al-Yaman to the mosque, demolishing fabric that stood in the way. Opposition to the plan was spear-headed by Selma al-Radi, a respected archeologist and conservator; it quickly gained high-level support. Saleh was finally persuaded that an addition to one of the oldest mosques in Islam, in the heart of a World Heritage site, was not possible or desirable. Instead, he commissioned an enormous new mosque on the southern outskirts of the city, which was inaugurated in 2008.

After years of conflict with GOPHCY, the Ministry of Endowments took up the banner of heritage in the early 2000s.[100] Officials began to describe their projects as "restorations," although they often conflicted with GOPHCY's standards. Architects at GOPHCY complained that the Ministry had no understanding of historic value: it was interested only in tenants, and would make any concession to assure future income.[101] Ministry officials conceded that projects did not always meet conservation standards, but this was the fault of the administrators and patrons (*mutabarr'in*) of *waqf* properties: it was *they* who did not appreciate historic value. "They can't be faulted," said a high-level official. "They are doing what they think is right." In an interesting twist on international discourse, he explained that there are two kinds of conservation: historic conservation (*hifadh tarikhi*), which is concerned with the outside of buildings; and archeological conservation (*hifadh athari*), which is concerned with the inside. "Each person works according to his conscience," he concluded. His statement suggests that the public face of the building is "historic" (*tarikhi*), and can be considered independent of utility; only a structure that has value as an antiquity (*athar*) should be conserved on the inside.

This official—who works in the Investment Sector and attends meetings of the High Commission for the Preservation of Old Sanaa on behalf of the Minister—told me a different

98 The main failure of the *waqf* system, according to some historians, was its failure to account for inflation, which in turn led to various abuses. Arab and European professionals say that *Awqaf* authorities are a problem everywhere in the region: they lack basic understanding of the aims of conservation; the easiest solution is to bypass them. At a 2005 conference in Sanaa, a Tunisian architect remarked: "We solved the problem in Tunisia by making all *waqf* property of the state."

99 A. al-Habashi, "*Athar* to monuments: the intervention of the *Comité de Conservation des Monuments de l'Art Arabe*," unpublished doctoral dissertation, University of Pennsylvania, 2001, Chapter 1.

100 For example, it published a color brochure celebrating its role as care-taker of Yemen's historic structures, and initiated an inventory of *waqf* properties.

101 The Ministry of Endowments is said to be one of the wealthiest ministries, and according to several informants, it is rife with corruption. Officials at the Ministry denied charges of corruption. They noted, however, that there was a great discrepancy between government rents (that is, those charged by the Ministry to other government agencies like GOPHCY) and those in the private sector. The rent from Samsarat al-Nahhas, for example, was no more than a "symbolic rent (*ijar ramzi*)," an inspector told me. There are obviously strong incentives to rent *waqf* properties on the open market.

version of what happened at Samsarat al- Baw'ani. The Ministry reached an agreement with the merchant who controlled the *samsara* (and thus paid rent to the Ministry): the merchant agreed to preserve the exterior of the building, but would renovate the interior. In this way, said the official, "the historical style (*al-tabi' al-tarikhi*) would be preserved, and the inside modernized (*mutahadiththa*)." The merchant had the right to do the work, he said: he "saved an historic structure" that was in poor condition. The official was not aware that GOPHCY had prepared drawings, nor was he aware of any stop-work order. He was surprised to hear the charge made by some that Ministry employees had received money from the merchant. Like others at the Ministry, he said that the main function of the *samsara* was to provide income for mosques: "The condition of the *samsara* is not the goal."

Architects see Samsarat al-Baw'ani as a tragedy.[102] But the official at the Ministry of Endowments invoked a precept of modern conservation to defend his agency's actions. Preserving the exterior, he said, was the most important thing, since it embodies the "historical style." Treatment of the interior could be guided by use value; as such, it could be "modernized." Apparently the Ministry had failed to understand—or had not been informed—that the structure was considered a "monument" and thus subject to "archeological" treatment. GOPHCY had a clear understanding of the class of the building, as one of the oldest and most historically important buildings in the city. Yet no agreement had been reached with the Ministry, which was clearly not prepared to carry out a careful restoration project without assurance of income. GOPHCY could have agreed, for example, to rent the *samsara* from the Ministry as it did with Samsarat al-Nahhas, the building designated as a training center for handicrafts—thus removing it from the market. Yet such a solution may not have been acceptable to the Ministry, since as a government agency GOPHCY's rent would be a fraction of what could be obtained on the open market.[103] The case of Samsarat al-Baw'ani underlines the difficulty of constraining private interests to undertake careful and costly restorations. According to one official, the tenant feared that his subtenants would not be able to afford to stay in the *samsara* after it was renovated. At the same time, the familiar distinction between "outside" and "inside" was here pressed into service to justify the work and to cover one or more lucrative financial deals.

Defining Heritage

The differing interpretations of historic value would be in part resolved by a classification system. But some foreign consultants and local professionals have been reluctant to implement such a system: they feel that historic value lies not in monuments or "class A" buildings, but in the fabric of the city which is composed of less "significant" structures. A classification system, they fear, would open the door to unwanted change; better to prohibit change entirely, knowing that it will happen anyway. This approach, frankly acknowledged by GOPHCY staff, was echoed by an official at the World Heritage Center: "We know that regulations will not be observed," she told me, "but we have to maintain them as

102 This is how architect 'Abdullah al-Hadrami, who worked with Barbato, described the end result. The GOPHCY project architect who modified Barbato's drawings said: "I never go into the *samsara*. I bear no responsibility for it. It is the responsibility of the officials [who approved the work]."

103 GOPHCY architects note that *Awqaf* doesn't like the arrangement, because it means the *samsara* is not in their control—that is, they cannot collect money off the books. Yet some charge GOPHCY with a similar, double-book strategy at the *samsara*.

ideals." But in prohibiting change, one misses the opportunity to guide change in desirable ways—and this may have unfortunate consequences."

In Europe and the United States, where historic districts have become desirable places to live, conservation regulations often abet gentrification. Many residents and business owners in gentrifying neighborhoods cannot afford to restore buildings according to exacting standards. They may be displaced, or if they own property they may sell at a profit. In both cases, conservation leads to the transformation of the community, and thus the place, it purports to conserve. The opposite situation exists in the Middle East and North Africa, where elites and aspiring middle classes continue to favor modern districts, and historic areas are progressively "ghettoized."[104] Conservation may exacerbate decay of the fabric by creating the impression of an "untouchable" place, protected by guidelines that are never fully clarified. As an architect who worked on the UNESCO Campaign for Fez notes:

> [In Morocco] urban documents and master plans [from the 1970s onward] designated the *medinas* as 'preservation zones,' but without providing any defined guidelines or norms. The most striking case is that of Fez. Indiscriminate preservation of such a large and complex unit has condemned it to decay. Experts have hesitated to tackle its challenges since it has been an untouchable site under the label of an undefined 'preservation.' Thus it has suffered from neglect and intense decay, and the sustainability of its socio-economic and cultural networks, which are in fact its *raison-d'etre*, has been disregarded.[105]

In Sanaa and elsewhere in Yemen, local professionals have struggled to come to terms with international practice and its relevance for the local context. In the early years of the Campaign, a small group of professionals was secunded to EOPOCS (as the conservation office was then called) from several ministries. They spent much of their time wading through reports on Sanaa and working out a Campaign strategy in conjunction with UNESCO consultants. They also studied the conservation laws of other Arab countries—Egypt, Iraq, Syria, and Oman—in an effort to develop a local model. They quickly realized that certain practices and legal instruments developed in Europe, and even in other Middle Eastern countries, were not easily applied in North Yemen. Over the years professionals have attempted to "yemenize" existing models, incorporating principles from their own practice in the field, *shari'a*, civil law codes, and the World Heritage Convention and other international charters.

Sanaa was put on the map of World Heritage not because of individual monuments, but because it is a spectacular "vernacular" settlement composed of some 5,000 houses in addition to numerous religious and public buildings. This massive quantity of built heritage presents a huge financial burden, especially since most donors will not finance restorations of privately owned buildings. Many national conservation laws require the state to purchase properties if they are not maintained by their owners. But in Yemen, the state is relatively

104 The phenomenon of ghettoization, noted in the early 1990s, has not been reversed. See S. Abdulac, "An Introduction to Urban Conservation problems today in the Arab Region," in *Symposium on the Integrated Urban Policy for the Conservation of the old City of Sanaa, Yemen*, 18–19 December 1991, UNESCO Working Documents, 1992. On the contrast between historic cities in the West and the Middle East/North Africa region, see also M. Lamprakos, "The Idea of the Historic City," especially pp. 18–19.

105 H. Radoine, "Conservation-Based Cultural, Environmental, and Economic Development: the Case of the Walled City of Fez," pp. 461–2.

weak and has limited resources, and a strong tradition of private property is enshrined in both Islamic and customary law. The authors of draft legislation proposed in 2004 carefully considered this issue.[106] They framed house maintenance in terms of the legal principle that "general interest precedes private interest": owners are required to maintain their houses, but the state will provide financial and technical assistance. The law proposed the creation of a small loan fund and free technical assistance and supervision by GOPHCY staff. At the same time, it strengthened the institutional capacity of GOPHCY by making salaries comparable to those at government research institutes.

Most of the regional laws consulted by professionals defined a monument as any building over 50 years of age, in some cases as little as 30. By this criterion most buildings in Yemen would be considered a monument, and would thus be subject to the highest degree of protection. In their discussions in the mid-1980s, the group proposed to deal with the issue of "living heritage" in two ways. First, they argued that historic cities should not be managed by the Antiquities Department. This represented a departure from practice in most Arab countries, and converged with the desire of foreign consultants and local policymakers to create an independent institutional structure.[107] They also proposed a distinction between two basic classes of buildings, based on their experience in the field. They defined an "archeological monument" (*ma'lam athari*) as a building, or the remains of a building, that no longer has a function and no longer accommodates, or is surrounded by, life; an "historic monument" (*ma'lam tarikhi*) is a building that is lived in, or can be lived in. As one professional noted, the historical layering of Sanaa means that even within a single building, the "archeological" and the "historic" may be mixed, as in a house built on an archeological site.

The terms "archeological" and "historic" are commonly used at GOPHCY, although one professional says there is still no consistent distinction between them at the technical level. In practice, however, there does seem to be a distinction. While the term *athari*, "archeological," originally meant a "dead" monument, it is now commonly used to indicate an exceptional work that should be a accorded a higher level of protection—that is, simply, a "monument." On a survey form for the old city developed in 2005 by an Italian team and GOPHCY officials, a building of the highest architectural quality—what the Italian authors call a "monument"—is translated as *ma'lam athari*. Other classes of buildings, ranked by quality—"excellent (*mumtaz*)," "good (*jayyid*)," "ordinary (*'adi*)"—are seen as part of the historic fabric of the city; the only negative attribution on the list is "inconsistent (*ghayr mula'im*) [with the context]." This understanding of *athari* and *tarikhi* corresponds to that of the official at the Ministry of Endowments: a *ma'lam athari* is a building whose *interior* cannot be changed, because it has a higher level of protection; in the case of a *ma'lam tarikhi*, only the exterior must be conserved since it is representative of the historic style (*al-tabi' al-tarikhi*).

The terms *athari* and *tarikhi* are thus generally understood to indicate a building's relative age and importance. In practice, both types can accommodate and sustain life. The usages represent an interesting twist on international terminology. Riegl defined an "historic monument" as *any* building over 60 years of age, since it can be viewed through the lens of historical distance; the building may or may not have use value, but its primary

106 This section is indebted to conversations with GOPHCY's legal counsel, Tarek al-Hammadi, and the "Draft Law for the Conservation of Historic Cities, Areas, and Monuments, 2004."

107 See Chapter 3.

role is as a bearer of “age value.” In the Yemeni usage historic (*tarikhi*) indicates age, but specified the *continuation* of use and the support of life. As we will see in the following chapters, this understanding of *tarikhi* is shared by builders; it has significant implications not only for the designation of buildings, but how buildings are conserved.

Chapter 5
Old and New Professionals

A number of recent studies have shown that local or "traditional" experts—builders, craftsmen, and artisans—are often marginalized by the heritage enterprise. They are seen as bearers of a noble tradition, but they are also excluded from the benefits of modernity, and from new global networks of production and expertise.[1] As we will see, a similar process has occurred in Yemen. But benefits to new groups—in terms of economic and cultural capital—have been limited by political circumstances and the vagaries of donor aid. Professionals who began their careers under the UNESCO Campaign lived through the restructuring of the conservation office in 1990–1991, the economic crisis that followed the first Gulf War, and the 1994 civil war; the modest renaissance of heritage activities in the late 1990s is now stalled, in the wake of the 2011 uprising. As such, heritage has never quite taken off in Yemen. At the same time, a relatively young architectural profession has not fully displaced the prestige of the master builders, or *usta*s. Different kinds of knowledge and expertise coexist, and intersect in interesting ways.

In this chapter I explore *turath* from the perspective of these "old" and "new" professionals. In first part of this chapter, I describe the experience of architects, planners, and other new professionals relative to the typical process of heritage training and management. The contrast derives in part from the fact that conservation was transferred to Yemen in the framework of development assistance. It also derives from specific conditions in Yemen, especially the persistence of traditional building practice. In the second section, I discuss the *usta*s who, independently or under the supervision of architects, do the actual making. I attempt to paint a portrait of builders' practice as they inherited it, and the pressures exerted by the new conception of *turath*. These pressures are material and social, even psychic—challenging the identity of builders as expert persons and historical agents. At the same time, the new context of *turath* provides an opportunity for builders to influence the ongoing production of culture and conceptions of the past.

The New Professionals

Because of their training and their mandate, conservation professionals are on the "front lines" in the negotiation of international and local practice. They interface with international agencies like UNESCO and the World Bank, and may work with the programs of donor nations like Germany and the Netherlands.[2] They are aware of the unique circumstances in which conservation has been introduced into Yemen—particularly its association with

1 On this "double-edged sword" of tradition, see M. Herzfeld, *The Body Impolitic*, pp. 4–5. See also T. Marchand, "Process over Product"; and C. Joy, *The Politics of Heritage Management in Mali*, especially, pp. 157–75.

2 The German development agency GIZ (Gesellschaft für Internationale Zusammenarbeit) has been particularly active in cultural heritage in Yemen. Yemen was a focus area for the Netherlands in the Middle East; but the main Dutch initiative, the Urban Cultural Heritage Project (UCHP), stopped in 2004.

development which, for some, changes the nature of conservation. "Cultural heritage is a Western term," said 'Ali 'Oshaish, one of the founders of the Campaign office. "It has limited application in a place where the built environment continues to fill the need for shelter and social and spiritual sustenance."

Many of the conservation professionals cited in this work are graduates of professional schools—that is, they are architects, engineers, or planners, and in one case, a lawyer. A number of individuals were sent abroad on government scholarships to universities in Europe, the United States, the former Soviet Union, Pakistan, and the Gulf. During the Campaign there was a kind of internal hierarchy in the office: one's position was determined by class and origin (*saada* vs. non-*saada*; Upper vs. Lower Yemen), but also by the place one studied (Western vs. Eastern Bloc or South Asia) and by one's command of English. This hierarchy has to some extent dissipated, since those with the most prestigious degrees and the best English have left to work at the Social Fund for Development (SFD). In the mid-2000s, the majority of professionals at GOPHCY were graduates of the University of Sanaa: some had majored in architecture, while others had studied history, archeology, sociology, philosophy, journalism, and business administration.[3] Still others were graduates of local vocational or technical schools (*ma'ahid fanniyya*), including one school specialized in the fine arts.

A small number of professionals received training in conservation at the Institute for Housing and Development Studies (IHS) in Rotterdam through scholarships from the Dutch government, or in brief training courses elsewhere in Europe or in Sanaa. But the majority of those working in conservation have no specialized training: they learn their work from their colleagues in the field, most of whom also lack training. A few inspectors, who are daily called upon to authorize or reject work by residents, have developed a keen sensitivity to the local architecture. Yet most lack formal training in architecture, conservation, and the history of local building. In the mid-2000s the lack of specialized training, and the absence of resources in Arabic on Sanaa's architecture and history, were seen as pressing problems at both GOPHCY and the Social Fund.[4] Although most international projects have included training components, these were often short, or reduced to a token "study trip." On-site training opportunities were limited, at least in the early years, by what most see as an over-reliance on foreign expertise. As we saw in Chapter 3, the problem was compounded by the internal politics at EOPOCS, with leadership restricting access to foreigners.

Veterans of the UNESCO Campaign admit that conservation was a "new idea … we had to convince the people of its value." At the same time, certain aspects of conservation seem strangely familiar: it affirms certain "historical" practices that in many cases have not died out. "Conservation and sustainability (*istidama*) are old concepts," said architect Yassin Ghalib. Sanaa was built through practices like maintenance, reuse, and the recycling of water and solid wastes. "Why do we always have to look to Europe for concepts that we already have?" he asked. Tarek al-Hammadi, legal counsel for GOPHCY, pointed out that

3 The Architecture Department at the University of Sanaa graduated its first class in 1992.

4 This was partially remedied by 'Abd al-Raqib Tahir and Yusuf 'Abdullah's book, part of the Organization of Islamic Capitals and Cities series on architecture and planning in the Islamic world: *Dirasat Tahliliya 'ala al-'Asima Sanaa, al-Jumhuriya al-Yamaniyya*, Sanaa: Markaz al-Tahir lil-Istisharat al-Handasiyya, 2005. In the late 2000s the SFD initiated a project to translate Serjeant and Lewcock's book, some three decades after its appearance in English.

conservation emulates certain principles of customary law (*'urf*) and practices that have been neglected in recent decades.[5]

As we saw in the previous chapter, professionals' concern with preserving the "historic character" of the city often translates into practical planning concerns—density, the height of buildings, and so forth. The standard to which they refer is not a written manual, but the old city itself—which was the training ground for many builders who are now in their fifties and sixties. Builders learned the "rules" of the old city during their apprenticeships. I asked *usta* Ahmad al-Rawdhi, who specializes in traditional construction, how he goes about designing an addition. He replied:

> I look at the surrounding neighborhood to see what it wants (*ashuf aysh yisthi al-jiwar al-mujawir*) and I look at the house itself, to see what it wants. I build a kind of dam around the house, so the adjacent houses are not affected. I look at the style of the surrounding houses. I build what will be appropriate to the environment (*jaw*, literally "atmosphere") the house is in; it should fit like the fingers of a hand.

Conservation professionals accept or reject new work in the old city according to whether or not it is consistent (*mula'im*) with the surrounding context, in its form and materials—what al-Rawdhi called the *jaw*. The addition of stories depends on the height of surrounding buildings. If a house is shorter than those around it, an inspector told me, floors may be added to it. Those not trained in architecture sometimes use revealing metaphors to describe notions like "context." At the 2004 UNESCO-sponsored training seminar, a GOPHCY employee asked why a certain addition, built of concrete, was classified as a violation (*mukhalif*). "It would be fine outside the old city," said one of his colleagues. "It's like going to the village wearing pants. If you did that, you would be inconsistent (*ghayr mula'im*) [with the context]." The old city requires an architecture that is more "traditional," then, but also more familiar and domestic—like the clothing one dons when one returns to the village. This was a natural metaphor, since many professionals grew up in villages and return there regularly for the 'Id holiday and other vacations. For some, the village is a primary reference for *turath*. Ghalib has a unique, even poetic, appreciation for *turath*. When I asked him how he had developed this appreciation, he replied: "I grew up in the village. The smells, the textures, the light; the songs my mother sang ... I drank of *turath* (*shiribt min al-turath*) throughout my childhood."

For many professionals, the contrast between the Yemen they knew as children and the Yemen of today is a kind of "cultural shock." Rapid modernization seems to confirm the idea of historical rupture that is embedded in conservation discourse. Like many other educated Yemenis, professionals see themselves on the other side of an historical divide. "Change has happened so quickly," said al-Hammadi. "Everything is so new, that we now long for everything old." For some, the sense of historical rupture is felt as an alienation

5 *'Urf* may favor a kind of "preservation" based on cultural, rather than historical-aesthetic, norms. An aim of *'urf* is to maintain the *hima*, the protected perimeter of the house, which ensures that women will not be seen by unrelated men. According to an entry in Lewcock's field notebooks from the early 1970s, a judge in Shibam-Hadhramaut directed the owners of a collapsed house to rebuild it in the same form, with the same layout of windows and openings, to ensure that the *hima* of neighboring houses was protected. On the *hima* see P. Bonnenfant, *Sanaa*, pp. 70–71.

from one's own culture. "Even our beauties are not beautiful," said Amat al-Razzaq, a female architect who directed the women's handicrafts project. She explained:

> The [modern] window has no relationship to life, or to the building. We have no idea where it came from. It's like a slap in the face ... The same is true of clothing. The black *abaya* has no relationship to the life or culture or the spirit of the Yemeni woman.[6]

For professionals, conservation is perhaps a means to preserve part of their own identity—now conceived as a national identity, that all Yemenis can claim. This perhaps explains why they found the Campaign's aims—to protect not only the physical fabric, but a traditional way of life—so appealing. In the words of another architect:

> Conservation (*hifath*) means preserving not only forms, but the social life (*al-hayat al-ijtima'iyya*). There are people [in the old city] who remember how it was. If the new generations come in, we will lose that. Yes, there have been changes in the social life [he said in response to my question]. Relations in the quarter (*hara*) are not as tight as they used to be. But [the old city] is still closer to the past than to the future.

While this architect is from an old Sanaani family, most professionals in the conservation office are not from Sanaa. As elsewhere in the public (and private) sector, most technical and managerial positions at GOPHCY are filled by individuals from Lower Yemen and the south, where there is a tradition of higher education. They may view the old city of Sanaa as a beautiful relic of urban highland culture, which is nevertheless backward and alien. While they feel that conservation should preserve social forms, they in fact consider some of these forms primitive, especially attitudes toward women.

The idea that a World Heritage site is both a collective masterpiece and a unified whole suggests that it evolved over time, adjusting slowly in response to changing conditions. But the "historical time line" sets an endpoint to this evolution. "Can we stop history?" asks one (male) professional.

> We can't ask people to live as they did generations ago—people will leave. When I visited houses [during a 1996 survey by a German-led team], I saw girls in blue jeans. They were modern girls; why are we asking them to live as their ancestors did? ... Bir al-'Azab is different from the old city; every period should talk about its history. Time will pass and [the new] will become traditional.[7]

Since local professionals have a similar background to mine, I thought that most would agree that the physical fabric of the old city should continue to evolve. But I was mistaken: while they acknowledge the need to accommodate changes in lifestyle and use, most feel that such changes should have minimal impact on the physical fabric. Their language often seems to echo the international charters. "Yes, history is evolution," said one architect,

6 The black *abaya* (worn in Saudi Arabia and other Gulf countries) is now worn by virtually all Yemeni women in cities and towns, having replaced regional and local forms—many of which were very colorful.

7 This man is from Aden, evident from his dark skin. He was allowed into the houses because he was an outsider; an unrelated Sanaani man would never have been allowed in, he explained.

"but now we've decided that evolution should stop." Today, evolution means erasing all difference and distinction:

> In the past evolution occurred through changes in successive floors [of a house]; there were no major changes in the lower ones. So you can read the history of the house in its stories. Now, with the current changes, everything will become the same.

Modernity—which marks the end of "history"—is identified with sameness. Even in the old city, the facades of new construction and additions usually have identical window layouts and openings. This reflects, on the one hand, changes in taste. "Today everyone wants large windows," said al-Rawdhi, who struggles to satisfy his clients' desires within the framework of the Sanaani style (Figure 1.3). But the wide, identical window openings also reflect the changing use of space within the tower house: in contrast to the old communal living arrangement, nuclear families often occupy separate floors.

As might be expected, the most ardent advocates of international practice are architects and graduates of technical or fine arts programs: they have deep knowledge of and appreciation for the special qualities (*khussusiyyat*) of local architecture and want to conserve them. "There are fifty Sanaas in Sanaa," said Ghalib. He asked his Italian colleagues how such diversity can be conserved:

> It's not enough to specify that stone be cut by hand, rather than by machine. There is a kind of cut stone unique to Sanaa, produced by a special tool. If workers come from outside, they will bring different tools and the feeling will be different.

For Ghalib and others, the old city is all they have left. "It's like a knife that my father gives me," he said. "It will not be repeated." (The reference to a knife, *janbiyya*, is not casual: an object of deep social and symbolic value, its price varies with age and provenance; the best ones fetch extraordinarily high prices in the *suq*.)[8] The uniqueness of the old city is linked to a local aesthetic sense that is very much alive, and which, for many, must guide and limit intervention. "The architectural style of Sanaa cannot be changed," said 'Abdullah al-Hadrami, veteran of the UNESCO Campaign and a noted restoration architect. "Sometimes if an owner wants something new, I will do it very carefully. But most do not know the style, so they end up ruining the facade."

Builders express the same notion, in remarkably similar terms. I pointed out to one builder that the architecture of Sanaa was not static, but rather had changed over time; should it not continue to evolve? "Change can be allowed," he replied, "but it must be within the framework (*nitaq*) of the Sanaani style." Both these statements seem to uphold international strictures against change. Yet as we will see in the following section, architects and builders do not always agree on the practical application of the Sanaani aesthetic. Their differences are rooted in different conceptions of *turath*.

8 There is also active artisanal production of *janbiyya*s for an expanding market; see Mermier, "The Suqs of Sanaa," pp. 158–9.

The Changing Role of Builders

As we saw in Chapter 2, the social and economic changes that followed the revolution produced an upheaval in traditional practice, particularly in the capital city. New materials and technologies, in particular reinforced concrete, were initially brought north by Adani contractors after the revolution, and diffused via a rapidly expanding road network. By the 1970s these techniques were in general use, not only for new commercial and institutional typologies but also for houses.[9] While the buildings incorporated certain local materials, these were increasingly reduced to cladding. The master builder (*usta*; often, but not always, a mason) initially benefited from the construction boom and experimented with the new technologies. Increasingly, however, projects came to be directed by new professionals—engineers, architects, and construction managers—who had skills that were not part of the *usta*'s training. Some *usta*s acquired new skills and became prosperous contractors. But in many cases highly trained *usta*s ended up working in concrete construction for contracting companies, typically owned by men who had had experience with new construction technologies in the Gulf. No longer the head of a prestigious profession, the *usta* was now subordinated to a building process that valued different skills.[10] "Skilled artists have been transformed into unskilled workers," said architect Sultan Sallam, professor at Sanaa University and author of a remarkable work on local building practice. His observation highlights an aspect of modern building practice—the deskilling of labor—that is seldom acknowledged by architects and conservators in the developed world.[11]

The building arts have traditionally been highly valued in Yemen, as testified by the remarkable variety and quality of so-called "vernacular" architecture; even in Sanaa, many excellent examples are of fairly recent origin (Figure 5.1). As Eickelman notes for Morocco, building skills are passed on according to a genealogical model of authority, which may or may not correspond to blood lines.[12] In Sanaa, the ideal appears to be to pass building skills on to one's sons, although the rigors—even abuses—of apprenticeship may be incompatible with kinship, particularly agnatic ties.[13] Some families in Sanaa, as elsewhere in Yemen, acquired reputations as great *usta*s: builders often say that they learned the trade from their fathers, who learned it from their own fathers—reaching back many generations. 'Abdullah al-Rawdhi (Ahmad's brother) described the transmission of skills in his family: "My great-uncle passed the trade on to my uncle 'Abdullah, when it was about to become extinct *(tawarrath al-'aml ila al-'amm 'Abdullah, wa kanat sa tanqaridh)*.

9 By the mid-1970s about 20 per cent of new houses in Sanaa were built in concrete (F. Varanda, "Tradition and Change in the Built Space of Yemen," p. 158).

10 F. Varanda, "Tradition and Change in the Built Space of Yemen," pp. 156–8.

11 Sallam's work is an unpublished Master's thesis, *Al-hiraf al-taqlidiyya al-islamiyya fi al-'amara al-yamaniyya* ("Traditional Islamic Crafts in Yemeni Architecture"), Department of Fine Arts, University of Cairo, 1988. The anthropological literature on deskilling is substantial, but most studies have focused on the industrial context; see T. Gieser, "Enskillment inhibited: 'industrial gardening' in Britain," *Journal of the Royal Anthropological Institute*, vol. 20, 2014, pp. 131–49. On deskilling see also Herzfeld, *The Body Impolitic*, from p. 81.

12 *Knowledge and Power in Morocco: The Education of a Twentieth Century Notable*, Princeton: Princeton University Press, 1989, p. 308 (cited in Marchand, *Minaret Building*, p. 10).

13 T. Marchand, *Minaret Building*, pp. 138–9. Goody attributes the training of unrelated family members to the development of a market economy and the increasing division of labor ("Learning, Apprenticeship, and Division of Labor," in M.W. Coy, ed., *Apprenticeship: from Theory to Method and Back Again*, Albany: SUNY Press, 1989).

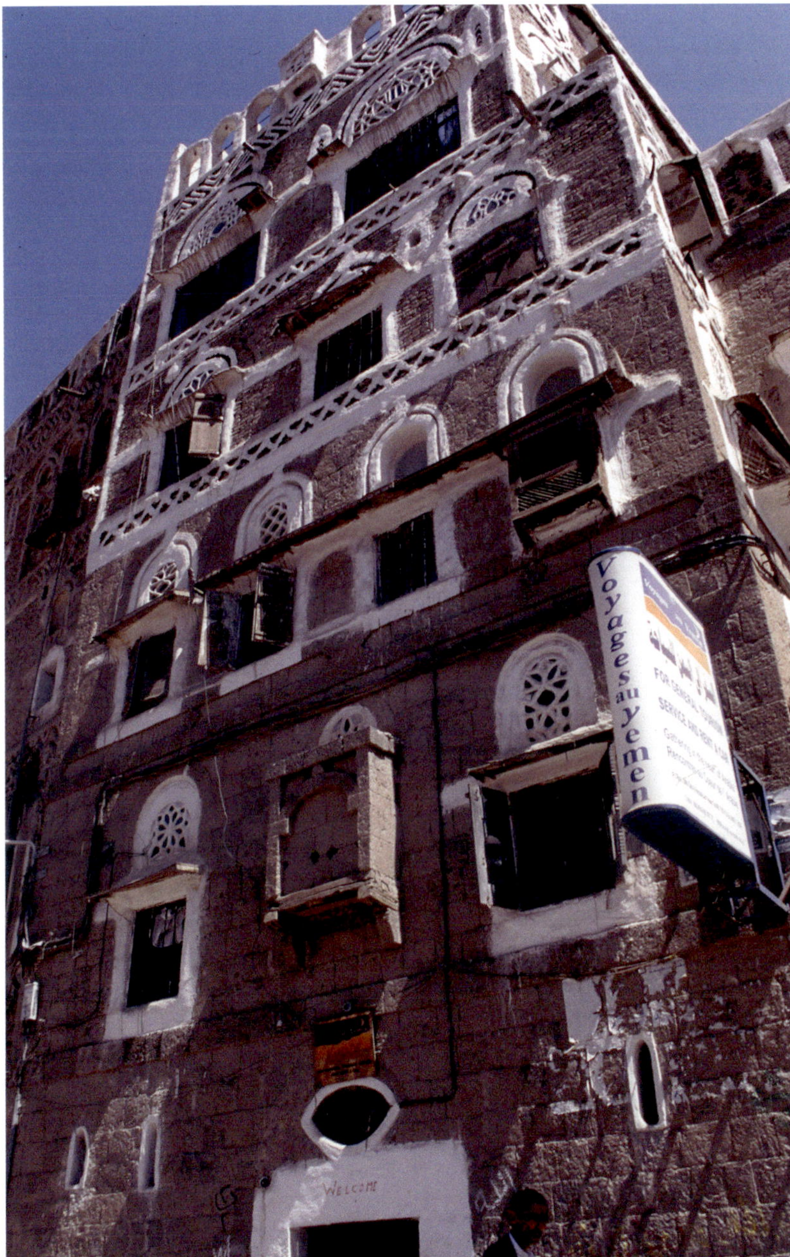

Figure 5.1 **The upper stories of this house on the Wadi Sa'ila were built by the father of Muhammad al-Rawdhi, lead *usta* for the UNESCO Campaign**

They are the ones from whom I inherited the trade (*tawarrathat al-mihna minhum*)." It is significant that he used a word that derives from the root *w-r-th*, "to inherit": the usage may be a modern one, perhaps influenced by the new currency of *turath* which, as we have seen, derives from the same root. Builders often maintain that they will pass the trade onto their sons, even though this seems to be less often a reality. But according to some *usta*s, even in the old days a great building house (*bayt*) rarely lasted more than four generations: by the fourth generation its glory had expired, and sons had migrated to new fields.[14] In recent decades, the process has accelerated—due primarily to the faltering social status of builders, which has encouraged their sons to seek careers in new professions that carry greater prestige and monetary rewards.

Conservation has, to some extent, renewed the prestige of *usta*s trained in what they call the "Sanaani style" (*al-tabi' al-sanaani*). *Usta*s have benefited from increased patronage in old-style work—mostly additions and restorations—in the old city and suburbs, as well as in neo-traditional architecture throughout the city. Some builders who began their apprenticeships after the revolution were converted back to the old style and now specialize in it. Although they continue to work in concrete construction—their bread and butter—they are advocates of traditional methods and materials, which they believe produce better and more durable architecture. "Concrete may last a generation," an *usta* told me, "but a building of stone and brick (*hajar wa yajur*) will last hundreds of years, and more." Some builders try to persuade clients to use traditional materials. If finances permit most opt for concrete and steel, which evoke modernity and permanence. But almost invariably, clients choose to clad concrete in a neotraditional veneer.

For some builders, conservation has validated the work of their fathers and grandfathers: they see the old buildings as a training ground for them, challenging them to perfect their own skills. "When I first built next to my father's work," said one builder,

> I felt ashamed [because my work was inferior]. But gradually I learned from his work, and came to feel proud. There is nothing like the old work, before it or after it (*al-shughl al-qadim ma bish ba'dih battatan*). When I began to work at Bab al-Yaman [the old city gate] and I saw the work, I was amazed (*baharni*) and I asked myself, where are the *usta*s who built this? The work was perfect (*nadhif*).

Despite his assertion that there is "nothing like the old work," the builder described himself (on tape) as a "great builder (*usta kabir*), specialized in the old city … I learned this work from my father, who learned it from his father." As Andrew Shryock notes, the belief that an age has ended goes hand in hand with the desire to portray oneself as heir to that age.[15] While builders are proud of their work in the "old style," they are equally proud of their work in a neo-traditional style—what one builder describes as "the renewal of heritage (*tajdid al-turath*)." These commissions include exterior cladding and ornamentation of private villas, office buildings, and government institutions; and religious buildings like mosques, minarets, and *madrasa*s, usually financed by private

14 Marchand was told this by builders in the mid-1990s, to explain why many of the well-known building families of the 1970s and '80s were no longer practicing. He notes the uncanny parallel to Ibn Khaldun's comment in the *Muqaddima* (14th century): "[Prestige] reaches its end in a single family within four successive generations …" (cited in *Minaret Building*, p. 139).

15 *Nationalism and the Genealogical Imagination: Oral History and Textual Authority in Tribal Jordan*, Berkeley: University of California Press, 1997, p. 14.

patrons as good works (*sadaqa*) and signs of their social status.[16] In some cases, builders have managed to become contractors themselves. In the case of minarets, described by Marchand, *usta*s continue to contract directly with patrons, and exert control over the creative process and its execution. In contrast, work on other types of buildings is often performed under subcontract to a general contractor; the builder may have a say in the design, depending on the contractor or architect. In commissions for private houses, owners may defer to the authority of the *usta*, who freely changes the architects' plans and elevations on site; while these new houses have a very different layout, they are often designed with traditional features.[17] But especially in larger projects, including the more opulent villas, traditional motifs—translated into new materials—are reduced to cladding and are often executed in a formulaic way. As Sallam puts it, "details have become engineering."

The popularity of traditional motifs is not only due to conservation—nor is it, as has been suggested, due to the cheaper cost of traditional work, which has equaled or surpassed that of modern construction.[18] The taste for the local was in part a reaction against Egyptian modernism, which was perceived as an alien and imported style. Pride in local architecture was reinforced by European and American consultants, and canonized in laws and codes which, despite the absence of enforcement mechanisms, are widely observed. This conformance is largely due to a shared perception of local forms and motifs, and their associated values, as "traditional" (*taqlidi*).[19] As we have seen, much neotraditional construction is disturbing to professionals, both local and foreign—evidence that the old, refined sensibility of builders has dissipated under the influence of rapid change. For many builders and their clients, new work in the "Sanaani style" continues and validates past practice. But some builders, too, lament the lack of refinement in modern work. At a *qat* chew in a house in the old city, ʻAbdullah al-Rawdhi pointed out the window and said wistfully:

> As you see, the experience of fifteen hundred years—even in the quality of the brick—has been lost. If there is no shepherd, the goat gets lost. We builders are goats without a shepherd: each works the way he wants.

Ironically, conservation may contribute to the lack of continuity. Because of regulations in the old city, it is difficult for building practice to evolve there. Innovation is confined to new districts where construction and typologies are radically different, and there is little context—and no regulations—to guide change. As Wim Denslagen observed two decades

---

16 T. Marchand, "Process over Product," p.123.

17 J.-L. Arnaud, "Formation de l'Architecture Contemporaine à Sanaa, 1965–90," in *Sanaa Hors les Murs*, pp. 169–70. See also interview recorded by Hatim al-Sibahi in the 1980s, transcribed in *Sanaa, Yemen: Tradition and Modernity in Sanani Architecture*, Eschborn: Deutsche Gesellschaft für Technische Zusammenarbeit (GTZ), 1996. According to architect Nadia al-Kawkabani, the functional layout of new houses often resembles that in old houses—although spaces are grouped on a single floor. While new houses may be inspired by Jordanian or other Arab models, they invariably have a *diwan*.

18 Marchand wrote that traditional construction was cheaper, but most architects (and some *usta*s) told me that this was not the case. The cost of traditional materials, they said—especially structural wood and brick used in old work—has continued to increase because they are scarce.

19 Marchand, *Minaret Building*, p. 217. According to ʻAli ʻOshaish, in 1972 a law was passed to preserve the "Yemeni architectural style" ("Toward a strategy for the preservation of the urban cultural heritage of Yemen").

ago, "the preservation of monuments aims to prolong the life of the monuments, while killing the architectural climate that has produced them."[20]

It is often said that builders are fundamental to the conservation effort: it is they who possess the skills needed to restore and maintain old buildings. But in Yemen, as elsewhere, the attitudes of conservation professionals toward builders are mixed. A few professionals acknowledge that they have learned much of what they know from builders, who should be privileged partners in conservation. At the other extreme are those who say that the role of builders in conservation is provisional: builders would ultimately be superseded by a well-trained class of professionals. Most fall somewhere between these two positions: they acknowledge that builders are crucial to the conservation process, but that their skills must be elevated and refined in order to work on sensitive historic monuments. In this, they echo the sentiments of consultants dating back to the Quaroni Bonifica studies—which assumed that the *usta*s' skills had declined or had been lost, and had to be "recovered."[21] "The local knowledge [of builders] is part of our experience," said one professional, "and we must find a way to integrate it." "Once you start a dialogue, *usta*s will react quickly," observed a project manager at the SFD. "They will propose solutions." But this requires changes to traditional practice as the builders know it:

> This is the paradox of conservation: traditional builders are most qualified to work on old buildings, but they are not sensitive to its historical aspects; they are interested only in pleasing the architect. They are separated from the architect by a cultural gap (*fajwa hadhariyya*).

Many builders have assimilated the notion of the old city as a special and protected area, where innovation should not be allowed. While affirming these principles and the superiority of their forebears' work, *usta*s often try to improve the buildings, in aesthetic or functional ways. This approach to the past conflicts with that of conservation professionals. Most have been schooled, directly or indirectly, in international methods and standards—although as we will see in the following section, they often disagree on their application or choose not to observe them. They have internalized the notions of the "historical time line" and the "monument" (a designation they attribute to buildings in the old city in varying degrees). But since there is no classification system—only a blanket guideline prohibiting change—judgment is often left to individual inspectors, many of whom have no specialized training. The basis for such judgments is not always clear to builders (or to residents). Professionals and builders learn each other's conceptions of the "historic" on site—for example, when builders try to make changes that will improve the building. Selma al-Radi, who directed restoration at the 'Amiriyya *madrasa* in Rada, was a vocal defender of the *usta*'s role in conservation. With her authorization and assistance, the chief *usta* at the 'Amiriyya substantially rebuilt the foundations and other parts of the building (the extent of rebuilding provoked criticism from some foreign observers). The chief *usta*, recalled al-Radi, often tried to improvise, adding or changing details which he felt would improve the building. She had to persuade him that in the case of an historic monument, the restoration had to be faithful to the original.

For centuries repair work, reconstruction, and modification to meet new needs were essentially part of the same process. Now, there are clear distinctions between these

---

20 "Restoration Theories, East and West," *Transactions*, vol. 18, 1993, p. 6.

21 See Chapter 6.

operations.[22] The "cultural gap" between builders and conservation professionals produces competition over professional authority, and the right to interpret the past. But at a deeper level, the gap derives from competing notions of historicity, which are rooted in different types of knowledge.

'Abdullah al-Rawdhi won the contract for the repair of an historic bridge at Damt—a challenging project which, in his words, required "all the experience of my father and grandfather and great-grandfather." He had successfully bid on the job, making calculations on the basis of SFD drawings. But he was nervous when managers from the Social Fund visited the site, just after he had started work:

> The committee judges between *turath* and me (*al-lajna tahkhum bayn al-turath wa ana*) ... [When I began the job] I asked myself, how can my wall go with a wall that is four hundred years old? ... How am I to deal with *turath*?

Al-Rawdhi's words reveal an ambivalence toward *turath* and the new arbiters of the past—and a new uncertainty in a profession that has traditionally had a strong self-image. Builders are told that the work of their predecessors represents a pinnacle of refinement: now considered "historic," the work should not be changed—lest it be contaminated by new materials, methods, and tastes. Buildings are now valued according to temporal and aesthetic criteria defined by others. They are viewed in isolation from their performative context, which is intimately related to a complex of cultural norms—including the *usta*'s expertise and public status.[23] At the same time, the new values resonate with aspects of builders' discourse. *Usta*s agree, for example, that "there is nothing like the old work, before it or after it"; yet they can no longer engage creatively with their predecessors' work. The builder cited earlier, who had been ashamed at the quality of his work in comparison to that of his father, recalled the early days of Campaign:

> In the beginning I had problems with al-Haddad [former director of the conservation office]. He hired me to do renovations in the old city; I would do the work in new ways—for example, using cement with brick, different kinds of stone and stone that was machine cut. I would open a shop in the ground floor [of a building] and run away ... Yes [he said, in reply to my question], conservation was a strange idea for us, it was a mistake [sic]. When we began to work in the old ways, we realized we had done things that were not good, because they disfigured the old Yemeni style and also the old city and the old work in it (*shawwahat bil namudhaj al-yamani al-qadim wa aydhan bi Sanaa al-qadima wa al-shugl al-sanaani al-qadim*). When I first started to work in the old style my conscience bothered me (*ta'annab dhamiri)* because I had done things that violated the old buildings.

Once again, we hear a mixture of pride and shame in his words—and an attempt to translate the new view of his craft into familiar terms.

---

22 W. Denslagen, citing Niels Gutschow's account of the "cultural gap" between architects and builders in Nepal ("Restoration Theories, East and West," p. 7).

23 T. Marchand, "Process over Product," pp. 137–9.

Local Building Culture

It is difficult to walk through the old city of Sanaa—and countless other old towns in Yemen—and remain unmoved by the creative vision that produced it. Foreign consultants maintain that the World Heritage initiative has been invaluable in fostering an appreciation for remarkable human masterworks—especially since the forces of globalization would otherwise eradicate them. Yet decades after the World Heritage Convention, we are perhaps no closer to understanding the creativity that produced places like the old city of Sanaa—and hence the true "value" of a World Heritage site. "Yemeni conservators would do better in Florence," a senior architect told me. "That's where our understanding of architecture comes from. As for our local architecture, we still have no idea what it is."

Local architecture resists categorization and analysis according to established art historical methods. In Europe, too, notes Lewcock, the academic, evolutionary model—based on notions of "period" and "style"—has obscured local variations and the persistence of technologies and motifs over time. It has also tended to obscure the relationship between architecture and making. Indeed, historians and other scholars have often imposed rather abstract, and subjective, notions of meaning on architecture. It is not accidental that art historical methods developed alongside the new architectural profession, distinct from builders' practice, which defined itself in terms of a new specialized knowledge and an historical self-consciousness.[24] In the practice of builders in Yemen, we have the opportunity to observe a different kind of knowledge and way of constructing meaning that perhaps resembles that of builders in Europe, before architecture was "professionalized" in its modern form. The approach of Yemeni builders comes into sharper focus as one observes their interaction with new discursive practices and new hierarchies.

Sitting with builders at *qat* chews, and at meetings of foreign and local experts, I was struck by these disparate worlds. I wondered why foreign experts continue to be brought to Sanaa to evaluate and plan: not only is there a core group of local professionals with decades of experience, but there is a rich and vital building tradition practiced by men in the prime of their careers, some of whom are third- or fourth-generation builders. Indeed, *usta*s are keenly aware that they are the ones who possess knowledge and skills required to conserve the old buildings: they know that they are caretakers not only of local, but of *world*, heritage (*turath 'alami*). Yet they are rarely, if ever, consulted in discussions of what conservation is, or should be, in Yemen. Rather, they serve a system whose goals appear to be defined in the boardrooms of local and foreign agencies.[25] Why has this important resource not been tapped? The explanation perhaps lies in the competition between old and new professions—and ultimately, between different forms of knowledge.

As Trevor Marchand has argued, building is a form of knowledge that is constituted in the acts of performance and transmission; the latter typically occurs through some type of apprenticeship system. Apprenticeship is aimed not only at inculcating skills, but at creating a professional person and moral agent with particular behaviors and values.[26] Training is accomplished indirectly, through observation and imitation—as builders say, "just by

---

24 In the formation of modern professions, notes Larson, "[h]istorical continuity is not only implied; it is deliberately and actively sought in the attempts by organized professions to give themselves a culture with roots in a classic past" (*The Rise of Professionalism*, p. xv).

25 This point was made by A.G. Krishna Menon regarding builders in India ("Rethinking the Venice Charter.", pp. 51–2).

26 Marchand, "Process over Product," p. 139. The following is indebted to this article and to his *Minaret Building*; and to my own site visits and conversations with builders.

looking (*bi mujarrad al-nadhar*)." When asked how they learned their craft, builders relate an almost identical sequence:

> I worked first with mud (*khulab*), I mixed earth with water (*kunt akhallib*) … After that I worked throwing the mud to the *usta* (that is, as a *munawwil*). Then I worked on the inside of walls (*fil mathna'*, that is, as a *thanna'*). Little by little I got to work on the facade—first in stone, then in brick. After that I learned to work with the string (*mizan*), working next to the stone-cutter (*muwaqqis*); we use the string for measuring, too [rather than a meter]. We also learned to cut stone (*al-qassa*). Then we went up on the facade (*tala'na fil wajiha*) and worked there. All this I learned with my father, who learned it from his father …

In building and other trades, apprenticeship is often characterized by an almost total absence of verbal communication—normally thought to be essential in the learning process. This is due in part to the initiatory aspect of apprenticeship and the desire to safeguard specialized knowledge.[27] But it is also related to the type of knowledge that is being transmitted. Drawing on recent studies of language and cognition, Marchand argues that while building is a non-lexical form of knowledge, it nevertheless consists of concepts that emerge in the process of making. These concepts inform theoretical knowledge and in turn impact making: "knowing that" becomes "knowing how."[28] Knowledge is constituted through the engagement of the body and the material, in a performance that is professional and public—and is thus closely tied to the builder's status and claims of professional authority. The *usta*'s authority is vested in a specialized and secret knowledge, which is inculcated in stages during the course of apprenticeship.

Marchand sees the builder's knowledge as a form of "traditional discourse" which focuses on specific events and contexts rather than on theoretical concepts.[29] The builder's statements are assumed to have "truth value" not because the content is true, but because "the utterance [is] an event … produced by a customized person who embodies the supposed historicity and expert knowledge."[30] This focus on events recalls the term *tarikh* (history)—literally, "recounting." It closely parallels Dresch's and Shryock's accounts of local forms of historiography, both oral and written, which also emphasize events and persons: narrative and sequence are subordinate or absent, in favor of a continuity that affirms the identity of the group.[31] While scholars often assume that "history" has a positivistic sense—a "'chronicle of actual events' as opposed to 'myth,' 'legend,' or

27 Delbos and Jorian describe apprenticeship as a process of socialization which is as much about what is *not* said, as what is said (*La Transmission des Savoirs*, Paris: Editions de la Maison des sciences de l'homme de Paris, 1984, p. 239).

28 Marchand, "A possible explanation for the lack of explanation: or, 'Why the Master Builder can't explain what he knows'," *SOAS Working Papers in Linguistics*, vol. 10, 2000, p. 313. See also Connerton's discussion of "inscribing" versus "incorporating" practices, *How Societies Remember*, Cambridge: Cambridge University Press, 1989, Chapter 3.

29 Marchand, *Minaret-Building*, pp. 137–8, building on Pascal Boyer's work (*Tradition as Truth and Communication: a Cognitive Description of Traditional Discourse*, Cambridge, Cambridge University Press, 1990).

30 Marchand, *Minaret-Building*, pp. 218–19.

31 Dresch, *Tribes and Government*, p. 182; Shryock, *Nationalism and the Genealogical Imagination*, pp. 21–2. See also Connerton, *How Societies Remember*, p. 3.

'fable'"—in Yemen *tarikh* is often used to describe a past that is relativistic, and often frankly ideological.[32]

Marchand's study of minaret-building does not attempt to explain the content of builders' knowledge, since the rendition of one form of knowledge into another can only be partial. Rather, his detailed account of the building process and his own apprenticeship attempts to show how this knowledge is constituted, and the social context of its production and reproduction.[33] Sultan Sallam chose to study building practice in a way closer to that of the eighteenth century *encyclopédistes*: in 1988 he spent several months on building sites, observing how *usta*s make decisions, and drawing their techniques and construction details. "I express (*u'abbir*) for the *usta*s, in words and drawings, what they cannot say for themselves," he told me. "I am their ambassador (*safir*) to the architecture students, who are removed from [the *usta*s'] practice and from the old city."

I at first felt frustrated that the *usta* "cannot explain what he knows."[34] Yet as a practitioner and teacher of architecture, I realized that it is difficult to articulate knowledge about practice and aesthetics. Later, while reading Marchand's studies, I suddenly realized that the anecdotal stories and casual analogies I had been hearing from builders *were* their discourse. They do not explain the "knowing how": rather, they shed light on the richness of builders' knowledge within an intensely local context.

'Abdullah al-Rawdhi told me about his initiation in the 1970s. The story provides a glimpse into the internal workings of a vital profession:

> I used to work with my father; I made him lose money. He would joke with me and sing, 'Ask my history, "I am Yemeni"' [a song] ... He used to tell me to go to the house to bring the *mizan* (level). I would go [but I would] sit, and imitate what I saw my father doing on the wall. I would be late returning, and my father would hit me ...
>
> I used to work with other *usta*s too—they were the best, real perfectionists (*'ahl al-dhimma*). They did Bayt Mi'yad, Bayt al-Raqqas, Bayt al-Haymi, and other houses. Once I worked in Bir al-Bahma [a neighborhood in greater Bir al-'Azab]. That place is a place of scholars of the trade (*'ulama' al-mihna*) like Ahmad Miqbal Mi'yad—he is the summit of Yemen (*qimmat al-Yaman*)—al- 'Izzi al-Haymi, and others. They were all old men, and spent most of their time in the mosque. My father gave me a building to work on in front of the mosque. I had a *janbiyya saifani* [a knife of the highest quality]; I sold it and used the money for the work. When the *usta*s walked in the street, they would look at the building and say, 'That boy is talented ... his hand is good (*yaddih haliya*).'
>
> In those days 'Abdullah 'Ali al-Rawdhi was a big *usta* in Bir al-'Azab.[35] I was afraid of the *usta*s, afraid that they would criticize my work; in those days there was a lot of criticism. The old *usta*s used to ask, 'Whose son are you?' If an *usta*'s work wasn't good,

32 Michael Herzfeld suggested the relativistic character of *tarikh* at an early stage of my fieldwork. For a pararallel to Jordan, see Shryock, *Nationalism and the Genealogical Imagination*, p. 22.

33 "A possible explanation for the lack of explanation," p. 309.

34 The inarticulacy of the craftsman is the subject of Marchand's piece, "A possible explanation for the lack of explanation." The phenomenon has been noted by Herzfeld in Crete (*The Body Impolitic*, pp. 81–2) and others.

35 This man was from another family of *usta*s with the same name, al-Rawdhi. He built the Ministry of Health (now the Mayor's Office) on Tahrir Square. Its exceptional facade draws on a tradition of polychrome stonework, and also appears to translate decorative brick patterns into stone on

> he would hide it when an old *usta* walked by, so he wouldn't criticize it. [But] the old *ustas* encouraged me. Al-Rawdhi came by, and I liked him a lot. He called to me while I was making a *hizam* [the decorative brickwork that divides one floor from another] and asked me if I knew its name. I said that I didn't. He said, 'You are a talented boy' ....He entered the mosque, and the *ustas* spoke about me. Al-Rawdhi said, 'This boy has a great future' (*yaddih mubashira bil khayr*; literally, 'his hand is headed for great things'). Al-Rawdhi said, 'Come to me.' And I used to go to him. He taught me a lot. He told the *ustas* that I was great; but they said, 'Don't judge now: first let's let him build a *hizam* on Bayt Sunaydar (*yihazzim Bayt Sunaydar*), and then we'll judge.'
>
> The *usta* who was working on that house let me build the *hizam*. I worked on it, but I built a *hizam* of a different kind; I wanted to develop (*kunt ashti atawwir*) [the form], but the work was not good. I went down to the mosque to look at it, and I saw a hateful thing (*shay' karih*). I finished the work and went away. The *ustas* said to al-Rawdhi, 'You see, we told you to wait before judging.' After some time al-Rawdhi took me to his house. He was very, very mad at me because I made an ugly *hizam*, after he had told the *ustas* I was talented. He told me to go away. I was unable to sleep for three days. After that, I got a job: Bayt al-Bashiri in Bir al-'Azab, a building with stone cladding. As I worked on it, I remembered what al-Rawdhi had said [and] because of this, I worked from my heart (*min dakhil qalbi*). I made a *hizam* of stone and called al-Rawdhi to see it, but he refused. 'Go away,' he said, 'you're no good.' Finally he agreed to go and look. After he saw the *hizam* he said, 'I dare all the people in Bir al-'Azab to say that you are not an artist.' He gave me a certificate (*shahada*, literally 'diploma') ....[36]

The themes of the story were echoed in my rounds and talks with builders. Ahmad al-Rawdhi told me that an *usta* who claims to be expert in *turath* should know the many types and names of *hizams*, which had almost been lost to memory. As his brother's story suggests, the choice and execution of *hizam* is an important test of a builder's proficiency. An architect can immediately see that the *hizam* at each floor level—its height and form—establishes the main proportions of the facade. According to Sallam, who documented the various types of *hizams*, *ustas* design the facade by eye, from the ground, compensating for the distorting effects of perspective. Typically, the higher a *hizam* on the facade, the wider it is. But in my discussions with builders, it became evident that proportion is only one consideration: the *hizam* is an important aesthetic component, and in some sense establishes the character of the facade—not unlike, one suspects, the classical orders in ancient buildings.[37] For *ustas*, said Sallam, the *hizam* is like the belt in which a man carries his knife (*janbiyya*; *hizam* in fact means "belt"), or the scarf that he wraps around his head.

The *usta*'s body and perception are intimately engaged in the design process, which is akin to sculpture: a visual proportion is pleasing, said one builder, when it "fills my eye."[38] When a builder designs a *mafraj*, Sallam told me, he sits on the roof to determine the correct

---

this Ottoman-inspired, institutional building. See P. Bonnenfant, *Les Maisons Tours de Sanaa*, Paris: Presses de CNRS, 1989, pp. 102–3.

36 Other builders also told me they had received a *shahada* from an established *usta*.

37 The technique of manipulating perspective to please the viewing subject is also well-documented phenomenon in ancient Greek architecture.

38 Quoted in T. Marchand, *Minaret Building*, p. 178.

height of window sills. The conceptualization of space may begin, as Marchand suggests, in the body's engagement with the module of the brick. During the course of apprenticeship

> [c]ontinuous building practice, meaning the regular involvement in the processes of *making*, served to develop a knowledge of the primary building components such as the central column, stairs, and walls, and the manner in which they are assembled to create *space*. At this scale of *making*, one's own body became the principal source of measurement, and by inhabiting the process, reflection upon the body in relation to the physical environment became the principal means of planning spatial relations.[39]

The engagement of body and material may explain, in part, builders' anthropomorphic conception of buildings: like a person, a building pleases (or not) by virtue of its stature, proportions, and "clothing."[40] When I asked Ahmad al-Rawdhi how many floors were appropriate to a house, he said that it depends on the height of the surrounding houses, but also on the nature of the house itself:

> If a house is five meters wide, maybe it permits only four floors, like the house we are in … If it's nine meters wide, it could have six floors in order to have good proportions. A short house is like a person in *sha'bi* (popular) dress, with a *janbiyya*. A tall house with a *mafraj* is like a man in a suit and tie.

On another occasion I walked with an *usta* past a garden (*maqshama*), surrounded by a new wall that GOPHCY architects had adapted from the *tijwab*, the arched parapet wall on the upper floor of a house. The builder grimaced: the wall, he said, is like "a man who is so short that he can't breathe." Describing the technique of *fals*—the careful reconstruction of a wall by removing and replacing stones—*usta* Yahya al-Dhahbani said that a wall was bulging out 20 centimeters, "like a pregnant woman" (tracing a bulge in his belly). "We rebuilt it so it was the same as the others (*mizan wahid*)" (Figure 5.2).

The proportions of the window are, for builders, a particularly important characteristic of Sanaani *turath*. In order to be pleasing, Ahmad al-Rawdhi told me, the arch must sit on a "neck" (*raqaba*), a vertical extension of the sides of the frame. Even in wider windows, the arch should not sit directly on the window, but on a *raqaba*. As such, the 1:1 proportion of window to lite is retained (Figure 5.3). Without the *raqaba*, said al-Rawdhi (hunching his shoulders so as to appear neck-less) the arch is "like a head without a neck"; Sallam told me that builders poke fun at architects who design ugly *hizam*s and arched windows with no "neck." But as windows become wider, it is hard to maintain the neck—which should be proportional to the width of the window and the radius of the arch. The arch of a very wide window will bump into the ceiling, explained al-Rawdhi, so the neck is sometimes eliminated. But this is a mistake, he said: the *raqaba*, like the *hizam*, should be retained in their traditional proportions. He and others *usta*s cited poorly designed *hizam*s as one of the biggest flaws in new construction.

---

39 T. Marchand, *Minaret Building*, pp. 177–8.

40 Architect José-Marie Bel used anthropomorphic terms in describing Yemeni architecture (*Architecture et Peuple du Yemen*, Paris: Conseil international de la langue française avec la collaboration de l'Institut international d'architecture mediterranéenne, 1988). For Europe, see especially J. Rykwert, *The Dancing Column: On Order in Architecture*, Cambridge, M.I.T. Press, 1996.

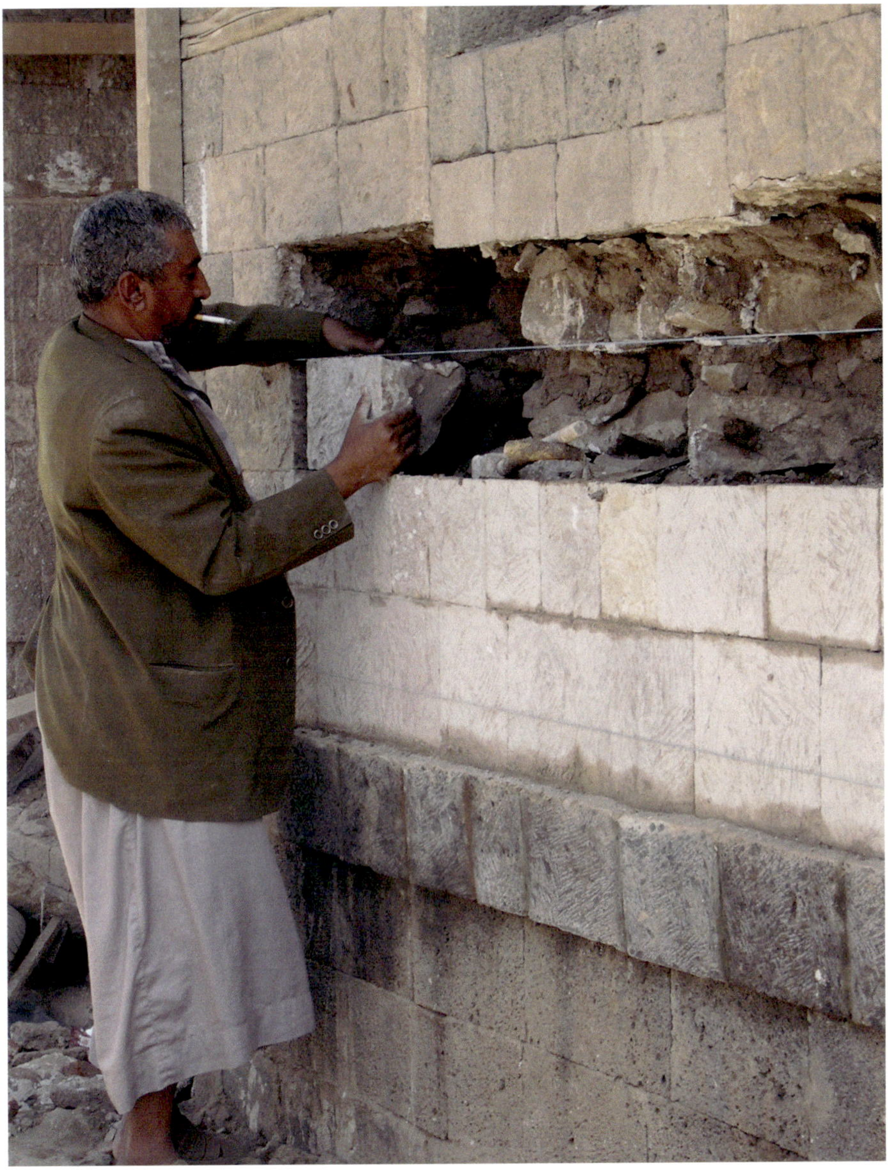

Figure 5.2 Yahya al-Dhahbani demonstrates the technique of *fals*—the careful reconstruction of a wall by removing and replacing stones

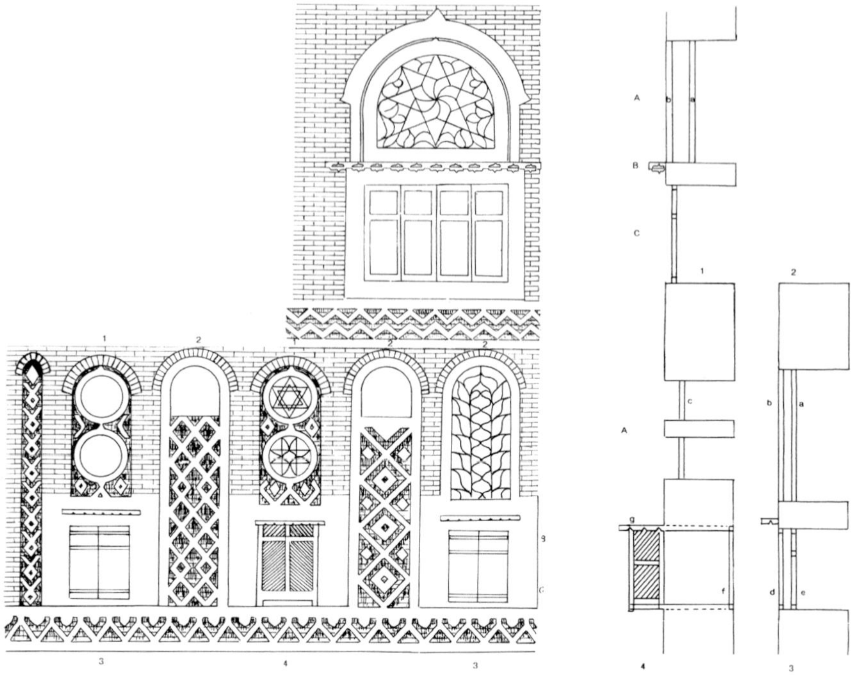

Figure 5.3 Historical evolution of window types, proposed by Paul Bonnenfant, Jean-François Bréton, and Christian Darles

At lower left, seventeenth–eighteenth centuries: a small window with shutters is surmounted by two alabaster oculi, in an arched recess in the brick. The height of the arched recess is twice that of the window. From the nineteenth century, the alabaster oculi are replaced by small windows with colored glass (center bottom). Early in the twentieth century, the two oculi give way to a single arched window (lower right). In the late nineteenth century wider windows of three or more shutters appear; the arch above has the same width and height as the window (top right).

Source: Image and citation courtesy of Guillemette Bonnenfant-Outrebon/Paul Bonnenfant, editor, *Sanaa, Architecture domestique et société*, Paris, CNRS-Editions, 1995, Chapter 21, "Le livre des jours et des modes," figure 32, pp. 482–3.

Interestingly, some of the same terms occur in various art forms—suggesting the existence of certain "affinities" in local arts and aesthetics. Saba al-Suleihi suggests that these affinities can be used to study the arts from a culture's own perspective, rather than through theories external to it.[41] A noteworthy example is the use of *hizam*-like illumination to separate chapters in Qur'an manuscripts—visible, for example, on some very early fragments in GOAMM's manuscript collection.[42] Parallel structures and terms have been

41 S. al-Suleihi, "The Lyrical Facades of Sana'a," p. 84. In addition to poetry, al-Suleihi explores aesthetic and structural parallels between architecture, music, and textiles (pp. 18–26).

42 The fragments were discovered in the ceiling of Sanaa's Great Mosque in 1972, and were restored under a joint Yemeni-German project. They are held at the at Dar al-Makhtutat, adjacent to the Great Mosque.

noted in music and architecture. Jean Lambert suggests that architectural terms allow aspects of music to be conceived and visualized metaphorically. Both musicians and masons, he notes, are practitioners rather than theoreticians: "if one finds structures, they come after the fact, having been sanctioned by usage, rather than organizing practice *a priori*."[43] In poetry, a line is called a *bayt* (literally, "house"), while a song has a melodic-rhythmic structure called a *qa'ida* (base or model). A musical interpretation should follow a conventional sequence, writes Lambert, consisting of three rhythmic cycles or suites (sing. *qawma*):

> According to musicians [the *qawma*] must be like a building, *mabna'*, a structure in the architectural sense, *maqam*. The metaphor of the building clearly indicates the desire for perfection and cohesion of the entire sequence, in which the melodies are freely chosen by the musician. Moreover this does not exclude the existence of certain fixed sequences, doubtlessly because they had their hour of success: such is the case of a well-known suite called *al-qaws* (*arc*, but also arcade in architectural vocabulary) because it is made of several sections, *madraj*, particularly well assembled, *murakkaba*.[44]

Qawma and *maqam* derive from *q-w-m*; the many variations of this root include the meanings "to be erected" or "to erect." Lambert suggests that musicians' use of the term *maqam*—that is, a building—differs from the use of the term in Near Eastern music, where it means a mode in the tonal system.[45] Several builders I interviewed, however, use the term *maqam* in a way that seems close to the sense of "mode." When I asked Ahmad al-Rawdhi if he would agree to add two stories to a particular house, he said yes, "but I will make the windows in the same *maqam*; I will not go outside the *maqam* (*kharij al-maqam*)." When I asked him to explain what he meant by *maqam*, he did not answer directly, but said that *maqam* applies to the entire house (*al-bayt al-kamil*). Later (a couple of hours into *qat*-chewing) I asked him what types of houses are found in Sanaa. "There are no types," he replied. "All of Sanaa is of one piece (*qut'a*); the houses of Sanaa are of one *maqam*. Houses may differ by size but they are not 'types.' The 'types' are arches, windows, and ornament (*naqsh*)."

The last term, *naqsh*, is a significant one: builders use the term *naqqasha*, "to decorate," to describe the ornamentation of a facade, much as a musician embellishes a *maqam*. This term and others like it appear in Arabic literature and a variety of artistic and artisanal practices. As the eminent historian Oleg Grabar wrote:

> Decoration seems to complete an object, a wall, or a person, by providing it with quality. In classical Arabic *adab* literature, literary theory or practice, words like *naqqasha* ('to cover with decoration' in many different techniques), *zawaqa* ('to embellish'), and especially *isti'ara* ('to use metaphorically') have different meanings dealing with artistic or esthetic practice or description; all are positive in their judgment and imply effective completion and even transfer of meanings from one form to another.[46]

---

43 J. Lambert, "La Maison Tour et la Musique: Harmonies et Dissonances," in P. Bonnenfant, ed., *Sanaa*, pp. 165–73. Al-Suleihi makes a similar argument regarding the genesis of grammatical rules in Arabic, and rules of composition in poetry (*The Lyrical Facades of Sanaa*, p. 85).

44 J. Lambert. "La Maison Tour et la Musique," p. 167.

45 J. Lambert. "La Maison Tour et la Musique," p. 575, fn. 16. The term *maqam* in architecture indicates a commemorative structure, usually a mosque or saint's tomb—literally "a place where one rests." See A-R Ghaleb, *Encyclopédie de l'Architecture Islamique*, Beirut: Jarrous Press, 1988, p. 397.

46 *The Mediation of Ornament*, Princeton: Princeton University Press, 1992, pp. 25–6.

Decoration may also be a metaphor for contemplation.[47] At the same *qat* chew, a researcher asked al-Rawdhi about the relationship between the architecture and dance of Sanaa. "Like Sanaani dance," he replied, "the facade [of the house] goes from right to left: the eye dances on it, jumping from one beautiful thing to another." In Sanaani music, notes Lambert, the embellishment of a melody is described in terms borrowed from artisanry. In addition to *naqsh*, the term *harsha*—the embellishment of the main melody with the voice—also refers to the decorative motifs of the window-maker (*harrash*). *Tadfir* is used to indicate the interpenetration of several elements.[48] During a walk in the old city, one *usta* pointed to a vertical brick panel on a facade: it was displeasing, he said, because the undulating lines did not interpenetrate.

The detailing and decoration of the tower house has been presented by other authors.[49] I suggest here only a few thoughts regarding ornament as a product of builders' discourse—a critical subject, requiring extensive ethnographic and historical work. In addition to polychrome stonework, the main ornamentation of a facade consists of brick patterns and panels, parts of which are painted with a white plaster called *nura*. *Nura* is also applied around window and door openings, perhaps as water-proofing. Certainly functional reasons were long ago subordinated to a taste for decoration that surpasses that of other highland towns—although builders note that ornamentation in Sanaa has become excessive, even in the old city. Doors and window openings are places where evil spirits may penetrate, and so are often adorned with talismans, sculpted in plaster or carved in wood.[50] One wonders whether *nura* itself plays such a protective role, since the verb to plaster with *nura*, *nawwara*, also means to "bring light."[51] Families traditionally renewed plaster on special and happy occasions, like weddings or births. Plaster decoration is part of a decorative and iconographic program that extends into the interior of the house, in the ornament of important rooms. This program includes symbols and inscriptions that express gratitude to God and invoke divine protection, through a combination of Islamic formulae and pre-Islamic motifs.[52] The current tendency in academia to study the architecture of Islamic societies in isolation from religious and spiritual concerns seems inappropriate for Yemen. "The philosophy of the *usta*s," Sallam told me, "is that God is most beautiful." Builders are grounded in Islam, notes Marchand: it provides the basis of their socialization and hence of their individual and collective identities. In religion, as in building practice, their "grasp was of the *particulars*, each practice as a discreet object of knowledge, exercised automatically in response to a formalized set of circumstances."[53]

Usta 'Abdullah al-Rawdhi is a master oud player. He also happens to be illiterate. He likened the bridge at Damt to a song: in order to work on it, he needed to know the tones

47 Grabar cites a Jewish saying from the Book of Ben Sirach: "A mind settled on an intelligent thought is like the stucco decoration on the wall of a colonnade." Here, writes Grabar, "contemplation of something intelligent is compared to what makes a wall beautiful, that is something plastered on it, in a technique prevalent in Iraq whence the saying probably came" (*The Mediation of Ornament*, p. 25).

48 J. Lambert. "La Maison Tour et la Musique," pp. 167–8.

49 In addition to Sallam's work, which includes extensive documentation of doors, windows, and brick and plaster ornamentation, see parts four and five of P. Bonnenfant, ed., *Sanaa.*

50 P. Bonnenfant, "La Maison Microcosme," pp. 565–6.

51 I am grateful to Engseng Ho for suggesting the metaphysical aspect of this verb.

52 See P. Bonnenfant, Chapter 23, "La Symbolique de la Demeure," on decorative motifs and their symbolism; and chapter 22, "Les Inscriptions dans la Maison."

53 Marchand, *Minaret Building*, pp. 107–9. *Usta*s, as well as Muslim colleagues—even those who were not particularly observant—found it difficult to understand how Marchand could exclude God from his analysis (p. 26).

(*maqamat*) on which the song was built. "How can my voice go with [the voice of the *usta*] who built the bridge?" he asked. "I needed to know how my *maqam* would affect the old." His artistic impulses were reigned in by managers at the Social Fund, who tried to instill in him a different approach to the work. 'Abdullah al-Dailami, then director of the Cultural Heritage unit, saw the irony. "During the course of the project, al-Rawdhi learned about restoration," he told me. "He wanted to beautify the bridge. We wanted to make it a monument." But al-Rawdhi's proposal for repairing the bridge, he noted, did not differ substantially from that of the consulting civil engineer (Figure 5.4).

In order to get the job, al-Rawdhi had to read drawings produced by the SFD and submit an estimate. He was concerned, because he knew that the SFD would hold him to the scope of work as described in drawings. "The drawings were theoretical," he said, "not like reality … I knew it was a test." He tried to explain to them that the computer is different from his own method of "measuring":

> Anybody can play oud. But music doesn't come from the right hand; it comes from the left [the hand that controls the neck]. Feeling is the important thing. The right hand gives me the tone (*maqam*), but how do I embellish it (*kayfa unaqqishu*)? When a poet speaks from a window, he speaks words of poetry. When I go to build I sing in a different way. Each [art form] has its effect on the other. The artist, poet, *usta* need to be together in a *maqyal* (*qat* chew).
>
> Not all have talent. I'm illiterate; I understand my work through *wiratha* (inheritance). They say they're preserving *turath*, but they're treating it like a frozen thing. Architects shouldn't get their degrees until they are affected by Sanaa (*yata'ththaru bi Sanaa*)—until they have experience (*tajruba*) in old Sanaa. The architect should be the conductor [he motions with his hands]. GOPHCY should know [the name of] the *usta* who built each house, and who the family was; they should go to the documents (*watha'iq*) …

Architect Yassin Ghalib, who was present at the *qat* chew, added that each family of ustas had a particular signature or "stamp" (*basma*). "First I have to be good in the *turath* of my country," said al-Rawdhi. "Then I can leave a new stamp (*basma jadida*)."

As noted above, the authority of builders is vested in carefully guarded knowledge. Muhammad al-Rawdhi, Ahmad and 'Abdullah's father, was a lead *usta* for the conservation office during the UNESCO Campaign. 'Ali 'Oshaish, then Technical Director, recalled that they asked him to rebuild the *qutb* or central column of a building, using the technique of *fals*. (The *qutb*, which supports the central stair core, is considered the core of the house and is usually erected first.) They also asked if they could film the operation. The senior al-Rawdhi said yes—but then came on Friday and completed the work, when no one was around. When 'Oshaish asked al-Rawdhi for an explanation, the latter replied that *fals* was special knowledge that could not be shared with others. Indeed *fals*, more than any other operation, appears to embody the mastery and prestige of the profession. *Fals* involves the removal of stones and the insertion of wood (or iron) supports or wedges, so that the area below can be rebuilt (Figure 5.5). The stone courses above appear to be without support—and builders often use the word "hanging" (*mu'allaq*) to describe portions of a wall that seem to be almost magically suspended. *Fals* can be complex and dangerous, as in the above example when the *qutb* was rebuilt, since it supports the whole house. Some *fals* operations are remembered in builders' lore as superhuman feats—endowing the *usta* with a mythical prowess:

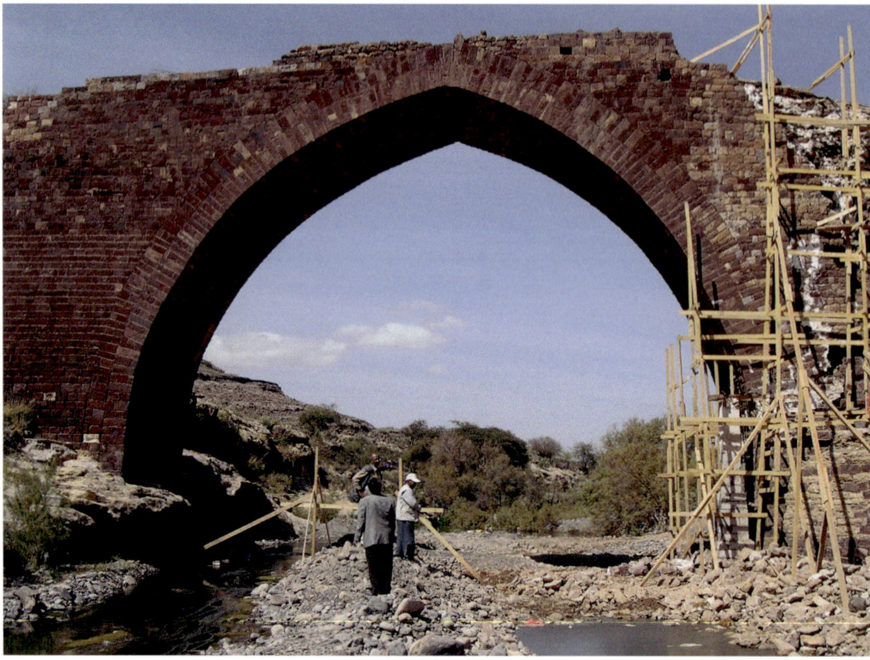

Figure 5.4 ***Usta* 'Abdullah al-Rawdhi restored the Rasulid bridge at Damt for the Social Fund for Development. In the foreground is a wooden centering device he built, to determine the radius of the arch**

> *Fals* is normal for us. The most important thing is the support. We can take out the [lower section of the] *qutb*, and hang it (*na'allaqu*) from the second floor ....My father could remove the lower four floors of a building and leave the *mafraj* hanging—suspended in the air!

Muhammad al-Rawdhi was specialized in restoration or repair (*tarmim*). "He was a surgeon," said his son Ahmad. "He could take apart (*yanqudh*, 'destroy') a building and rebuild it from ground up, just as it was."

Builders and Conservation

Modern professions like architecture and conservation must create and control a market for their expertise. They accomplish this, in part, by transforming the "situated" knowledge of builders and craftsmen into abstract or managerial knowledge.[54] The struggle between architects and builders (and later, between conservators and builders) in the developed

54 On situated learning and practice, see especially J. Lave and E. Wenger, *Situated Learning: Legitimate Peripheral Participation*, Cambridge: Cambridge University Press, 1991, pp. 26–7; and Gieser, "Enskillment Inhibited," pp. 134–5.

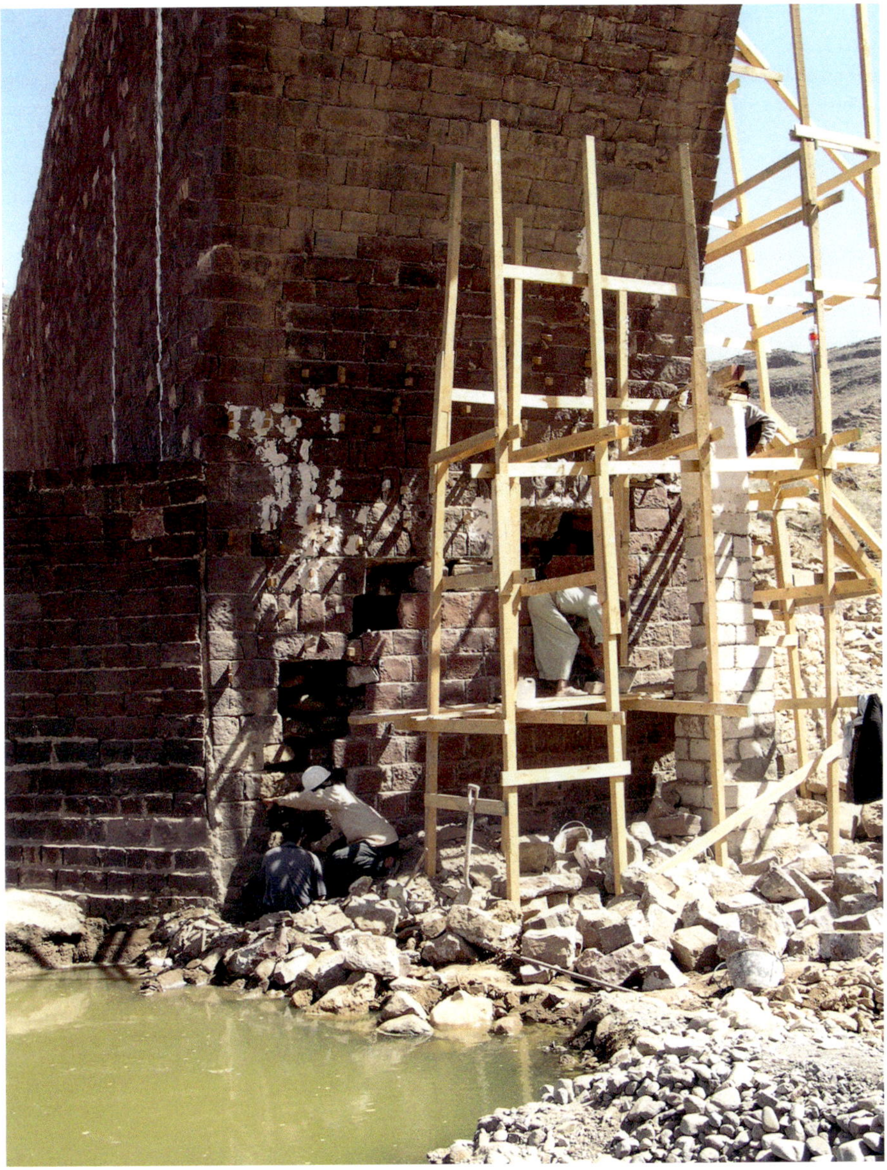

Figure 5.5 The operation of *fals* at the pier of the bridge

world is well-known, though rarely discussed in professional circles. But in Yemen, these professions have not yet attained a privileged position—perhaps because they have not been fully embraced by the political and economic elite. The builder's knowledge is not easily assimilated to a purely cognitive framework.[55] At the same time, other kinds of knowledge may threaten the still fragile identity of the new professions. Despite affirmations of the importance of "traditional builders" for conservation, they are excluded from discussions of policy and practice, and are economically and professionally marginalized. *Usta*s find this disturbing and confusing. While they speak fondly of certain professionals who respect their skills, they wonder why their status is not valued and rewarded—especially in public conservation projects that rely on general contractors. During my fieldwork, the Social Fund was restoring several small mosques in the old city. This involved, in part, undoing restoration work carried out decades earlier by the Ministry of the Endowments. An *usta* noted with irony that he was removing the same cement plaster that he had been hired to apply in the 1970s. He had known then that it was a mistake, but no one had asked him. He continued:

> Do you water a tree with petrol or water? If you keep watering it with petrol, it will dry up and burn. The *usta*s are the ones who know how to water. We know how to maintain the old city, just like our fathers did, yet we get no benefit from it. The benefit goes to the contractors, who have money but no experience.

The contradiction is a bitter one. The government and the international community celebrate the old city as "world heritage," yet they marginalize the heirs of the great tradition that built it.[56] "If they do not first preserve the craftsmen (*al-hirafiyyin*)," asked the *usta*, "how can they preserve the old city?"

Since the beginning of the conservation effort, architects and consultants in Yemen have lamented the disappearance or decline of traditional building skills. Some proposed training courses or schools for *usta*s, which would bolster their skills and sensitize them to the special qualities of historic buildings. Marchand believes that worries about the disappearance of traditional building skills have been overstated, and may be disingenuous. There may have been a shortage of masons in the 1980s, when many *usta*s and tradesmen were in the Gulf; but the situation changed after their return in 1991.[57] The reality may lie somewhere in the middle. The *usta*s' trade and numbers have perhaps been more affected in the capital than elsewhere, as traditional building methods have given way to concrete and other modern techniques. It is certain, for example, that *usta*s' sons tend to seek work in other professions, especially in Sanaa and other cities where such opportunities are more numerous. In this, material factors play an important role. In Shibam-Hadhramaut, for example, mud remains the preferred material for much new work. As a result, *usta*s have retained their status and authority—and "traditional" practices and techniques have

55 Marchand, *Minaret Building*, pp. 181–2; and "Process over Product," p. 138. On the emergence of modern professions see M. Larson, *The Rise of Professionalism*, pp. xiv–xvi and p. 11; and S. Bozdogan on builders and architects in the U.S.: "The Lincoln Cathedral versus the Bicycle Shed," *Journal of Architectural Education*, vol. 36, no. 4, 1983, pp. 10–15.

56 For an analogous situation in Crete, see Herzfeld, *The Body Impolitic*, p. 16.

57 *Minaret Building*, pp. 181–3 and endnotes 112–13, pp. 265–6; and pp. 236–9. On the competition between architects and builders see also J.-L. Arnaud, "Formation de l'Architecture contemporaine a Sanaa," pp. 169–70.

continued to evolve. This strength was used to advantage in an innovative conservation project in Shibam that supported the apprenticeship system (to be discussed in the following chapter). But this kind of approach has been rare.

The building process, notes Marchand, has many aspects that cannot be conserved through training courses. It includes the social roles and status of builders; their technical knowledge as embodied in practice and performance; the transmission of expert knowledge; and their social and economic relations with suppliers and clients.[58] Unfortunately, these aspects of the building process are rarely considered in conservation projects. Virtually all publically funded projects—including projects affiliated with UNESCO, the World Bank, and other international agencies—require the use of general contractors. *Usta*s usually cannot qualify as contractors, because they may not have bank accounts or money to put up as security. They also have difficulty with, or resist, competitive tendering, which involves certain procedures that are alien to traditional practice. An *usta* is accustomed to working directly with clients, determining with them the design and scope of work. He might charge a fixed fee for an entire job, or he might charge by the day.[59] But under the competitive tendering system, the *usta* must bid on a design conceived and drawn by someone else, usually an architect. He may be unable to read the drawings and specifications, and to estimate costs by area and quantity. Unless an *usta* can overcome these obstacles, the project is awarded to a general contractor—who may have administrative abilities and financial resources, but not necessarily skill in building.

Many conservation projects require contractors to hire *usta*s as subcontractors. As the employee of a general contractor, the *usta* suffers not only financially, but in terms of work control and social prestige. Yassin Ghalib, who managed the restoration of Samsarat al-Nahhas (the first restoration project conducted under the auspices of the Campaign), said this problem was evident from the beginning. "The *usta*s do all the work, bear most of the responsibility, yet they get a fraction of the contract sum," he said. "These huge sums should go to them. We have to encourage *usta*s, and their sons, or they will leave the profession." Another architect suggested that public agencies should contract directly with specialized building trades. "That way we will preserve the *usta*," he said, "not just the product of his work."

During my discussions with builders, the idea of a builder's society (*jam'iyya*) occasionally came up—perhaps inspired by other local societies (NGO's) rather than a traditional builder's guild, which curiously never existed in Sanaa.[60] One *usta* told me he had proposed the creation of a builders' society to the Director of GOPHCY, presumably in the hope of obtaining official support. The builder argued that such a society would set standards for work in the old city and would be accountable for the work of its members; it would include the various trades concerned with *turath*. A manager at the SFD noted that the existence of such a society would be helpful in many respects: it would, among other things, provide the local counterpart required in donor-funded projects. While some *usta*s with whom I spoke liked the idea, one said that control of such a society would be

58 "Process over Product," in *Managing Change*, p. 139; and *Minaret Building*, p. 182.

59 Lewcock and Serjeant encountered these two kinds of contracts in the 1970s: a fixed sum contract for a complete work (*'amal muqata'a*), and a contract for work by the day (*al-'amal bil yawmiyya*; "The Houses of San'a," in *San'a*, p. 468).

60 While craft guilds played an important role in Sanaa there is no record of a builders' guild. On this point see Marchand, *Minaret Building*, p. 72.

a stumbling block: alliances between several branches of his own family had ended in disputes and dissolution, because "everyone wanted to be in charge."

The deskilling of the craftsman and the lack of continuity from father to son, are, in Sultan Sallam's view, a "tragedy." It is a critical issue for conservation, he said, but one that has not been addressed. Moreover,

> with the advent of concrete and steel, we have lost the special qualities (*khasa'is*) that each part of Yemen had before the revolution. These new materials also bring problems of climate adjustment. So you see, there is no conservation of the human being and his environment (*la yujad hifath al-insan wa bi'atu*).

This brings us back to a point raised earlier, and which we will take up again in the following chapter: the tension between physical and social structures, or the question of *what* is to be conserved. In discussions with professionals, builders, and residents, I detected a certain dissonance: if conservation aims to preserve and affirm *turath*, why does it end by destroying traditional structures? As a veteran of the Campaign noted, conservation is a *modernizing* phenomenon. "GOPHCY, the Social Fund, Public Works, and other agencies all abide by new rules," he said. "They use competitive tenders rather than traditional mechanisms. So you see, even in conserving, you are destroying aspects of the traditional system."

In the mid-2000s the Social Fund for Development had begun to address these issues on an experimental basis, as cultural heritage became more important as a program area. SFD managers realized that although *usta*s were best qualified to work on historic buildings, they were disqualified by procedures and guidelines required by donor agencies. They quietly began to develop ways to facilitate the participation of *usta*s and strengthen their autonomy. In villages and small towns, they experimented with a system called "community contracting": local committees acted as general contractor and employed *usta*s directly; progress of work and contract payments were overseen by SFD supervisors. They had rarely tried this in cities, however (one suspects because multiple layers of bureaucracy all want a "cut," making the system untenable). Most SFD projects in Sanaa were awarded to general contractors, who were in turn required to employ qualified *usta*s. But in a project that restored a series of house facades on the Wadi Sa'ila (to be discussed in the following chapter), the SFD did try out a variant of community contracting. The Local Council acted as the client, and SFD helped *usta*s qualify as contractors. Architects and managers are also aware of the problems inherent in drawings: not only may the *usta* be unable to read them, but they may restrict his creativity. SFD project drawings were sometimes left intentionally schematic—allowing the *usta* to execute details which, as one manager said, "he can do better anyway."

The marginalization of the builder is by no means unique to Yemen. Theorists like Alois Riegl and Paul Philippot posited a rupture between past and present—and thus between traditional and modern building practice. Conservation requires the indefinite preservation of original material and traces of age. As the source of the monument's authenticity, they must be entrusted to a new kind of expert.[61] Modern conservation then, is not as a creative art, but rather a science of the past: as an historical document, a building must be restored according to concrete evidence rather than conjecture or interpretation. Modern

61 For Phillippot's views see above, Chapter 1; for Riegl's views, see his *Project for the Legislative Organization of the Protection of Monuments in Austria*, Italian translation by S. Scarrocchia, *Alois Riegl: teoria e prassi della conservazione dei monumenti*, Bologna: CLUEB, 1995, p. 212.

interventions and rebuilding are strongly discouraged; where new work is required, it must be clearly distinguished from the old.

Like residents, officials, and architects, *usta*s sometimes use the term "historic" (*tarikhi*): one builder described his work to me as an "historical treatment" or "cure" (*'ilaj tarikhi*). They also occasionally use the term "traditional"—*taqlidi*—which derives from the root *q-l-d*, "to imitate." While these terms appear to echo conservation discourse, builders use them to describe a process that is fluid and interpretive, rather than documentary. The Venice Charter's strictures against intervention have little meaning for *usta*s: rebuilding, for example, must seem natural to them, since they continue to practice the old techniques. Nor would it occur to them to distinguish their work from the old. As we have seen, the continuity of building techniques makes it difficult, if not impossible, to date Sanaani houses by means of style and details. This continuity has carried into the modern practice of conservation: indeed, builders pride themselves in the fact that their often substantial interventions cannot be distinguished from the original building fabric.[62] Interestingly, several builders specializing in *turath* seemed to have embraced the idea of reusing original materials—not only for practical reasons like economy and scarcity, but in the interest of what conservators would call "authenticity." The reuse of original materials renders the joints between old and new virtually invisible, thoroughly disguising their interventions.

While affirming the superiority of their forebears' work, *usta*s see themselves as heirs to their legacy. They maintain that their work is faithful to the "original"—even as they make changes that will improve the building, or adapt it to present conditions. On one occasion, a builder proudly showed me photographs of a house facade that he had rebuilt "exactly as it was," reusing the original materials. Yet upon further discussion, it became clear that he had in fact made significant changes to the building:

> The lower level was too short—like an old man hunched over—so I made it taller. The window sills also needed to be slightly higher, because the facade is *qibli* (north-facing); when the sill is higher, a child sleeping next to it will stay warm.

The apparent paradox dissolves when we realize that the *usta*'s statements are based on a different view of history—one that is embodied not in the object, but in practices that he has inherited and continues to perform in the present. Anthropologist Steve Caton had a similar experience studying oral poetry in the Khawlan, a tribal region to the east of Sanaa. Although history is extremely important to tribesmen, he was unable to elicit the recitation of ancient works:

> Midway in my fieldwork it dawned on me that I had not been thinking about this tradition in the right terms. I had been assuming that it is the text per se that is important to remember, the verbal object we associate with printed poetry, when in fact it is either the *process of composition* that is to be remembered … or *the finished product of that process as it relates to a historically remembered situation.*[63]

In building practice in Yemen, as in other art forms, the past is validated by its continuing relevance in the present. Builders may describe their work as imitating the past, but it is

62 Menon made the same observation regarding Indian builders ("Rethinking the Venice Charter," p. 53).

63 S. Caton, *Peaks of Yemen I Summon*, Chapter 2, note 3, p. 307 (italics in original).

creative and interpretive—in effect, *improving* on the past. This ideological evocation of the past is not so different from modern conservation, which insists on faithfulness to an original model. Theorists and practitioners have acknowledged that conservation involves change, material as well as social. But procedures and policies are generally aimed at masking or minimizing the apparent effects of change. While the discourse is internally consistent, it is not that different from the use of the term "tradition" in other societies and eras: the ideological usefulness of the term appears to lie in its ability to mask change. The language of conservation is thus familiar to builders, and easily adopted to express their own attitudes toward the past: like tradition, conservation involves change, but achieves its force through the rhetoric of the unchanging.

Chapter 6
The Evolution of Local Practice

In 1993, at an international symposium on the old city of Sanaa, two young Yemeni architects, Yassin Ghalib and ʻAbd al-Hakim al-Sayaghi, presented a paper that reflected and exacerbated the crisis at the conservation office. At issue was control of the decision-making process, especially regarding the formulation of policy and standards.[1] In their paper the architects compared two recent projects: the restoration of Bayt Sariʻ, part of a larger project by Quaroni Bonifica to conserve a "model neighborhood"; and the restoration of Samsarat al-Mansura, a caravansaray at the edge of the *suq*, financed by the government of Germany. For Ghalib and al-Sayaghi, the projects reflected two different models for conservation in Yemen, based on different attitudes toward local architecture and craftsmanship.

The architects felt that the Italian intervention was heavy-handed, and even at the time some foreign experts agreed. Quaroni Bonifica's preparatory studies had asserted that traditional building methods should be encouraged and improved according to sound building standards; consolidation measures "must support the logic of traditional construction practice."[2] Yet these measures should also

> compensate for deficiencies in traditional building and technical methods, to restore or strengthen the function of the single structural components, but always maintaining the structural and formal 'character.' Appropriately developed and coordinated, these solutions will serve to create within the structures to be preserved a truly structural 'skeleton', providing old buildings with new and greater capabilities to resist stresses. The solutions suggested are compatible with professional capabilities of the local ustad who will be called upon to learn and employ the new techniques and to pass these on to the young recruits so that local building traditions will be perpetuated.[3]

At the time, consultants working with UNESCO had confirmed the soundness of traditional building practice.[4] But Quaroni Bonifica felt that traditional building methods should be improved. Later, during the Implementation Phase, Bayt Sariʻ would later be the test case

---

1 See Chapter 3. Ghalib was project architect for Samsarat al-Nahhas, and al-Sayaghi for Samsarat al-Mansura. They described the paper to me, but were unable to locate it. It was presented at a "Sheraton conference"—probably the Symposium on the Integrated Urban Policy for the Conservation of the Old City of Sanaa, Yemen, held at the Sheraton Hotel, Sanaa, 15–19 December, 1991.

2 Studio Quaroni S.r.l. and Bonifica S.p.A., Proposal Phase, volume IIIA, "The Urban Texture," p. IIIi; and "Action Proposal: Implementation of a Conservation Reinstatement Measure for a Sample Nucleus of the Historical Centre of Sanaa and the Establishment of a Restoration Workshop/School," January 1986, p. 27.

3 Proposal Phase, IIIA, "The Urban Texture," item 18, 1981. The "deficiencies" of local building practice are enumerated in the preceding sections of this document, and are illustrated in drawings submitted as part of the "Implementation Phase" (1989–1990), to be discussed below.

4 For example, Jacques Heymann, Professor of Civil Engineering at Cambridge University, reviewed building techniques in both Sanaa and Shibam in the early 1980s at the request of Ron Lewcock. According to Lewcock, Heymann found traditional building satisfactory in terms of both

for this strategy: they created a rigid shell within the old structure—pouring slabs on top of the wood floor assemblies, and stiffening the stair core and walls with steel reinforcing, wire mesh, and concrete. Such measures were sometimes taken where earthquakes were a major threat, but Quaroni Bonfica did not explain the work in these terms.[5] Their conditions assessment found the house to be in an "initial state of degradation."[6] But local professionals recall that the building was in reasonably good condition. The cracks that could be found in this and other structures were due to water-laden soil—a direct result of the recent introduction of piped water. (Water-laden soil is still cited as the main reason for cracks and structural problems, by both architects and builders.)

For Ghalib and al-Sayaghi, the new structural skeleton within Bayt Sari' was "a strange object in a traditional body." They contrasted this approach to the restoration of Samsarat al-Mansura which, they believed, was a more appropriate model for Yemen. The project relied on traditional materials and techniques—a principle which, they noted, was reiterated in all the international charters. In a publication released after the completion of the project, one of the German conservators described the approach:

> Execution of the work ... was exclusively in the hands of Yemeni craftsmen, with the Yemen Construction and Development Company as the general contractor. In view of the extraordinary significance of Samsarat al-Mansura as evidence of craft traditions that are still alive today, special value was placed on the use of traditional materials and on skilled craftsmanlike practices, as well as on the preservation and repair of the existing building fabric ... Modern materials were used only in critical structural areas, for the renovation of the foundation with concrete ... and for mechanical installations.[7]

Traditional materials were not only desirable, but were compatible with the building's structural characteristics. Moreover, the use of local materials and techniques would keep costs low and provide a model of restoration that could be replicated.[8] These local techniques created a unique Yemeni heritage (*turath*) which is celebrated by UNESCO, Ghalib observed. "Why should we belittle our own experience and knowledge, and bring in outside experts—when we have techniques that are hundreds of years old?"

In the years leading up to and following the announcement of the UNESCO Campaign, local architects and engineers worked alongside consultants from a variety of countries. They often found the experience confusing. In contrast to the singular and monolithic "international" approach implied by the charters, foreign consultants often had different attitudes and approaches to conservation, depending on their country of origin and their training. At the time virtually all of these consultants were from Europe, or were trained in Europe. Conservation is a relatively young profession, and its precepts were,

---

materials and construction methods. The competition between UNESCO and Quaroni Bonifica should also be borne in mind (see Chapter 3).

5 Sanaa had seen two major earthquakes, in 1644 and 1908; the 1982 earthquake at Dhamar was devastating, but the effects in Sanaa were relatively minor.

6 "Reinstatement Plan" for Mu'adh Quarter, drawing A4; Quaroni Bonifica, Implementation Phase, 1990.

7 M. Petzet, "The Restoration of Samsarat al-Mansurah," in M. Petzet and W. Koenigs, eds, *Sana'a: the Restoration of Samsarat al-Mansurah*, Munich: Bayerisches Landesamt für Denkmalplege, 1995, p. 12.

8 J.M. Klessing, "Planning and Execution of the Restoration," in M. Petzet and W. Koenigs, eds, *Sana'a: the Restoration of Samsarat al-Mansurah*, p. 56.

and are, everywhere in flux. "There are as many approaches to conservation as there are conservators," remarked Lewcock. While there are generally accepted principles, interventions are designed according to the specific conditions of each case, and thus rely on the judgment of a highly skilled conservator. The experience was also confusing because the treatment of historic cities and "vernacular" fabric has not been clearly distinguished from the treatment of monuments. As we saw in the first chapter, the designation of a World Heritage site imposes conservation standards that were designed for monuments. Failure to comply with these standards may result in the removal of a site from the World Heritage List. Indeed, UNESCO has periodically threatened to remove the old cities of Sanaa and Zabid from the list because of ongoing "violations," like lack of maintenance and intrusive new construction.

Many foreign professionals are keenly aware of the problems inherent in cultural heritage management, although they may not express these doubts in public forums. In their role as international consultants, most take the view that there is a clear, agreed-upon body of knowledge and practice that can be taught to professionals in other parts of the world. In principle they acknowledge that it must be adapted to local conditions, according to the current approach of "cultural relativism." But in practice, few foreign consultants who come to Sanaa on short-term missions or for conferences seem interested in working with local professionals, or learning about their experience. "Most Western consultants treat their colleagues in the Middle East and other parts of the world as if they were idiots," a consultant from Latin America told me. "That's what happens when you think what you're doing is objective science." Foreign consultants usually interact with officials, who are themselves removed from the realities of the field. Working at this level, they are continually frustrated by the inefficiency of local institutions, which—as Yemeni professionals themselves complain—are unable to document past work, manage ongoing efforts, and plan for the future.

Were these foreigners to probe more deeply, they would find a more interesting and complex reality. Many local professionals have had decades of experience in decision-making positions and in the field, working alongside foreign consultants and builders and interacting with residents. In the early years there seems to have been much dynamism and enthusiasm, which has sadly been blunted by the weight of bureaucracy and political uncertainty. But local professionals remain reflective and critical: they have assimilated aspects of international practice, evaluating them relative to the local context. Like builders, they may affirm aspects of conservation discourse while making decisions that contravene it. A local approach—or approaches—to conservation can be discerned in the design and management of projects; it can also be gleaned from formal and informal discussions, where fundamental issues are raised: what, in fact, is conservation? Why do we conserve? For whom do we conserve, and how? How do we reconcile development and conservation—especially when the latter is a low priority for policymakers? What is the value of conservation for a poor country, where the basics like food and education are beyond the reach of so many, and where democracy remains elusive? Is it simply a means to attract tourists, whose numbers are so limited? Is "World Heritage" a means to validate the local, or simply a kind of political and cultural domination? "The implications of World Heritage are mainly legal ones," one professional told me. "It permits external interference—it gives foreigners the same kind of domination (*imtilak*) they had in Egypt." Yet on other occasions, as we will see, local professionals invoke the authority of UNESCO and the World Heritage Convention against what they see as misguided undertakings by their own leadership.

This chapter will trace the evolution of local practice through a series of conservation projects, which represent only a small portion of the work conducted over the last few decades. These projects were selected because they focus on components of the "living historic city"—the *suq*, the residential fabric, and other "minor" architecture—and because they demonstrate various types of interaction between professionals, foreign consultants, builders, and residents. While the projects are colored by the particular dynamics of Sanaa, they highlight issues that are at stake in conservation in other parts of the world: for example, the tension between the "living" on the "historic," and the contradiction between conserving a city for its inhabitants, and financing costly restorations.

The first part of the chapter looks at several projects conducted under the auspices of the UNESCO Campaign, which were cited by professionals as formative in the development of their thinking and practice. The restoration of Samsarat al-Nahhas (1988–1989) was the first restoration project conducted under the Campaign. Sponsored by the government of Norway, the work was largely in the hands of local professionals, who learned about conservation on site and debated appropriate treatments. While the restoration is generally seen as a successful project, the institution it was to house—a National Center for the Revival of Yemeni Handicrafts—is widely regarded as a failure. I will then discuss the rehabilitation of a "model neighborhood" by Quaroni Bonifica (1989–1992), which included the restoration of Bayt Sari'. Despite the careful and thoughtful analysis of Quaroni Bonifica's early studies, the project was conceived according to a vision that seems to have had little to do with local realities; the work was executed and supervised largely by foreigners.

The second part of the chapter will look at a number of recent projects undertaken by the Social Fund for Development in cooperation with GOPHCY, in the modest renaissance in conservation activities that occurred when donor interest resumed in the late 1990s and early 2000s. While still following a plan of action drafted by UNESCO and EOPOCS in the 1980s, these projects demonstrate an interesting synthesis of international and local ideas and practices. The first of these projects addresses the long-standing issue of house maintenance and restoration. An innovative project in Shibam-Hadhramaut, also a World Heritage site, will be discussed by way of introduction. Co-sponsored by the SFD and the GIZ (Deutsche Gesellschaft für Internationale Zusammenarbeit), it is unique in Yemen because it focuses on local builders as agents of conservation.[9] I will compare this project to a more modest effort in Sanaa to restore a series of house facades on the banks of the Wadi Sa'ila (2003), an area that is valued for its appeal to tourists. I will then compare this small and carefully designed project to the mass white-washing project undertaken by the Ministry of Culture, in preparation for Sanaa's year as Cultural Capital of the Arab World (2004). The concluding section of this chapter describes the rehabilitation of the *marna'* or well enclosure of Talha (2002–2004) which illustrates the independent direction of conservation in Sanaa.

The Restoration of Two *Samsaras*

The first restoration projects executed under the UNESCO Campaign were commercial buildings called *samsaras* (Arabic pl. *samasir*), usually translated as "caravansarays" in

9 The agency was formerly called GTZ—Gesellschaft für Technische Zusammenarbeit, the German Development Agency.

English.[10] Most of the existing *samsara*s are located around the edge of the central *suq*, Suq al-Milh, and probably date from the seventeenth and eighteenth centuries. Most *samsara*s are *waqf*, endowed for the benefit of a mosque or other charitable structure, but Samsarat al-Mansura is property of the state (*mulk al-dawla*).[11] The *Qanun Sanaa*, a legal document that outlines market regulations, says that *samsara*s were used as warehouses. But they probably also functioned as inns, with merchants sleeping on the upper floors and the ground floor reserved for goods and animals.[12] Most *samsara*s linked the wholesale and retail trades, but some also served regulatory functions, like the weighing of merchandise and the collection of duties; one was even used as a bank.[13] They were usually rented to an individual who managed the property (*samsari*), and who rented rooms to other merchants for the storage of goods or as retail space.[14] While some *samsara*s are simple, one- or two-story structures organized around a courtyard, the most stately ones have four or more floors organized around covered, central halls. The latter have imposing facades, echoing certain elements of domestic architecture (Figure 6.1).[15] Most rooms are accessed from within, but some *samsara*s—like al-Mansura—had shops that opened onto the street, and were read as part of the *suq*.

The restoration of selected buildings was to be carried out in a second phase of the Campaign, following infrastructure work. But these projects were in the works well before infrastructure was complete—because donor nations were anxious to sponsor projects, and EOPOCS was eager to begin them. Drawings of various *samsara*s had been executed by Serjeant and Lewcock's team and by Quaroni Bonifica, and from the mid-1980s detailed studies of several *samsara*s were carried out by EOPOCS staff and foreign consultants. The Campaign's promotional brochure, published in 1989, outlined plans for many of the 26 documented *samsara*s, as well as other market structures. *Samsara*s were seen as ideal buildings for restoration and reuse because they were public buildings (although most were controlled by the Ministry of Endowments). Unlike religious buildings, they could accommodate a variety of new uses. Most importantly, the *samsara*s were seen as key to the revitalization of the *suq*. They had been marginalized along with old production networks;

10 *Samsara* derives from the quadriliteral verb *s-m-s-r*, which means "to act as a broker or middleman." It may be a later word, since in the ninth century Ibn Rustah uses the term *khan* (a Persian word that is commonly used in Arabic); *khan* was sometimes used in Sanaa in the twentieth century (R. Lewcock, "The Buildings of the Suq," in Serjeant and Lewcock, eds, *Sanaa*, p. 277). The following comments on the history of the *samsara* draw on this article, pp. 277–90.

11 "Conclusions on the Restoration and Refunctioning of Samsarat al-Mansuriyya," EOPOCS proposal to the Foreign Office of the Federal Republic of Germany, 18 August 1988, p. 2.

12 R. Lewcock, "The Buildings of the Suq," p. 278; at the time he was writing, some of the smaller *samsara*s were still used as stables and inns (p. 284). The last edition of *Qanun Sana'a* had appeared by 1836 (N. Lowick, "The Mint of San'a," in Serjeant and Lewcock, eds, *San'a*, p. 309).

13 The *samsara* of Muhammad Ibn al-Hasan ibn al-Qasim (seventeenth c.) was known as *Dar al-Mal*, that is, as a "bank": a service charge was paid to the *amin* (official) on money deposited there (R. Lewcock, "The Buildings of the Suq," p. 278). Other *samsara*s with regulatory functions were Samsarat al-Ghumruq (the old customs house; commonly called Samsarat al-Zabib, for it houses a raisin market); and Samsarat al-Mizan (*samsara* of the scale).

14 Like other service-oriented trades, that of the *samsari* is considered a dishonorable profession, practiced by *muzayyin* class (Lewcock, "The Buildings of the Suq," p. 278).

15 W. Koenigs, "Samsarat al-Mansurah—Description of a Historic Building," in M. Petzet and W. Koenigs, eds, *Sana'a: the Restoration of Samsarat al-Mansurah*, p. 33. A few *samsara*s have a double courtyard scheme, such as Samsarat al-Majjah, next to Samsarat al-Mansura.

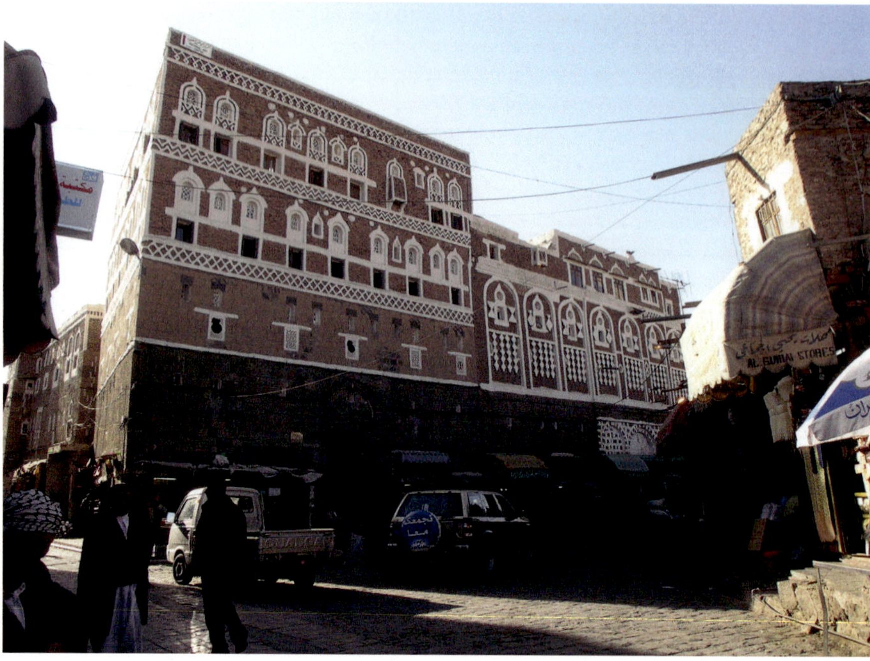

Figure 6.1 The *samsaras* of al-Mansura (left) and al-Majjah (right)

by this date, many new transportation routes and facilities bypassed the old city entirely.[16] Yet many *samsaras* remained in use, with merchants renting space as they had in the past.

Virtually all foreign consultants saw the *suq* as the economic backbone of the old city, and hence a vital part of the conservation effort. But their approaches varied—suggesting different conceptions of the market and its role in the wider city. The Quaroni Bonifica studies, which devoted an entire volume to the *suq*, argued for its modernization. At the time, the *suq* was in poor condition: roads were unpaved and gutted, congested by traffic, and strewn with garbage. Quaroni Bonifica saw the one-story shops as dilapidated and unsuited for modern commerce, in terms of space, access, and communications. They proposed to demolish the shops, and create a two-story modern commercial center in their place, with parking underground (Figure 3.1). This modern facility would tie together restored *samsaras*, gardens, and other elements to retain the *suq*'s historic image. Economic activities in the new *suq*, they argued, should be as numerous and diverse as possible.[17]

Other consultants, and most Yemeni professionals and policymakers, saw the destruction and rebuilding of the *suq* as a radical and unacceptable solution. They may have also objected to a modern role for the *suq*, which brings with it alien building types and

16 G. Barbato, "The Methodological Approach for the Restoration of the Samsarah al-Bowani and Samsarah al-Majjah," unpublished paper, Symposium on the Integrated Urban Policy for the Conservation of the Old City of Sanaa, 15–19 December, 1991.

17 Quaroni Bonifica, "Safeguarding the Historical Center of Sana'a," PIV, Introduction, 1.1–16 and 3.3.

materials, and new functions that undermine social and economic relations. Lewcock, for example, saw the old city as a regional market, specialized in traditional goods and serving a local clientele; the latter included both rural folk and Westernized Sanaanis who came to the *suq* for certain specialized products. He recommended conservation of the physical fabric of the *suq* and its institutional structure, noting that "[t]he government is in favor of trying to preserve the market as it is to conserve the way of life that it contains and that surrounds it."[18]

Lewcock's view was important in the development of the Campaign strategy, which also envisioned the development of compatible new activities as a means to revitalize the *suq*.[19] These activities, he argued, should be aimed at a local market; tourist activity in the old city should be limited. Along with other consultants and local professionals, he wanted to install services that would be useful to the local population, like a post office and a bank. But for many working on the Campaign—both foreign and local—tourists were the natural market. Yemeni policymakers saw tourist revenues as essential: they would persuade residents to accept the conservation effort, and would help finance the numerous and costly restorations envisioned by the Campaign. Former Prime Minister 'Abd al-'Aziz 'Abd el-Ghani, who chaired the Executive Committee, recalled that most members wanted to transform old houses into hotels. But most UNESCO consultants and some EOPOCS staff opposed this: they feared adverse effects on the communal life of the old city.[20]

These different attitudes influenced planning for the *samsara*s. In Lewcock's mission reports to UNESCO, he observed that these buildings were in need of extensive conservation work, but were not productive enough to pay for themselves; he envisioned their reuse for the local wholesale trade.[21] But Yemeni policymakers and professionals saw the buildings as ideally suited for tourist activities, and this approach was endorsed in the revised Campaign strategy.[22] The Campaign's promotional brochure, co-published by UNESCO and the UNDP, outlined new uses for *samsara*s that included museums, cultural facilities, and the production and exhibition of handicrafts. Some projects would have

18 R. Lewcock, introduction to B. Lane, *Pilot Restoration Projects*, p. 9. He describes the role of the *suq* as follows: "30. The essential strength of the suq in the old city seems to lie in its role as regional market to a large hinterland of traditionally oriented farmers and villagers. Many products are unique to the suq. It is to capitalize on this potential, as well as on its role in the city itself, that action needs to be taken. 31. It is felt by a number of authorities that judicious help for certain trades and crafts from the Government would rapidly revive the prosperity of the suq. 32. Principles which might guide the revival of the suq are: i) the traditional fabric needs to be renewed. If the old structure is renewed, this will control the revival of commerce into the traditional social pattern ...." (UNESCO mission report no. 3, 1985, pp. 6–7).

19 One of the Campaign's objectives was "to reinforce and maintain the traditional economic activities of the old city and to develop compatible new activities to ensure a viable economic and commercial base for life in Sanaa" ("Campaign Strategy and Action Plan," 1988–1997, August 1988).

20 Lewcock told me that he and other UNESCO consultants asked embassies to discourage foreign nationals from living in the old city. The Third International Working Group stressed that the utmost care should be taken "to preserve communal structure of the old city ... therefore conversion of houses into hotels should be done only after careful study ...." (Recommendations of meetings held 24–26 January, 1989).

21 Mission report, 9–16 April, 1982, item 66; and report no. 3, 1985.

22 The 20 projects developed under the 1988–1997 Action Plan included "six first-class caravanserais which were to be adapted as tourist amenities containing information centres, cafeterias and other facilities or devoted to handicraft training"; as well as improvement and conservation projects for traditional markets, the Great Mosque, and service buildings like wells, bathhouses, etc. ("Draft Terminal Report/Restricted, UNDP/YEM/006," pp. 4–5).

benefited crafts in decline, like smithing and carpentry, through "soft" upgrading—street paving, the provision of water and utilities, roof repairs, and so forth. These projects would have reorganized and reoriented production by establishing pilot craft centers, which would revive traditional products and develop new ones.[23]

Most professionals agree that tourism was the main impetus behind the revival of handicrafts and other activities, and to some extent determined project priorities and locations.[24] Their feelings about this were, and still are, mixed. They realized that most of the functions envisioned for the *samsara*s would have little relevance for the community, but they would promote employment, traditional skills, and the development of new skills and products. These products could be tailored to new markets. "The National Center for Handicrafts was not intended only to revive handicrafts, but to develop them," said Yassin Ghalib, project manager for Samsarat al-Nahhas, who also briefly directed the Center. "Handicrafts as they were traditionally produced would not be able to survive; we could put these skills and materials to new and creative uses." New products based on old models and made with traditional materials—lighting, furnishings, fabrics, woodwork—could be used by architects in new and conserved structures. According to a former EOPOCS official, the revitalization of handicrafts was also as a way to introduce new attitudes in the old city, part of a wider "cultural revival."[25] But the use of a *samsara* or other endowed building must be compatible with the intentions of the patron—the same criterion that was invoked by residents of the old city in the dispute over Bustan al-ʻAmri. There is little reference to the constraints of *waqf* properties in the Campaign literature, although consultants were certainly aware of them.[26]

---

23 As of 2005, only two *samsara*s had been restored: al-Nahhas was restored in 1988–1989, and al-Mansura in 1991–1993. The Campaign brochure described proposals for several *suq*s and *samsara*s. The UNDP project financed studies of Samsarat al-Bawʻani, al-Majjah, and Muhammad Ibn al-Hasan ibn al-Qasim by G. Barbato. A map on file at CATS, drawn circa 1987 (title block of the General Organization of Antiquities and Libraries; labeled in English), identifies the extant *samsara*s, their condition and current use, and proposed plans for reuse. Virtually all the proposals envisioned tourist-oriented or cultural activities. See B. Lane, *Pilot Restoration Projects*; and Lane's unpublished paper, "Samsarat Muhammad b. Hasan b. Qasim, Suq al-Baz Bank and Trade Center," presented at the Symposium on the Integrated Urban Policy for the Conservation of the Old City of Sanaa, Yemen, 15–19 December, 1991. See also two reports by Barbato: "Training and Exhibition Centre for Vanishing Handicrafts," UNESCO/UNDP, Paris 1991; and "The Methodological Approach for the Restoration of the Samsarah al-Bowani and Samsarah al-Majjah."

24 The early projects, in particular, were selected for their appeal and visibility to tourists. The first paving project was along a major tourist route (see above, Chapter 3). Samsarat al-Nahhas was chosen to house the National Handicraft Center in part because of its location on the main road from Bab al-Yaman, also a major tourist route.

25 On the idea of "cultural revival" see Chapter 3.

26 E. Galdieri of the Quaroni Bonifica team wrote that due to the *waqf* status, one could not really speak of reuse of the *samsaras*, only of "provisional recovery (*provvissorio recupero*)" ("Rapporto sul tipo edilizio del samsarah," 1983, Fondo Maria Luisa Neri, Associazione Archivo Storico Olivetti, Ivrea). Throughout the centuries, many ways have been found to work around the restrictions of endowed properties (see for example, J. Peterson, *Yemen*, p. 19 and fn. 16); and the intentions of the patron are always subject to interpretation. In 2004–2005 a proposal was on the table to install a restaurant in Samsarat al-Nahhas. Yassin Ghalib, former project architect for the *samsara*, opposed the proposal because the building was endowed as a trade structure. But an inspector from the Ministry of Endowments told me that a restaurant was an appropriate use, consistent with the wishes of the patron. The restaurant, he said, would be small and "popular (*shaʻbi*)." Perhaps he qualified it in this way to indicate that it would not be oriented to tourists.

In the 1970s and '80s, many agencies were questioning the efficacy of tourism as a development strategy. Foreign consultants and donors tended to emphasize "community development" over tourist revenues. In the development world "microenterprises" were receiving new attention, but in many cases their production was oriented to an elite market interested in hand-made goods. The creation of new products and markets was seen as the only way local crafts could survive in an industrial and increasingly globalized world. Around the same time the term "cultural tourism" was gaining currency: in this loftier version of tourism, international tourists and representatives of local culture would come together in an equal encounter, each possessing a unique culture. A program under UNESCO's World Decade for Culture and Development (1988–1997) proposed to restore caravansarays and postal systems as symbols of cultural encounter and exchange; they would be used as venues for cultural tourism.[27] The approach to the *samsara*s of Sanaa, then, were part of this wider trend.

Planning for the restoration of several *samsara*s was going on during the first paving project, that is, in 1986. In 1985–1987, Norway and Germany were asked to help with these projects. A Norwegian mission in September 1986, led by the resident ambassador in Jeddah, developed the terms for a demonstration project that would raise national and international interest in Sanaa. The restored building should house functions that "will strengthen or develop the social fabric of the community; and support traditional occupations in the old city." The bilateral agreement required the conservation office to run a concurrent study "to document the community-related consequences of city conservation and development." Norway suggested three *samsaras*—al-Nahhas, al-Mansura, and al-Majjah—as possible candidates.[28] These three were also favored by the conservation office, because of their location and architectural quality. All three are located at the edge of Suq al-'Inab (the grape *suq*), the southwestern portion of the central *suq* near the Great Mosque (Figure 6.2).

In 1988, handicrafts were identified as a distinct sector in the revised Campaign strategy, and the creation of a national center for the revival of handicrafts was proposed. The project's vocational training and marketing aspects seemed ideally suited to Norway's criteria, and in the early summer of 1987, a Norwegian mission agreed that the restored building could have a craft-oriented function.[29] After evaluating the three *samsara*s, the office decided Samsarat al-Nahhas was the best choice for the project: it was located on the main artery from Bab al-Yaman to the *suq*; it was large (four stories, 6,000 square meters), in good condition, and of exceptional architectural quality (Figure 6.3).

According to a hand-written report by a Yemeni official at EOPOCS (in English), the *samsara* restoration and similar projects aimed to revive the economy of the old city, which depended on craft activities. The project would bring economic benefits to the residents of the old city, enabling them to "carry heavy costs of traditional building maintenance, and afford costs of services." In a surprisingly direct way, the document portrays tradition as a form of

---

27 *Rethinking Development: World Decade for Culture and Development, 1988–1997*, UNESCO, 1994. The World Decade for Culture and Development aimed to assert and enhance cultural identities as the basis for development; this would broaden participation in cultural life, and promote international cultural cooperation. As part of the initiative, UNESCO sponsored eight case studies on the links between culture, tourism, and development in the Arab World; one of these was on Yemen (H.M. 'Abdulla, "The [sic] Cultural Tourism and Development: the Case of Yemen," March 1996).

28 "Outline of Terms of Reference for the Norwegian Project," no date, apparently by the Norwegian foreign office; in a folder entitled *Dirasat 'Amma*, Planning Department, GOPHCY.

29 Hand-written document from EOPOCS, entitled "The Norwegian Project background," no date, probably the second half of 1987; with preliminary estimate in Arabic.

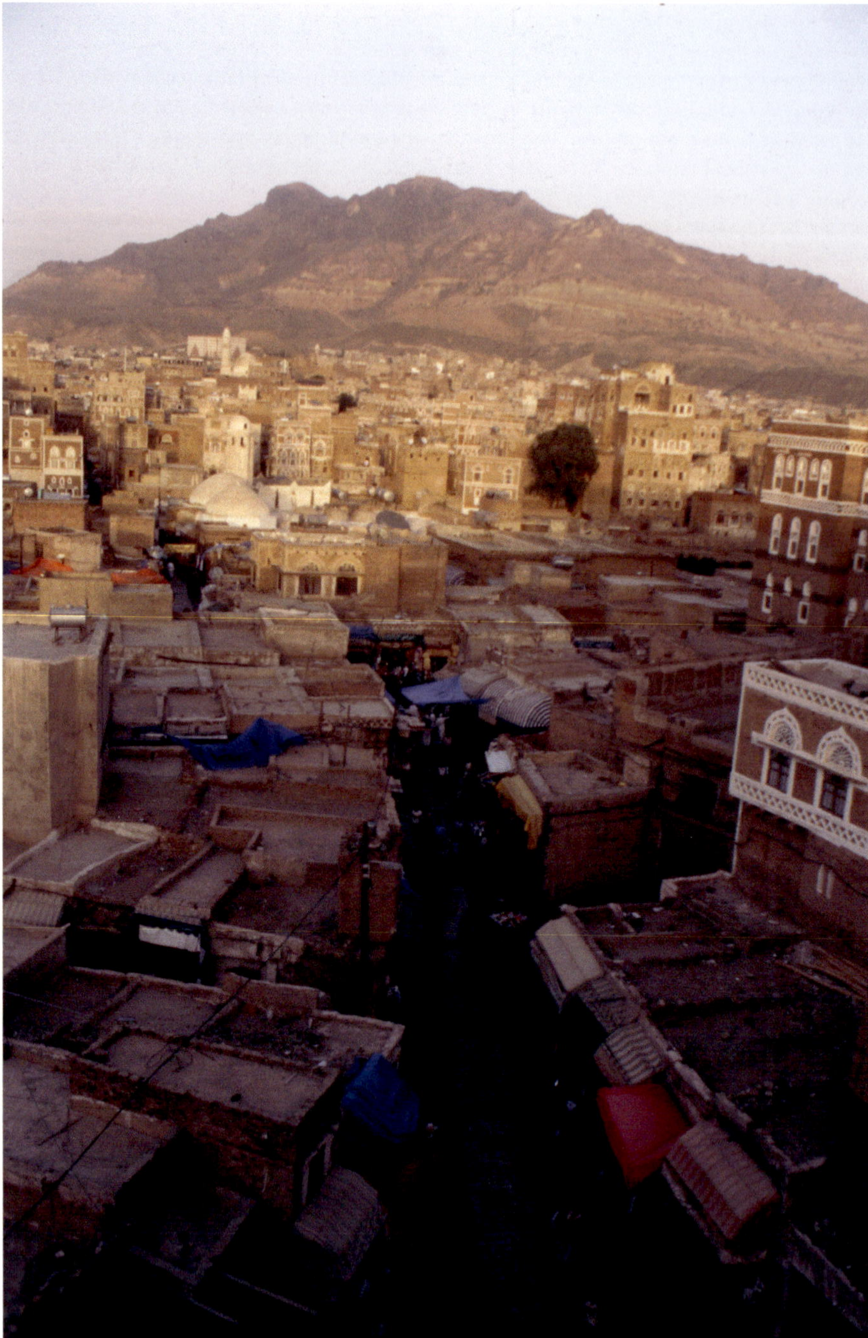

Figure 6.2 **View of Suq al-ʻInab, looking east to Jabal Nuqum**

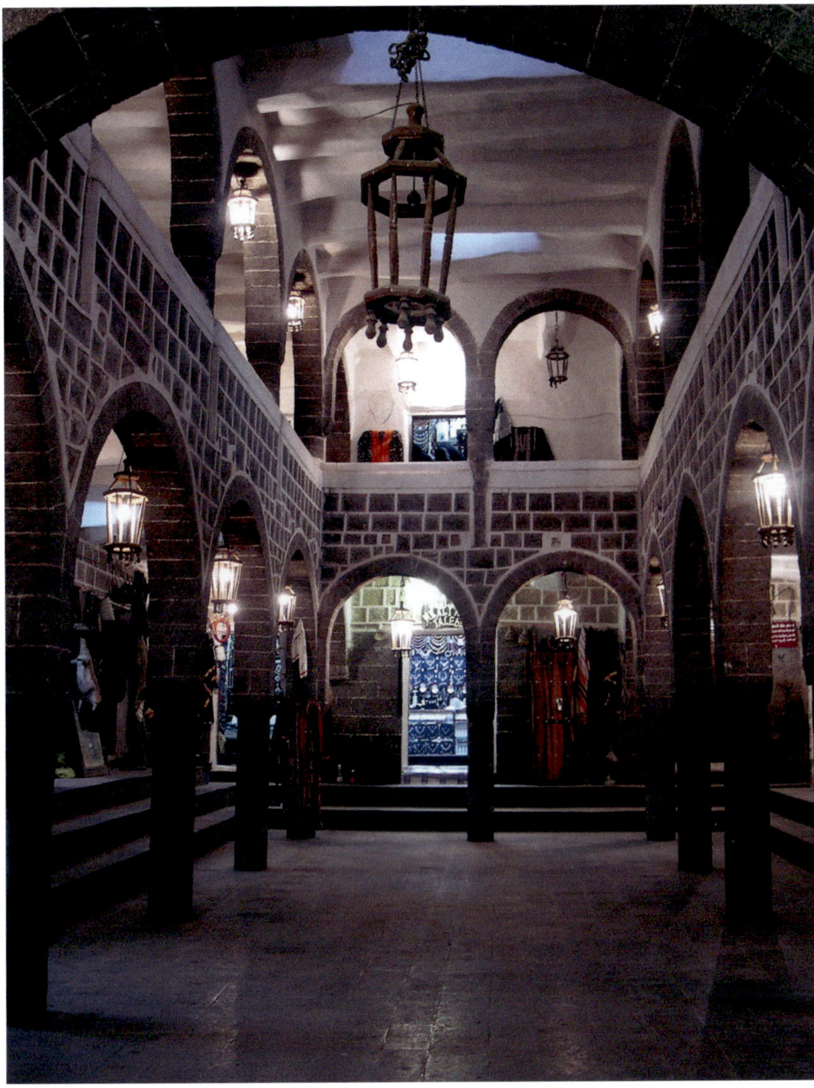

Figure 6.3 Samsarat al-Nahhas, view of central hall

marginality, produced and sustained by the West.[30] It is the "developed nation of Norway" that brings assistance to restore a "traditional" structure; "traditional building maintenance" is a burden which, implicitly, can only be carried with the help of tourist revenues (from developed nations).[31] The foreigner helps to revive and support the "traditional"; but it is

30 See M. Herzfeld, *The Body Impolitic*, pp. 1–5.

31 "The Norwegian Project Background." A document in Arabic on EOPOCS letterhead, with no date but apparently written around the same time, also describes the revitalization of the old city as

also the foreign tourist who drains Yemen of its indigenous crafts—one of the reasons a project for handicraft revival was necessary.[32]

Norway's development office (NORAD) had committed two million Norwegian Kroner (YR 4,982,184 or USD 320,000 in 1987), to be spread out over two to three years.[33] But almost as soon as the project began there were significant cost overruns, because of work required by "traditional building practice (*wifqan lil bina' al-taqlidi*)" that had not been accounted for in the original estimates.[34] The Norwegian mission had said that it might provide technical assistance, but experts were not sent until the end of the project, for a brief period; several EOPOCS professionals were also sent to Norway briefly for training. Norway hired a local company (staffed by Chinese engineers) as its representative in Sanaa, but these engineers had no training in conservation.[35] A Peace Corps volunteer drew plans and sections of the *samsara*; despite lack of detail, these were used as the project documents for the technical specifications and estimate, drawn up by EOPOCS staff.[36] Yemen Construction and Development Company (*al-sharika al-yamaniyya lil insha' wa al-ta'mir*; hereafter YCDC) was selected as general contractor; it was later awarded the contract for Samsarat al-Mansura and other projects, but also had no experience in conservation. Samsarat al-Nahhas was thus an exception to the tendency of EOPOCS to rely on foreign expertise. "Yemenis did everything on this project," recalled Ghalib. In Lewcock's assessment, the project suffered from the opposite problem: lack of experience, and a clear approach to the conservation of the building.[37]

The site was delivered to the contractor in May 1988—three months late, because the financial supervisor (*al-mukhassis al-mali*, presumably from the Ministry of Finance) did not release funds on schedule.[38] Work was again delayed because tenants in the *samsara* did not vacate their premises according to schedule. Formerly a *samsara* for copper goods, Samsarat al-Nahhas was used as a warehouse by various merchants—who sold dates, oil, butter, and other foodstuffs—and also by a number of retailers. They were subtenants

key to its self sufficiency (*himayat dhatiha*) in paying for repair and maintenance of buildings (*Dirasat 'Amma*, Planning Office, GOPHCY).

32 According to a professional who worked on the initial stages of the National Center for Handicrafts.

33 Financial report no. 3 (in Arabic), handwritten, no date; in Samsarat al-Nahhas project file, Projects Department, GOPHCY.

34 Document dated 8/23/87 by or for contractor, Samsarat al-Nahhas project file, Projects Department, GOPHCY; and Technical Report no. 2 (in Arabic), handwritten, no date, no signature; written after June 1987 by the current project manager, in folder entitled *Abhath wa dirasat*, Planning Department, GOPHCY. Cost overruns appears to have been paid for by NORAD and EOPOCS; according to a GOPHCY memo, the final project cost was YR 5,767,320 (in English, no date, Samsarat al-Nahhas project file, Projects Department, GOPHCY).

35 The local company was al-Sharq Contracting; the 7 per cent fee for this company was included in the grant (financial report no. 3, in Arabic; no date).

36 Drawings dated 12 July 1987, signed "JMS" (presumably Jamie Stone); and Technical Specifications (in Arabic), no date. Officials confirmed that these drawings represented the project documents.

37 The following section is based on interviews and site visits with Ghalib, Lewcock, and other architects and officials; as well GOPHCY files held at Samsarat al-Nahhas. Access to the files was arranged by Jamal Kushih, director of the Handicrafts Center, whom I interviewed regarding the programmatic aspects of the project.

38 *Bayanat al-mashari'*, Samsarat an-Nahhas, Projects Office, GOPHCY (no date).

of the well-to-do merchant who "controlled" the *samsara*.[39] In October the Minister of Endowments was asked to enforce the evictions; he agreed to do so, but noted that the Ministry had to approve any work in the building.[40] Around this time the square inside Bab al-Yaman, the main entrance to the old city, was being paved and upgraded (Figure 6.4). Numerous shops were torn down, and contrary to assurances from the conservation office, the owners were not compensated. As a result, a former EOPOCS official told me, the tenants at Samsarat al-Nahhas were reluctant to vacate, not trusting the conservation office to fulfill its promises.

The lack of experience in conservation on all sides—EOPOCS, the Chinese supervisors, and the contractor—slowed the decision-making process and the progress of work. Although the building was in relatively good condition there were structural problems, especially cracking on the north and west facades due to groundwater and settlement, and water infiltration at the roof. In a technical report written shortly after the commencement of work, Ghalib wrote:

> The restoration of old buildings is distinguished by technical, architectural, and constructive methods unique to it. I am convinced that the traditional *usta* must be well chosen [especially] at the beginning of restoration work: in view of his wide experience in this field, he is the foundation (*asas*) of the work.[41]

The contractor had been required to hire an *usta* to execute the work, but the work of the first builder was unsatisfactory. He was replaced by Muhammad al-Rawdhi, known as a master of *tarmim* (repair or restoration). Al-Rawdhi went on to work on Samsarat al-Mansura and other projects, and was retained as a consultant to EOPOCS through the early 1990s.[42]

Because the staff of EOPOCS had little practical or theoretical background in conservation, every step of the project involved negotiation—and there was no local experience on which to draw. The architects, for example, did not understand the properties of different types of local stone.[43] They relied heavily on the experience of al-Rawdhi, but differed with him on certain matters; and architects and engineers in the office differed among themselves. Some found certain items in the specifications objectionable: the use of

39 Survey of *samsaras* by Jamie Stone, April 1987. Officials were concerned that upper floors were not strong enough to bear the large loads of dates stored there. In October 1987 several storerooms had still not been vacated, and remained locked (memo to al-Haddad, 5 Oct. 1987, from Technical Director and Project Manager, Samsarat al-Nahhas project file, Projects Office, GOPHCY).

40 On the back side of this request from the *amin* of the *samsara* is an *aya* from the Qur'an with the following handwritten note, dated 12 July 1987: "No work can be done in the *samsara* without agreement from the Ministry; do not permit anyone to undertake work in it without our permission (*la yatimm ay 'amal fil samsara illa bi muwafaqat al-wizara. La tismahu liahad bi ay 'amal fiha illa bi muwafaqa minna*)" (Samsarat al-Nahhas project file, Projects Office, GOPHCY).

41 Letter from supervising architects of al-Sharq Construction and Civil Engineering Company Ltd., in English, dated April 5, 1988; and Technical Report no. 2. EOPOCS appears to have suggested laying a new foundation adjacent to the existing one, but the solution was rejected by the Chinese engineers.

42 His son, Ahmad al-Rawdhi, told me that his father's contract with EOPOCS was terminated when al-Haddad resigned. He died tragically in a motorcycle accident in 2000, around the age of 70.

43 "We thought that basalt (*hajar aswad*) and limestone (*balaq*) had the same properties," he said. "When we worked at Samsarat al-Mansura, we did some tests on machine-cut limestone (*balaq manshur*) and we realized the difference. This should have been basic information."

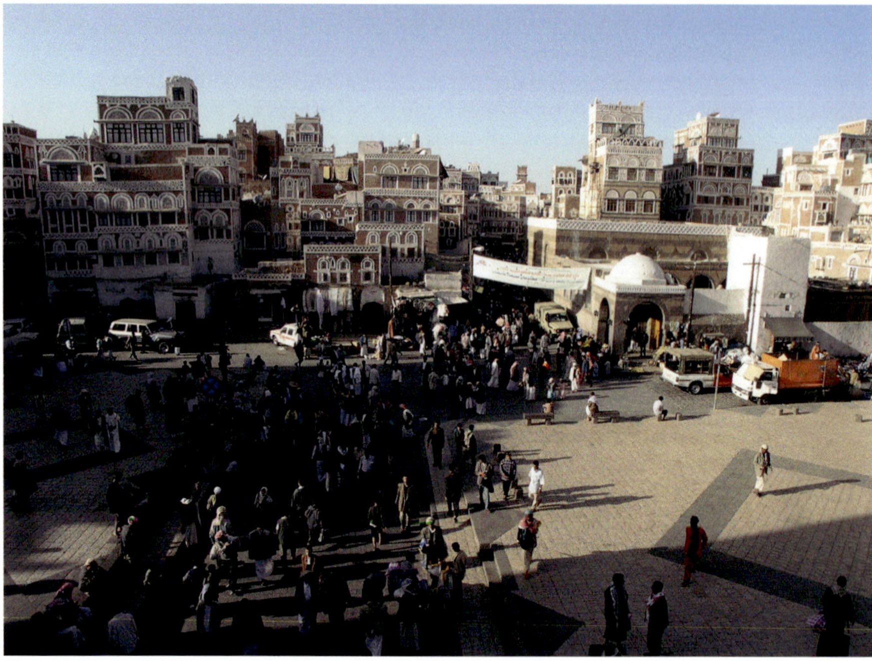

Figure 6.4 The square at Bab al-Yaman

cement mortar, the removal of existing plaster, terrazzo tiles, and plumbing and electrical installations that were not suited to the building's fabric or character.

Mortar, in particular, was a point of contention. Like most other buildings, the foundations and lower floors of Samsarat al-Nahhas are stone. While brick walls were constructed with a binding mortar (typically gypsum), stone walls were usually laid dry; masons filled in gaps with soft clay mortar and rubble. The wall was held together by the weight of the courses above, and tied at various points with timbers that allow for movement. Some EOPOCS staff wanted to use cement mortar to restore the foundations and walls of the *samsara*. Others argued that only traditional techniques should be used in the building: clay or mud (*tin*) should be used as mortar.[44] But by the time of the Campaign, most if not all *usta*s in Sanaa were using cement mortar: "They saw the strength of cement, and thought it was a secret material," recalled Saba al-Suleihi, an architect with EOPOCS during those years:

> Muhammad al-Rawdhi thought mud mortar was archaic. 'You're spending all this money on the project, why use mud?' he asked. He realized that there was a new international interest in Sanaa, that this kind of thing was part of a new trend.

44 While the term "soft clay mortar" is used throughout Petzet and Koenig's publication on Samsarat al-Mansura, architects often use the term "mud mortar" in English. Builders say simply *tin* (clay), without the term mortar (*mawna*).

Lewcock and others expressed concern over the use of cement mortar. Mortar is intended to fail before the masonry; if it is stronger than the masonry, it can destroy the wall.[45] But cement mortar also requires less skill in the cutting and laying of stone. In traditional construction, the strength of the wall depended not only on the mason, but on the skill of the stone-cutter (*waqis*). He carefully worked three faces of the stone, to assure that the bearing surfaces fit perfectly; the rough face was hidden in the wall.[46] Cement mortar requires less skill in stone-cutting, observed Ghalib:

> Now it's no longer important how good the stone is, or the stone-cutting; the only thing that has to be good is the exterior face. The traditional *usta*s also worked in new construction, they brought [new] things with them. So you see, even their beliefs do not follow the traditional system.

The western (rear) wall of Samsarat al-Nahhas stayed open for two months while professionals discussed the issue of mortar.[47] In the end cement mortar was used.

The project staff encountered many issues on site, and were not always sure how to deal with them. The interior bearing walls, for example, were not attached to the exterior walls. "Maybe the *usta*s built it this way because of differential settlement," Ghalib suggested. There were no guidelines at the conservation office to deal with such things; the architects relied on their own judgment, and especially on the experience of al-Rawdhi. He and his partner used the technique of *fals* where there were cracks, rebuilding walls from the foundation.[48] Because al-Rawdhi was accustomed to working by the day, Ghalib helped him prepare his initial estimate to the contractor, and later, his pay applications; these had to be calculated and reported by quantity (square meters). As Ghalib examined each line item, he realized that the *usta* was getting only 20 or 30 percent of the contractor's fee for that item.

> He was doing almost all of the work, yet he got a fraction of the money. The contractor's supervising engineer had no experience with restoration. They gave all the traditional work to the *usta*. The contractor did not acquire experience, and it's still that way. The contractor even put the responsibility for demolishing and rebuilding walls on the *usta*—so if [the conservation office] architects didn't like the work, he would have to bear the expense and redo it. From this day I suggested that we require the contractor to pay the *usta* by the day, rather than by quantity …

---

45 "It need not be stressed that cement is a most unsuitable building material for the repair of this kind of traditional building, being too hard and strong for the bricks, so that thermal movements cause cracking. In addition, cement introduces free salts which can crack the walls" (R. Lewcock, mission report no. 2, p. 13).

46 See diagram in Paul Bonnenfant's and Guillemette Bonnenfant-Outrebon's discussion of stonework ("La Pierre," in P. Bonnennfant, ed. *Sanaa*, p. 288). Timbers were also used to "stitch" together cracks, a practice that Jacques Heymann found acceptable (R. Lewcock, mission report no. 5, 1989, section 43).

47 Ghalib recalled that the foundation was only one meter deep on this facade, and only one meter in width. The kind of cement recommended by an engineer in the office, *ciment kibreeti* (used in sewage piping and other infrastructure work), was hard to obtain; this was one reason for the delay.

48 See above, Chapter 5.

As noted earlier, neither EOPOCS staff nor the contractor had accurately estimated the operations and materials required for the project. For example, the contractor estimated *qaddad*, the traditional waterproof plaster specified for the roof of the *samsara*, at a fraction of the actual cost.[49] At the time *qaddad* had almost died out; two decades later, professionals and craftsmen had managed to approximate the old *qaddad* work, and it was being used frequently in restoration projects (Figure 6.5).[50]

As noted in the previous chapter, government regulations require the use of a general contractor in donor- and publically-funded projects, but contractors do not have skill in conservation and restoration. During the Campaign they relied heavily on their *usta* subcontractors—and still do today.[51] The dilemma was noted by J.M. Klessing, one of the German technical supervisors at Samsarat al-Mansura, which began two years after Samsarat al-Nahhas was completed. The contract for al-Mansura was awarded to YCDC which, wrote Klessing, "could at least demonstrate some experience in the field of urban rehabilitation." Having completed Samsarat al-Nahhas, the company was then paving streets between the Great Mosque and al-Luqayh Square.[52] But Klessing noted:

> In accordance with stipulations for government projects of this size, only general contractors that could furnish evidence of the necessary capacities and securities and that had the know-how for modern book-keeping came under consideration for execution of the work. A dilemma seemed to be preprogrammed since firms of this type work almost exclusively in the booming branch of new construction and seldom have experience in the field of restoration, yet the restoration work that was to be undertaken could only be achieved by means of traditional building skills and craftsman's trades which mostly are organized on a family or local basis.

As at Samsarat al-Nahhas, YCDC's inexperience was remedied by requiring the company to "use only provenly [sic] qualified traditional craftsmen as subcontractors for each individual restoration measure, always in consultation with the project manager." At Samsarat al-Mansura and elsewhere, it was hoped that the use of traditional materials and techniques would keep costs down and provide a replicable model for restoration.[53] But the benefits of this approach have largely been cancelled out by the use of a general contractor: it drives up costs and subordinates the *usta*—undermining his status, and thus discouraging replicability.

49 The contractor had estimated the cost at YR 200/sm when the actual cost was nine times that amount. When bids were solicited for Samsarat al-Mansura three years later, the cost was YR 1800/sm (see J. Klessing, "Planning and Execution of the Restoration," p. 91, fn. 11). That amount was doubtlessly the cost of the general contractor; the *qaddad usta* who worked under subcontract would have gotten a percentage of this amount. The cost of *qaddad* in 2004 was YR 5,000–6,000/sm in Sanaa, or $26–31/sm. The tremendous labor input becomes clear when one recalls that a semi-skilled laborer in *qaddad* made around YR 800, or U.S. $4.20, per day.

50 On *qaddad* see S. al-Radi, "Qudad, the Traditional Yemeni Plaster," in *Yemen Update* (American Institute of Yemeni Studies), no. 34, winter/spring 1994.

51 By the mid-2000s some contractors had gained experience, but most still had little knowledge of traditional construction. They continued to rely on the *usta*s who worked under them. "A contractor is either a *qabili* (tribesman), or has a sheikh protecting him," an architect told me.

52 According to GOPHCY records, that area was paved from October 1988 to August 1991; the restoration of Samsarat al-Mansura began in October 1991.

53 J.M. Klessing, "Planning and Execution of the Restoration," pp. 54–6, 80.

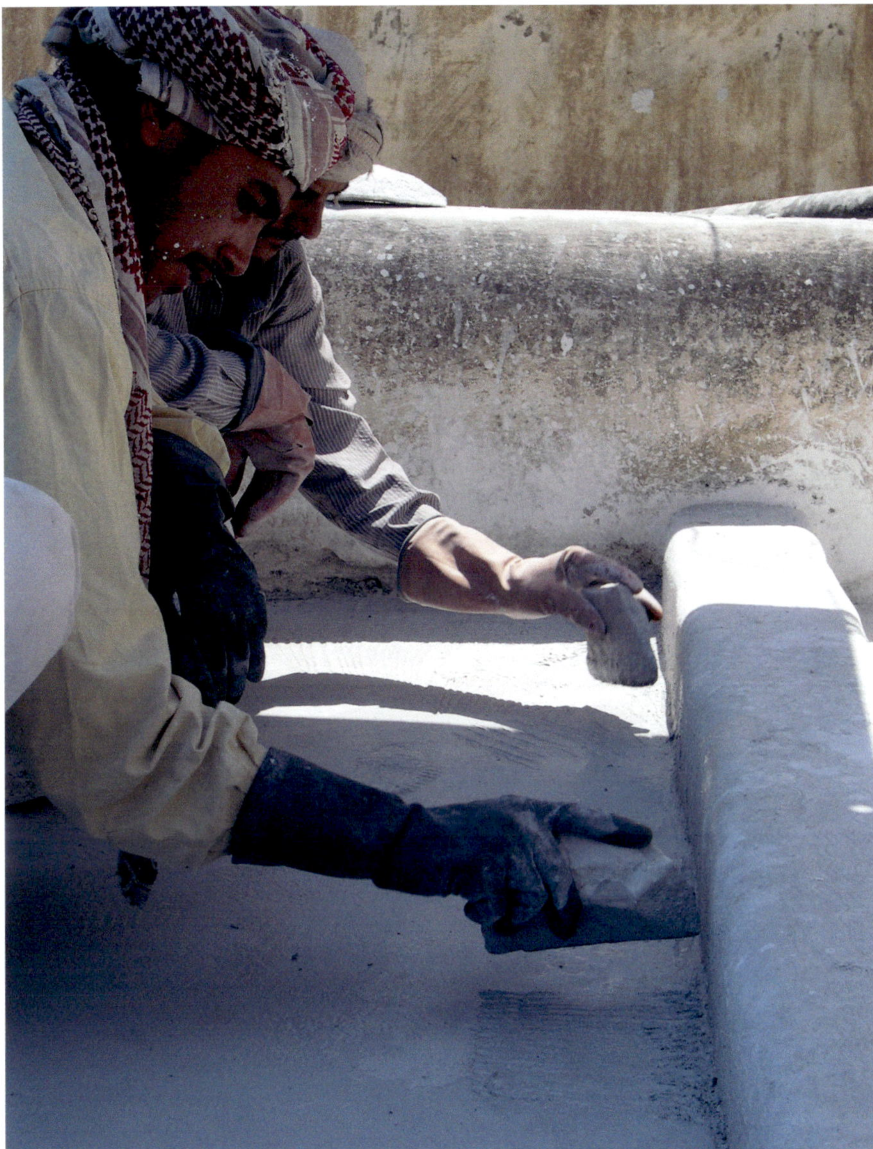

Figure 6.5 Craftsmen burnishing *qaddad*, a traditional and labor-intensive waterproof plaster, on the roof of a small mosque in Sanaa

The professionals working on Samsarat al-Nahhas had different views on the treatment of historic structures generally, and this building in particular. For some there was too much emphasis on aesthetics; but notions of aesthetics also differed. One argument involved the plaster finish on ceilings and beams. The Director and others at EOPOCS thought that plaster should be applied with a trowel, because they found a smooth surface more pleasing; Ghalib argued that it should be applied by hand, in the traditional way—and eventually he prevailed. He believes that the taste for smooth plaster, terrazzo tiles, and so forth was guided by a modern sensibility. The historic character of the *samsara* was compromised by decisions taken too quickly, without adequate thought, he says. For example, the rooms of the *samsara* had been filled with many built-in shelves and niches which were an important aspect of the interior character; these had been removed along with the original plaster, as directed by the specifications. Rather than using a wire brush to clean the surfaces of stone columns and arches, laborers were directed to chip away at the stone. "We lost an entire layer," recalled Ghalib. "So we lost our connection to the old building. We relate to old buildings through an exchange of our humanity, and by taking away this layer we lost this [exchange]."

Although Ghalib and others regret many mistakes at the *samsara*, they see it as a valuable learning experience—an opportunity for them to discover and test ideas of conservation in their local context. Lewcock, who was called back to Sanaa in March 1989, was brief but generous in his appraisal of the work. "The building has been satisfactorily conserved," he wrote in his mission report. "In the process a number of deficiencies have been discovered which it is hoped to rectify in the conservation of future buildings." Even with remedial efforts (new water pipes and drainage), Lewcock expected that the foundations in the old city would not settle down for 10 or 15 years. Cracks would continue to appear in the walls of the *samsara*, and should be carefully monitored. He stressed the importance of developing a clear methodology. "The preparation of a work plan for conservation is not enough in itself," he wrote. "It has to be preceded by a decision on the attitude which is to be taken by the conservers towards building." Did the architects want to restore the building to a hypothetically "original" state, or would they retain later changes and accretions? What parts of the building could be modified to allow for new uses and visitors, and what would be the character of these changes? The installation of modern amenities, like electricity and plumbing, should be carefully thought out and designed. Before conservation begins all aspects of traditional technology should be studied and materials tested; this would be facilitated by the laboratory to be provided under the Italian project. Serious attention should be given to the revival of building crafts, so that they could replicate the quality of the original workmanship. "The most serious deficiency in this respect," wrote Lewcock, "is the quality of carpentry in Sanaa, which has steadily fallen into decline during the last thirty years." In addition, great care should be taken to obtain materials suitable to the character and quality of the original building.[54]

Some of these technical and philosophical deficiencies were addressed during the restoration of Samsarat al-Mansura which, in contrast to Samsarat al-Nahhas, was carried out under the supervision of a resident German team, working with local architects. EOPOCS made an initial request to Germany for technical and financial assistance in 1987, shortly after the Norwegian mission had come to discuss a project. In June of the following year, Germany proposed a project that would contribute to the revitalization of the *suq* and the handicraft sector; and that would contribute "to the upgrading of restoration works on traditional buildings in Sanaa." Visits by an expert in Yemeni handicrafts and two experts in restoration were planned. The Germans proposed Samsarat al-Mansura for the

54 R. Lewcock, mission report no. 5, 1989, items 42–49. Ghalib also found the carpentry objectionable: "The doors should have been made in a more traditional form," he told me.

project, to be perhaps followed by Samsarat al-Majjah.[55] The project would be financed by a special commodity aid fund dedicated to the restoration of buildings in the old city, administered by the Central Planning Organization.[56] In July EOPOCS signed an agreement with the Bavarian State Conservation Office (BlfD), outlining the terms of reference for the project. Under the agreement BflD would carry out detailed documentation of the building and develop a restoration plan. Work on the building would be conducted according to "international standards of preservation," including:

> Preservation of the whole historical substance of the monument, i.e. architectural elements and surfaces;
> Where restoration work is required, it shall be done by traditional methods and materials—with preference of repair over renewal! ...
> There shall be no changes to the plan of the building (excluding some alterations of non-bearing walls);
> The facades shall be restored to their original appearance, with plasterwork, etc. ....[57]

A second agreement between EOPOCS and the German Ministry of Foreign Affairs reiterated that the project would be conducted according to the principles of monument protection, "that is, with the preservation of the original material of the building and the use of traditional techniques and materials."[58] Both agreements foresaw an "exchange of experiences"; the BflD's restoration concept would provide the basis of the work, "in cooperation with the Yemeni partner's concept of restoration."[59] BlfD would provide on-site supervision throughout the project, and develop and test "appropriate and economic techniques of restoration works on traditional buildings, in cooperation with EOPOCS and Yemeni masterbuilders." These techniques would include a more economical production of *qaddad*, and strategies for designing and installing plumbing and electrical works. The installation of utilities—electrical, plumbing, and telephone—would be planned and carried out in cooperation with the Vocational Training Center of Sanaa. Lighting would

55 "Proposal for the German Contribution to the International Campaign for the Preservation of the Old City of Sana'a (according to the request of EPOPCS, September 1987)"; handwritten (English), no signature; submitted to EOPOCS on 6/21/88 (according to the memo to which it is appended, from Werner Ligenau to EOPOCS, 6/29/88). The Yemeni handicrafts expert was Hamid al-Iryani (apparently residing outside the country); and two restoration experts, M. Petzet and W. Koenigs. Samasarat al-Majjah was not restored; Japan promised funds for it in 1991, but apparently the project did not go forward (G. Barbato, "The Methodological Approach for the Restoration of the Samsarah al-Bowani and Samsarah al-Majjah").

56 "Administrative Procedure for Managing the Counterpart Fund of the Commodity Aid provided by the Federal Republic of Germany," 12/21/87, appended to memo from Werner Ligenau to EOPOCS, June 6, 1988.

57 Text of BflD-EOPOCS agreement, reprinted in M. Petzet, "The Restoration of Samsarat al-Mansura," p. 13.

58 This is from the Arabic text of the agreement, item 5. The copy I obtained has spaces for signatures by representatives of the Yemeni and German governments, but is not signed or dated, and the first page is missing.

59 Text of BflD-EOPOCS agreement, reprinted in M. Petzet, "The Restoration of Samsarat al-Mansura," p. 13; "Conclusions on the Restoration and Refunctioning of Samsarat al-Mansuriyyah," signed by W. Ligenau, consultant to the German Ministry of Foreign Affairs, and 'A. 'Oshaish for EOPOCS, August 18, 1988, p. 4.

be designed by a German lighting designer, who would be brought to Sanaa on a short-term mission.[60]

At the beginning of discussions between EOPOCS and the Germans, a handicraft function was foreseen for the building. According to the agreement between EOPOCS and the BflD, the building "would house workshops and exhibition areas for artists and artisans." The August agreement with the German Foreign Ministry said that the *samsara* would be used by an artists' cooperative as workshops and exhibition space, and that their needs would be taken into account in the restoration. A statement by the cooperative, appended to the document and signed by eight artists, said that they would also hold classes in the building, and would set up a printing workshop and a library. The Arabic text of the German-Yemeni agreement stipulates that if the cooperative is unwilling or unable to make use of the building, it will be used by a similar society and/or for "representational cultural purposes (*liaghradh thiqafiyya mumaththila*)."[61]

Work on Samsarat al-Mansura was to begin in early 1989, but there were many delays.[62] Evictions appear not have been a major issue, since most of the *samsara* was vacant; the tenants of shops on the ground floor, facing the street, would be allowed to stay during most of the work, but a retail shop, installed in the entrance, was to be removed.[63] The project did not begin until October 1991, in difficult times: unification had occurred the year before, and the government had undergone massive restructuring; most donor funding had been cut off in retaliation for Yemen's position on the Gulf crisis. The conservation office, newly reorganized as GOPHCY—and now responsible for all the historic cities of Yemen—was in the throes of an internal crisis. Some senior staff members had been transferred or dismissed. The project for al-Mansura had been developed by several staff members who later fell into disfavor. They wanted more interaction between foreign experts and local architects. "We believed in the idea of national counterparts," recalled an EOPOCS staffmember. "The Germans insisted on this. They understood what was happening [at the conservation office]. What is the benefit of foreign consultants if there are no Yemeni counterparts?"

---

60 "Conclusions on the Restoration and Refunctioning of Samsarat al-Mansuriyyah," pp. 3–5. Lighting and plumbing were major concerns for architects, as they are for restorers in other parts of the world, because they require chases for wires and pipes which must be added to or subtracted from the walls. The choice of fixtures was also a dilemma because, as the agreement with the German Foreign Ministry notes, the selection of lighting fixtures in Yemen was very limited.

In a 1989 memo to the Director of Projects, Ghalib wrote that the contractor intended to import fixtures from abroad for both al-Nahhas and al-Mansura. These fixtures would be expensive, and would not serve their purpose. He suggested that they fabricate light fixtures locally, from materials available in the *suq*; these could be inspired by local fixtures, and would be more appropriate to the buildings (memo dated 1/22/89, handwritten, in Arabic; Samsarat al-Nahhas project file, Projects Department, GOPHCY). Both the directors of Projects and Technical Affairs signed their approval, but in the end imported light fixtures were used.

61 Arabic text, item 6. The artists' statement (in English and Arabic), appended to the English version was signed by Fu'ad Futaih, Mazhar Nizar, 'Abdullah Sallal (also of EOPOCS); 'Abdulkarim al-Iriani; Amin Nasher; 'Ali al-Dharhani; Fu'ad al-Haj; 'Aidaroos al-Jefri. See "Conclusions on the Restoration and Refunctioning of Samsarat al-Mansuriyyah."

62 Petzet attributes these delays to the difficulty of raising the DM 400,000 required for the project; the problem was resolved by allocating funds from the YAR's purchase of German commodities ("The Restoration of Samsarat al-Mansura," p. 12). But the commodity fund for restoration had already been established in December 1987.

63 The shop sold *janbiyya* belts. The owner was to be relocated; EOPOCS would compensate him, if required, but this compensation should come from the project funds ("Conclusions on the Restoration and Refunctioning of Samsarat al-Mansuriyyah," p. 2).

In contrast to the restoration of Samsarat al-Nahhas, detailed investigations of the building fabric were carried out before work was begun. Photogrammetric studies of the facade built on earlier surveys by EOPOCS staff and a Peace Corp volunteer in 1988–1990.[64] Engineer Richard Hughes, who had conducted other studies for the Campaign, tested clay mortar samples from al-Mansura. 'Abd al-Hakim al-Sayaghi, project manager for EOPOCS, recalled that he and other architects spent two long afternoons on the roof with Hughes and Lewcock, debating the use of clay mortar. The German team believed that clay mortar was compatible with the building's structural characteristics, which were distinguished by their "softness."

> The traditional use of soft clay mortar, bricks, and layers of wood within parts of the masonry permits a certain degree of plastic deformation of building components. The insertion of new, rigid building components or the use of materials that are too hard would severely disturb the existing load-bearing behavior in a critical case of stress (such as earthquake) and in the long run would cause greater damage than would the measures otherwise needed for repair.[65]

In the end, clay mortar was used in repairing and rebuilding walls; concrete was used only to stabilize the foundation.

Following international practice, the German team preferred "repair over renewal." Before making any decisions, they studied the structure of the building carefully and decided that it was essentially stable, despite its age. "They saw it as a perfect system, the most stable plan for load-bearing walls," recalled al-Suleihi, who worked on the project. "I remember Klessing's diagram: he drew the building as a tube, with an outer wall and an inner wall—two rings tied together." The Germans felt that structural problems in the *samsara* were due not to defects in construction but to weathering, lack of maintenance, and vibrations of vehicles. The exterior walls, and particularly the front wall, was pulling away from the structure; and the ceiling joists had begun to detach from the masonry. Left to his own devices, al-Rawdhi would have probably disassembled portions of the wall and rebuilt them from the foundation, using the technique of *fals*.[66] But the German team decided to tie the existing wall back with steel tie rods. Unlike al-Rawdhi, they saw the ceiling as the key to structural support. Klessing described the work as follows:

> Working floor by floor external or integrated wall anchors of various types ... were fastened with steel tie rods and stable screw joints at intervals onto the ceiling joists. By fastening the walls to the ceilings, which in principle function as a rigid plate through their layer of brushwood covered with a clay 'packing,' the process of detachment of the outer walls could no longer progress. In case of vibrations the building ought to be held together more firmly, as with a clamp, because of the restrained exterior walls.

The diamond shape of the wall anchors was inspired by the decorative brickwork on the building.[67] "Klessing and his colleague thought so carefully before putting in the ties,"

64 Plans, elevations, sections signed by S. al-Suleihi, B. Polluck, 'A. Sallal, no date (circa 1988), on file at Projects Department, GOPHCY.

65 J.M. Klessing, "Planning and Execution of the Restoration," p. 56.

66 See above, Chapter 5.

67 J.M. Klessing, "Planning and Execution of the Restoration," p. 56; 'Abd al-Hakim al-Sayaghi, interview.

recalled al-Suleihi, "because they were introducing a new feature into the building." Al-Sayaghi felt that it was an appropriate innovation: "There is no reason we can't adopt new things when they will improve the building." In other cases, wall areas that were deemed to have no bearing strength—due to groups of cracks or long vertical cracks—were "consolidated once again through careful, step-by-step exchange of the deteriorated elements." That is, the *usta* was permitted to demolish and rebuild. In addition, parts of the roof were rebuilt, reusing the original materials where possible.[68]

The dynamic of the project was complex, influenced by individuals with different backgrounds and skills: German conservation architects; local architects and engineers; and the *usta*s, craftsmen, and laborers. The Germans were clearly in a position of authority, but they "worked with the Yemeni colleagues as one team."[69] They respected the *usta*s and craftsmen, who solved difficult structural problems on site, and were clearly committed to the work.[70] But they felt that knowledge of traditional techniques, once of a uniformly high standard, was no longer complete:

> Indeed some things are known only by a few individual old master builders. The same is true for the practical ability to carry out traditional craft techniques. But for the preservation and repair of historic fabric this knowledge and the mastery of craft practices is imperative.[71]

The project was another opportunity for architects and *usta*s to negotiate their new relationship. I asked al-Sayaghi about the dynamic on site. "Sometimes the *usta*s take too many risks," he replied. "Al-Rawdhi wanted to disassemble large portions of the floor, but I preferred that he do a little at a time. I had to convince him that I respected his work, but that as project manager I was ultimately responsible."

Professionals who worked on Samsarat al-Nahhas and Samsarat al-Mansura generally remember these initial experiments as positive ones. Despite the lack of technical knowledge and a clear approach at Samsarat al-Nahhas, it was a learning experience and produced an acceptable restoration. Work with the German team on Samsarat al-Mansura seems to have increased the architects' self-confidence: they learned on site, and were also equal partners in the process. But the projects also established a new relationship—between architect, *usta*, and contractor. One *usta*, who worked on the restoration of the southern city wall (under a general contractor) said that several architects worked with him on site. "They learned many things, and then they went away"—that is, they took his knowledge, and then went on to prestigious positions in an internationally recognized endeavor. Some architects acknowledge their debt to the *usta*s and other craftsmen. Ghalib told me:

> From the beginning I wanted to give these projects to people with traditional skills. First, to encourage them; and second, to create a way for us to understand each other. Until now,

68 J.M. Klessing, "Planning and Execution of the Restoration," p. 58.

69 'Abd al-Hakim al-Sayaghi, "Samsarat al-Mansura Restoration Project: National Art Center Suq al-Ennab (Grape Market)," in M. Petzet and W. Koenigs, eds, *The Restoration of Samsarat al-Mansura*, p. 19.

70 Early on in the investigations—before al-Rawdhi had been hired—a Yemeni architect suggested that they bring him in to stabilize an arch on an upper floor that had no support; it was resting only on the wall of the neighboring *samsara*, al-Majjah. The German team was impressed by al-Rawdhi's work.

71 J.M. Klessing, "Planning and Execution of the Restoration," p. 80.

> it's too difficult for the architects to work alone; they still can't work on their own without skilled craftsmen.

The *samsara* restorations were less successful in terms of their social and use value. From the beginning, some EOPOCS staff had doubts about an art gallery. The fine arts were very new in Yemen: an art gallery would not be relevant to the residents of the old city, and perhaps not even to an elite Yemeni audience.[72] But the project was supposed to have a training component, which seemed to redeem it. In the end, the artists' cooperative was inactive. The contract was taken over by one of the artists, who used the space primarily to exhibit his own work; in recent years, the work of other artists has been exhibited on upper floors. Many professionals had had high hopes for the handicraft center: they hoped that it would have a direct impact on the economy of the *suq* and the revitalization of traditional crafts and trades. The National Center for the Revival of Handicrafts had its main venue at Samsarat al-Nahhas; it was overseen by a Handicrafts Department, created when the conservation office was reorganized as GOPHCY in 1990–1991. The *samsara* was to house the men's training center and an exhibition space for both men's and women's handicrafts—since the women's center (located in another building) could not receive male customers. Male and females professionals I interviewed saw the women's program as relatively successful: many women had been trained in sewing and embroidery and, according to the program staff, depended on the Center for income.[73] Most saw the men's program less successful.

The story of this project is a complex one, with many local and international players; only a relatively brief outline can be included here. As noted earlier, one of two UNDP projects to support the Campaign was for training and equipment for handicrafts; it was designed with UNESCO and financed in part by the ILO. The idea behind the project was to establish a national center for training, product development, and marketing; Sanaa would be the pilot area, and regional centers would later be linked to the national center. The project funded an initial survey of the current state of crafts throughout Yemen, which would determine the direction of the project.[74] "We had strong national feelings," recalled Jamal Kushih, who directed the Center. "We had counterparts from many countries, and we learned a lot from them." The new generation would learn traditional crafts and trades

72 "When we had our first exhibit of paintings," recalled one of the artists, "even the Minister of Culture didn't understand what it was. How can you promote the fine arts in such a country?"

73 The women's handicraft center was first located in Bayt Mutahir. It was moved in 2000 to the restored house of Sheikh Sinan Abu Luhum, in al-Abhar Square; there was also a separate weaving center for women. The men's center was first at Bayt al-'Amri on the Sa'ila; it was moved to Samsarat al-Nahhas circa 1990. In the early 2000s, I was told that GOPHCY provided insufficient support for the women's program. In addition to lack of funds, marketing was a problem because the exhibition/retail area was not easily visible from the street. In an attempt to remedy this, a group of foreign women rented the ground floor of Samsarat al-Dhamari in 2004, to exhibit products created by the women's handicraft center and other women's NGO's.

74 "Work plan YEM/87/008, Development of Handicraft Industries," no date (cover letter by project director Kevin O'Connor signed 8/30/88); in folder entitled "United Nations Project, 4" at Samsarat al-Nahhas. The book that resulted from the survey appears to have been a joint UNESCO-ILO publication, but I have been unable to locate a copy of it. The crafts that were covered in the initial survey were textiles (handloom weaving and embroidery), woodcarving, leatherware (including book-binding), jewelry (silver-smithing, gold-smithing, and semi-precious stones), and stoneware (stone pottery and alabaster; "Work Plan," Appendix II).

from the old; foreign consultants would bring techniques and equipment that would make production more efficient, and help develop new designs.[75]

Professionals and observers I interviewed, however, felt that the project was flawed in both conception and execution. It should have been conceived not as a training school, but as a center that offered support to craftsmen in the *suq* who were already practicing. These craftsmen would train apprentices at their shops, as they had in the past; the center would provide technical and marketing assistance. Instead, a formal school was established. Students received theoretical instruction in drawing and design (the position was described as an Islamic Design Instructor) and training in a craft. Various crafts were considered for the program; those finally selected were silverwork, leather, agate, and wood-carving, for men; and weaving and embroidery for women.[76] An unsigned memo, apparently from the UNDP office in Sanaa, said that "[p]roduction would focus mainly on traditional decorative crafts for the local, tourist, and export markets ... priority [will be given] to regional markets in the Middle East." Regarding instruction, the memo counseled that preference "be given to local expertise if it is available for some of the crafts groups."[77]

But the courses were largely run by regional consultants, including a woodworking expert from Azerbaijan; a Lebanese Armenian silversmith; and a Palestinian expert in embroidery.[78] In an interview, Kushih defended the use of foreign consultants, because local crafts were no longer intact: "There is a gap of history (*fajwat tarikh*) in Yemen ... [in the building trades and crafts] the chain of transmission (*silsila*) was not continuous." He said that Yemeni trainers worked alongside the consultants and benefited from their knowledge and labor-saving techniques:

> We were not organized or educated. We couldn't bring just anyone to teach. A craftsman (*hirafi*) doesn't know how to teach; few people would come [to study with him] ... Yemeni silverwork was the best in the world, in Sanaa and the Hadhramaut. But the traditional work took too much time, the materials were too expensive. We brought a Lebanese instructor to work alongside a Yemeni [master craftsman]. The Lebanese man showed us how to make lighter, cheaper things in the style of the old ... What a Yemeni craftsman took two or three days to make, the Lebanese could do in two hours, using electric tools. We had to modernize our crafts, otherwise our craftsmen couldn't compete.[79]

Observers, however, said that the new techniques and foreign influence changed the nature of Yemeni crafts: "Our embroidery looked Palestinian," remarked a GOPHCY staff member. Because the trainees had no relation to the *suq*, said another professional, they left Sanaa

75 An architect trained in Pakistan, Kushih headed the National Handicraft Center with Kevin O'Connor from 1988; the latter's mission was funded by the ILO.

76 According to Amal al-Shaiba, instruction in *dhahabi*—embellishment of garments with gold thread—was later added later; new courses in carpet-weaving and sewing were envisioned in 2004.

77 The undated memo was intended to brief handicraft advisor Kevin O'Connor on the first six months of his mission, which began in 1988; in folder entitled "United Nations Project, 4."

78 United Nations volunteers (UNV's) were considered for consultancies, but were too expensive. A letter dated 2/2/88 recommends that Middle Eastern and North African experts would be paid only $2,500/month, and so would be more affordable than the $132,000 envisioned for two UNV's and six man/months of consultancies (in English, no signature; in file entitled "Projects Reports 4"). The memo makes no mention of local experts.

79 Yemeni craftsmen involved in the project included a master silversmith and an 80-year-old master wood-carver from Zabid. Kushih recalled that the silversmith was given a test before he was hired; this is also mentioned in documents in the project file.

after their training to work elsewhere. Kushih maintained that students had to be sought outside the *suq* because craftsmen and merchants were suspicious of the project—although several sent their sons to train there after the center had been operating for a year or so.

In the mid-2000s a few trainers remained at the center—a woodcarver, a silversmith, and an agate-polisher—but they lacked students and funds. The women's center was locked in an ongoing struggle with GOPHCY over its budget. Most professionals agree that the project effectively stopped with UNDP funds, in 1991. Kushih acknowledged that since the civil war of 1994, the center has had virtually no budget. GOPHCY continued to rent the building from the Ministry of Endowments, and to sublet the shops to tourist-oriented businesses. But Kushih said that the sellers of agates and other gemstones were trained at the handicraft center. He believes that it has had an important impact on the *suq*: "The students have now become teachers of new students." When the project started, for example, there were only three master agate polishers left in the *suq*. Now thousands of people work in agate and other gemstones, assisted by machines; some have even found employment in the Gulf. While some traditional crafts are still produced, it is unclear whether this can be attributed to the handicraft center. Merchants in the *suq* who sell old style silverwork, made by artisans in Sanaa and other towns, are proud that customers cannot distinguish them from the old work.[80]

The *samsara*s are remarkable buildings; the two discussed here were carefully restored, albeit with different approaches. They were seen as nodes that would help revitalize the *suq*, the economic backbone of the old city, which would in turn keep the old city alive. But tourist-oriented functions predominated, and the *samsara*s were functionally and socially isolated. At the 1991 symposium Werner Ligenau, German attaché with the Department of Antiquities and promoter of the Campaign, expressed exasperation at the *samsara* restorations and other high-profile projects. The old city was being transformed for the benefit of outsiders under the banner of "cultural heritage." But these "spectacular, international … projects," he said,

> are divorced from the population they are intended to serve. How is the term culture to be understood? If we understand it in the sense of high culture, distinct from the culture of daily life, … the Madinah reduces itself easily to its fascinating silhouette of magnificent facades … an impressing spectacle in which a mixture of exotically dressed walker-ons and strange smells and noises entertain the culturally interested audience from around the world.
>
> But, the subject of observation changes completely as soon as one perceives and accepts that the Madinah with its market and traditional heritage of a—for Europeans today hardly understandable—Every-Day Culture which includes and integrates all ranges of life. This kind of Culture is not set off from the real structure of politics and economics, but is an integral part of all societal dimensions: the political order, the economic processes,

80 When I asked the owner of a well-known jewelry and craft shop where the new silverwork is made, he said with a smile "In our workshops"; he clearly did not want to reveal their location. It is often sold to unsuspecting customers as "Jewish silverwork," which implies that it is old silverwork produced before 1948. One merchant defended this usage: for him, "Jewish" indicates not age or authenticity, but quality. The new Yemeni silverwork that I saw in the mid-2000s seemed distinct from the ubiquitous (and lighter weight) Indian silverwork that was found throughout the *suq*.

> the used technologies; also from societal value conceptions, perceptual images, lifelong habits, etc. to the artistic genre of facade decorations.
>
> The preservation charge for the Madinah must be understood from the standpoint of the heritage to be preserved and not from that of our Culture Industry where culture is seen as a Commodity …
>
> What, then, is Culture? Does UNESCO refer to it in the sense of Every-Day Culture—as demonstrated in the living example of the Historic Centre of Sanaa …? If this is so … the physical and spatial stock [must be seen as] inseparably interwoven ….[81]

Ligenau, then, identified the two "kinds" of culture that informed some thinking at the conservation office. An authentic local or traditional culture would attract new and affluent audiences; revenues would, in turn, help the community bear the burden of conservation. At the same time conservation would promote the kind of "cultural revival" envisioned by the revolution—bringing an authentic Yemeni spirit into contact with new audiences. But tourism never increased to expected levels because of ongoing political instability, and because of the absence of tourist facilities outside the major cities.[82] Nor have local elites "bought into" the old city of Sanaa: it has not developed the kinds of "new uses" that revitalized historic centers in Europe and elsewhere.

The Restoration of a Model Neighborhood, 1989–1993

In addition to the *suq*, the heart of the "living historic city" was its houses—some 6,000 of them. As in Venice, the beauty and authenticity of the city was seen to derive from this remarkable fabric, rather than from individual monuments.[83] This brought up an important issue that had already been raised in Venice, Fez, and other cities: how was this fabric, virtually all of it in private hands, to be conserved? Consultants' reports produced in the late 1970s and throughout the 1980s warned that houses were falling into disrepair, largely due to outmigration. Not only were absentee landlords and rural migrants unable or unwilling to maintain their houses; traditional materials and builders were in short supply, making repairs unaffordable. New and cheap materials like cement were being used for repairs and additions.[84] By the late 1990s some consultants thought that maintenance had improved, perhaps as a result of increased property values and a conservation "ethic."[85] But the problem of house repair and maintenance has remained daunting—especially since privately owned property is generally ineligible for donor aid.

---

81 W. Ligenau, "Remarks on the correlation between cultural and economic revitalization in the case of the old city of Sana'a," paper presented at the Symposium on the Integrated Urban Policy for the Conservation of the Old City of Sanaa, Yemen, 15–19 December, 1991, pp. 46–7.

82 H.M. 'Abdulla, "The Cultural Tourism and Development: the Case of Yemen," March 1996.

83 This view, held by Lewcock and others, was reiterated in a number of promotional films produced by UNESCO. Of several films on file at UNESCO's documentary services, see especially *Programme 13: Venise, Sanaa*, UNESCO/Transtel, M0365, no date.

84 R. Lewcock, mission report no. 2, 1982, pp. 11–13.

85 Eckert and al-Maqrami, "Social Assessment Report," p. 25.

In several reports, Lewcock argued that the repair of houses through a loan fund was an essential part of the Campaign. In his 1989 mission report, he noted that although the conservation of several houses would soon begin,

> ... the Old City represents a unique problem in that there are approximately seven thousand buildings within it, most of which are private houses and homes. A strategy needs to be implemented quickly for encouraging their rehabilitation in a careful and sensitive way before it is too late. For this purpose, the establishment of a development bank for the Old City to provide loans at a low rate of interest is a *matter of the greatest urgency*, as is the study of, and introduction of, other suitable incentives.[86]

According to ‘Ali ‘Oshaish, the Campaign strategy envisioned the restoration of 30–100 hundred private houses as demonstration projects, and others would be financed through a loan bank. But plans for the loan bank were never implemented; in the mid-2000s only very limited funds were available to homeowners. Moreover, loans were rarely available from banks, because “traditional construction” is deemed ineligible.[87] Not all policymakers agreed with the idea of a loan bank: many felt that the restoration of private houses was not the responsibility of government. The Board of Trustees agreed that the whole city should be conserved, recalled Muhammad Tayyib, former Deputy Minister of Municipalities. But they felt that homeowners should maintain their own houses. The government would undertake infrastructure and large rehabilitation projects, like the paving and upgrading of the Wadi Sa’ila; as the city’s condition improved and its perceived value increased, homeowners would be encouraged to maintain their properties. A loan fund would not be practical, he said: neither homeowners nor institutions were accustomed to such a thing, and it would not be respected. His view points to the different attitudes inside and outside the Campaign office. But even Campaign officials acknowledged that the costs of house repair and maintenance would fall primarily on homeowners—one of the main reasons tourist revenues were needed.

During the Campaign, only one project dealt with the repair and restoration of houses. This was a demonstration project conceived by the Quaroni Bonifica team, as a practical application of its work in the old city. Outlined during the first phase of its studies, the project was formalized in a 1986 agreement between the governments of Italy and Yemen and Quaroni Bonifica. It read in part:

> ... the studies carried out thus far and the first results obtained from the international ... Campaign, in fact urge that a first practical example of urban recovery, implemented in line with the principles of scientific conservation, be undertaken without delay ...

86 Mission report no. 5, 1989, p. 43 (the emphasis is Lewcock’s). Earlier he had proposed that funds for the loan bank be borrowed on favorable terms from the Bank of Kuwait (mission report no. 4, 1987, p. 10). Lewcock’s sense of urgency was echoed by the Third International Working Group, in which he also participated: “the government should as a matter of urgency implement their plans for a Campaign loan bank to support and provide incentives to private owners to restore their buildings in the old city in accordance with standards set up by the Campaign Executive Office.” In addition, the Campaign office should have access to funds that would allow it to intervene in house repair or reconstruction, where because of inheritance disputes or other reasons, residents were unwilling to undertake the work themselves (“Recommendations,” 24–26 January, 1989).

87 Tarek al-Hammadi, legal counsel for GOPHCY. Such restrictions apply in some other countries, for example, in the oasis district of Morocco (Salima Naji, personal communication).

> The aim of the project is to salvage an urban neighborhood of Sanaa al-Qadima, chosen as a prototype for the commencement of a functional urban conservation policy.

The agreement anticipated a second grant from Italy, but the grant was delayed for several years.[88] There is little documentation on this "Implementation Phase" of the Quaroni Bonifica studies—in Sanaa, generally called the "Italian project." It was a controversial project, difficult to reconstruct because of the death of key protagonists (Quaroni, DeCarlo, and al-Haddad) and the dissolution of Bonifica S.p.A. The main issue was an ownership dispute: Bayt Sari', having been supposedly acquired by EOPOCS, was forcibly reoccupied by its owners while the house was under restoration. The incident provoked an official protest from the government of Italy. As a result, the Italian team did not hand over documentation, and some of the equipment promised for the restoration workshop was not delivered. The following account is based on interviews with former EOPOCS staff, including architects 'Abdullah al-Hadrami and Jamil Shamsan; Jamil Sari' and his family, who own the house; Maria Luisa Neri, architect with Studio Quaroni; and Giovanni Boccardi, resident architect for the Italian Foreign Ministry, who subsequently wrote about the project for *Chroniques Yemenites*, the journal of the French Institute in Sanaa (CEFAS).[89]

The southwestern quadrant of the city had been a focus of attention since the paving of al-Abhar Square in 1972; the streets leading from the Sa'ila to al-Abhar were paved in the first upgrading project paid for by the government under the banner of the Campaign, in 1984–1985. This major thoroughfare for both local visitors and tourists splits at al-Abhar Square: the northern branch leads past the Great Mosque to the *suq*, the southern branch to Bab al-Yaman. A number of stately houses are located on the square and in neighboring quarters. In the first phase of the Quaroni Bonifica studies, the authors had chosen a project site that was in the same area, extending west from the northern edge of al-Abhar Square. They gave several reasons for the choice: the neighborhood represented a complete range of housing types of a consistently high architectural quality; the houses exhibited representative types of structural damage; and the area was characteristic of old Sanaa's urban morphology.[90]

---

88 The project agreement refers to a bilateral cooperation agreement in the Italo-Yemeni protocol of December 1985. The contract, entitled "Conservation of a Sample Core of the Historic Center of Sanaa and the opening of a restoration laboratory/school—General Agreement," was signed by Francesco Pulcini, Ambassador to Italy; Riccardo Bichara, of Studio Quaroni and for Bonifica SpA; and Abd al-Rahman al-Haddad, Director of EOPOCS, Sanaa, 22 February, 1986. I was permitted to consult (but not to photocopy) this document, which is on file in the Director's Office, GOPHCY.

89 G. Boccardi, "Projet de la Coopération Technique Italienne pour la conservation du Centre Historique de Sanaa," and "Point de Vue," *Chroniques Yéménites*, 1993, p. 52, 54–5. Jamil Shaman directed the Department of Planning for EOPOCS; 'Abdullah al-Hadrami directed the Department of Restorations, and later, Architectural Studies and Technical Cooperation. I am grateful to Jamil Sari' and his family for granting me an interview and allowing me access to the house, which they have rarely done in the years since the restoration. According to informants, Bonifica S.p.A. was implicated in Tangentopoli, the investigations into wide-spread corruption that rocked Italy in the early 1990s. The Sanaa project may have been investigated as part of a wider inquiry into Bonifica's operations at several Middle East and North African sites.

90 Studio Quaroni S.r.1./Bonifica S.p.A, "Safeguarding of Sanaa Historical Center in Relation to the Urban Development of the City," Rome, 1981–1990, for the Yemen Arab Republic/General Organization for Antiquities and Libraries Proposal Phase, volume PIII, "The Urban Texture," 1983, p. 7. Harat Mu'adh had already been studied during the Reconnaisance Phase; it was now part of a larger "island," no. 4 (on the "island" concept see Chapter 3).

The project was an ambitious one: an example of integrated urban conservation that would provide a replicable model for other neighborhoods. It included an initial training seminar, and an "international publicity Campaign" aimed to promote the international role of the old city.[91] The project would restore several representative houses, selected on the basis of age, architectural features, and degradation, to serve as models for future work by the conservation office and by private owners. Buildings with structural problems would be consolidated with "materials and procedures compatible with traditional techniques, but with modern conservation principles."[92] The work would be carried out by local *usta*s: their skills would be upgraded as required by scientific conservation practice, and they would eventually become trainers of new recruits.[93] This training would take place at a restoration workshop and laboratory, a new facility to be built under the project. The workshop would serve as a school for *usta*s and "technicians"—presumably conservation office staff—and also as a research facility: planning issues would be investigated here, and there would be a specialized lab for the study of materials and techniques. The workshop would initially be run by Italians and Yemenis; ultimately, it would be administered by local technicians and *usta*s' cooperatives.[94] In addition, a ruined slaughterhouse at the northeastern edge of the project area would be rebuilt and upgraded as a "polyfunctional center" for the neighborhood. It would serve different purposes at different times (a kindergarten, meeting room, or clubhouse) depending on the needs of the community. A public park in the same area was also envisioned.[95]

An important part of the project was the "reordering" of what the authors called the "urban texture," respecting traditional precedents and while providing for contemporary needs. Traffic would not be eliminated entirely, but rather would be made to skirt around the edge of the neighborhood. Within the neighborhood, a system of pathways and "piazzas" would be reserved for pedestrian traffic, leisure activities, and parking.[96] This network would be consolidated by reclaiming small spaces that had been appropriated by residents, obscuring the original urban morphology of small squares (*sarhat*) and gentle widenings in streets. Residents had fenced off areas in front of their houses and built illegal structures; in some cases, alleyways had been closed off with gates of wood or sheet metal, and "other forms of 'modern' ugliness." These spaces would be "better suited for use by the community in a new collective sense." While each household would

91 Meeting minutes, 16–19 February, 1986, item 5f. The training seminar was held in January 1992, led by Gabriella Esposito, Quaroni's widow, through a translator. Documents on the seminar may be found in the Fondazaione Archivio Storico Olivetti/Fondo Maria Luisa Neri.

92 Project agreement, February 1986; 10 to 12 houses were to be selected, and of these two to four would be restored within the contract period.

93 The authors acknowledge that "[t]he time span for the gradual passage from the first stage to the second, i.e. the abandonment of old practices, and the adopting of new methods, will be rather long" (Quaroni Bonifica, "Safeguarding of Sanaa Historical Center," volume PIII, "The Urban Texture," 1983, pp. 2–4).

94 Meeting minutes, 16–19 February, 1986, items 5–6; Studio Quaroni S.r.l. and Bonifica S.p.A., "Action Proposal," January 1986, p. 22. This January 1986 study, carried out just prior to the signing of the project agreement, elaborated the idea of a workshop/laboratory which had been mentioned in the Proposal Phase.

95 Quaroni Bonifica, "Safeguarding of Sanaa Historical Center," volume PIII, "The Urban Texture," 1983, pp. 8–10. Quaroni Bonifica described the ruined building as a slaughterhouse (*macello*), but informants in Sanaa called it a *samsara*.

96 Neri recalls that parking was a particular concern of the Yemeni authorities; residents interviewed by the Quaroni Bonifica team expressed a desire to keep a car close to their house.

conserve its own private spaces, it could utilize contiguous spaces, possibly passing from one street to another (Figure 6.6). The Quaroni Bonifica team hoped that the conservation office would implement the legal measures, especially a property registration system to determine ownership, since the project required an exact determination of "public, semi-public, and private use." Eventually, outdoor spaces could be administered through a form of condominium ownership.[97]

Although the project was intended as an example of "urban recovery" (*recupero urbano*) street paving and infrastructure were not originally included; perhaps the Italian team expected that this would be a contribution of the Yemeni government. In 1986, at a meeting that preceded the signing of the project agreement, EOPOCS officials suggested that some project funds be shifted to infrastructure. They felt this was more practical, since access to the buildings and open spaces would be difficult to secure. The Italian team agreed, but said that the restoration of houses must remain the priority. Infrastructure works and landscaping would be expanded, connecting the project area to "at least one relevant access point to the Medina."[98] In the subsequent months, architects at Studio Quaroni developed designs for streets and open spaces, including paving patterns and water elements like public fountains (Figure 6.7). They also developed schemes for illumination.

Under the project agreement, EOPOCS officials would conduct a preparatory study and work with the Italian team to identify two or three houses for restoration. They would assure that these houses, as well as the land for the workshop/laboratory, be made fully available for work.[99] Perhaps because of difficulties in securing access to these properties, the project was delayed for several years. The delay may also have been due to the death of Quaroni and DeCarlo and the reorganization of the firm under Quaroni's wife. The project was finally inaugurated in the first half of 1989, with great fanfare: Italian VIP's –politicians, artists, scholars, filmmakers, journalists—were flown to Sanaa for several days of events at the university, celebrating Sanaa's status as a world heritage site and the historic ties between Yemen and Italy. Preparations for the work, including construction documents, were underway in the summer of the same year. Several houses had been discussed as candidates for restoration, but drawings were only done for two, Bayt Sari‘ and Bayt al-Thawr, both to the north of al-Abhar Square. The owners of the latter house did not agree to the restoration work, so only Bayt Sari‘ was restored—although this project, too, ended in conflict. Work began on the ground in 1990 and was completed in April 1993; the project opened officially on the same day as Samsarat al-Mansura. The contractor was an Italian firm based in Addis Ababa; its Sanaa office, staffed by Italian technicians, worked through a prominent Yemeni company called YCon.[100] Work was supervised by architects

97 Quaroni Bonifica, "Safeguarding of Sanaa Historical Center," volume PIII, "The Urban Texture," 1983, p. 2, 7–8; and volume PO, ""Foreword and Summary," pp. 14–16; M.L. Neri notes, "Gruppo Restauro," 1984, Fondo Maria Luisa Neri, Fondazaione Archivio Storico Olivetti.

98 The scope of infrastructure and urban upgrading—landscaping, walks, street lighting, and so forth—would only be determined after the number and cost of house restorations had been finalized (meeting minutes, 16–19 February, 1986, item 5d; and project agreement). A note in the margin by an EOPOCS official cites "garbage collection"—an activity that had not yet been addressed.

99 Meeting minutes, 16–19 February, item 3, 1986.

100 Owned by the Yemen Bank for Reconstruction and Development, YCon has undertaken many high-profile construction projects. I was unable to identify the Italian contractor, or to locate other project records. According to al-Hadrami, the project budget of L. 778,558,030 was to be disbursed between December 1990 and April 1993. He says that the final costs were around USD 6,000,000 (personal communication).

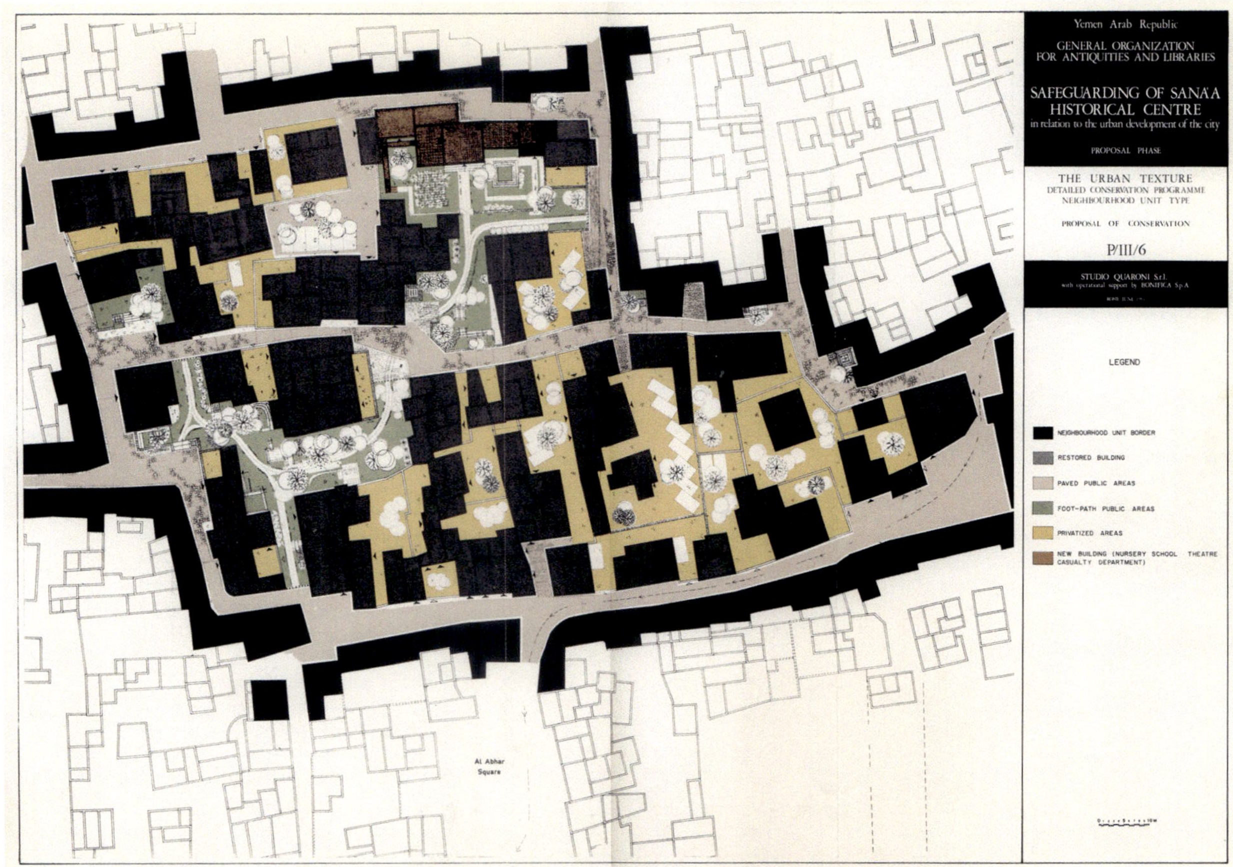

Figure 6.6 **Quaroni Bonifica's conservation plan for a neighborhood unit (Mu'adh quarter)**

Source: Studio Quaroni S.r.l./Bonifica S.p.A, "Safeguarding of Sanaa Historical Center in Relation to the Urban Development of the City," Proposal Phase, 1981; reprinted with permission of Maria Luisa Neri.

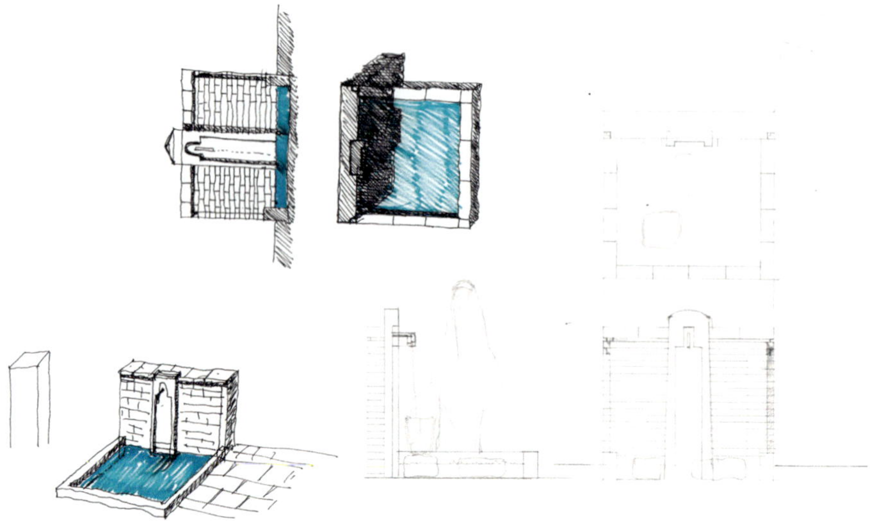

Figure 6.7 Sketches for a fountain in Mu'adh quarter by Studio Quaroni

Source: Image courtesy of Adriano Olivetti Foundation, Maria Luisa Neri Collection, Associazione Archivio Storico Olivetti, Ivrea.

representing the Italian Foreign Ministry, including Boccardi. According to local informants neither Ycon nor EOPOCS played a large role, although al-Hadrami worked on the project.

The last phase of the project was the restoration laboratory and workshop "in an abandoned and deteriorated zone of the same quarter [which will] accommodate and promote all the activities linked to the conservation of the old city …."[101] The workshop would train both architects and builders. The skills of the latter were assumed to have declined; the project thus accounted for

> the present state of preparation of the traditional workers (the 'ustad', responsible for the survival until today of the residential area of Sanaa al-Qadima) and the possible interest of recovering the ancient trade within an institutional context that can guarantee the conserving of the level and which will see to the training of young workers.[102]

The building—later named the Center for Architectural and Technical Studies (CATS)—was built on the site intended for the community center, which was not executed. The ruined building and land turned out not to be abandoned: as the project got underway, some 50 persons claimed ownership. Most were poor, and easily agreed to sell their shares; but one owner who owned a house on the site refused to sell. The new building was a concrete structure, in a design intended to be sympathetic to local architecture (Figure 6.8). But most local architects did not see it that way. The Italians sought a complete contrast between the

101 G. Boccardi, "Projet de la Coopération" (translation mine).

102 "Action Proposal," January 1986, p. 10.

new building and its surroundings, in keeping with the Venice Charter.[103] "I believe in this idea, but not here, because of the culture," said 'Abd al-Hakim al-Sayaghi. "It will harm the International Campaign. We are an official body, trying to guide the people. What we do will be taken as an example." Indeed, residents sometimes point to CATS in exasperation: "They tell us not to use concrete, and look what they did!" The programming of the building did not turn out as planned. A research lab and training center were installed on the ground floor, but the equipment was not delivered at the end of the project. Few workshops in restoration have been conducted there. Moreover, Quaroni Bonifica's vision of technicians and *usta*s running the center jointly was probably idealistic.[104] CATS opened in 1995 as a documentation center for GOPHCY, but was not operational.[105] In the mid-2000s it housed two other departments, Projects and Planning; a small auditorium is used for meetings and staff training.

Work at the urban scale included street paving, the upgrading of utilities, and minor facade repair. In his reports Lewcock had recommended that electrical and telephone lines, which criss-crossed house facades, be removed for aesthetic reasons and professionals at the conservation office agreed with him. The Italian project planned to do this, and laid conduit under the new paving for this purpose. But the wires were not moved underground, because it was not clear who should bear the cost of the work.[106] Street lighting was supposed to be part of the project, but was not done.[107] According to a foreign consultant, certain residents took advantage of the project. When the contractor's men went into houses, some residents asked them to make repairs, and the workmen complied on a good will basis. But some residents were demanding: when workmen refused to do what was asked of them, they got angry. In two cases residents fired on the workmen and the supervising architect—not an uncommon occurrence in Sanaa.[108]

---

103 Quaroni Bonifica's outline of the project invokes this principle of the Venice Charter: "... every innovative component, with regard to the material culture of that specific typology, must remain distinct in terms of technique and materials, avoiding any imitation of the characteristics thereof" ("Action Plan," January 1986, p. 30). The building originally consisted of a ground floor and two stories; an additional floor, designed by architects at GOPHCY, was added in the late 1990s. The building served an unintended purpose: it was used as a bunker during the civil war, since it was the only concrete building in the area.

104 At the meetings that preceded the signing of the agreement, the Italian team described the training of craftsmen at the center. An EOPOCS official wrote a question mark in the margin, crossed out craftsmen, and wrote "draftsmen." He wrote in the margin, "We prefer this."

105 When the Dutch fact-finding mission wrote its report in 1997, CATS was not functioning ("Preservation of Historic Cities in Yemen," p. 31).

106 Apparently the electric company expected payment for the work from the Italian project; the Italians thought it should be an input from the Yemeni government.

107 'Abdullah al-Hadrami, interview; G. Boccardi, "Projet de la Coopération Technique Italienne," p. 52. A small number of street lights were fabricated and installed in Mu'adh quarter after the design of a foreign consultant. Circa 1988 al-Hadrami proposed that a factory be created at Samsarat al-Dhamari, to produce street lights inspired by old fixtures in the *samsara*s. Funding was sought and obtained from the UK, and construction drawings were drawn for the *samsara*. Although the equipment for the factory was delivered, the project never got off the ground (*Mashru' masna' fawanis*, project drawings on file at GOPHCY's Planning and Projects Offices). Samsarat al-Dhamari was restored in the late 1990s, but some professionals disapproved of the work, saying that it was substantially rebuilt when it could have been restored. In 2004 it became a retail outlet for women's handicrafts.

108 A team of workers and SFD staff were shot at as recently as 2002, in connection with the restoration of a *maqshama*. Lewcock himself was caught in crossfire on one occasion, in a house at the edge of the *suq*.

Suspicion, and even hostility, were not new. But they may have been exacerbated by another aspect of the Italian plan: the attempt to reorganize outdoor spaces according to a new "collective" concept. Local professionals and some foreigners worried that the plan was insensitive to local culture. Vehicular access would, after all, be permitted within the neighborhood. One street would run along an alleyway on the west side of Bayt Sari', and would continue through a series of ruined courtyards (*ahwash*, sing. *hawsh*) to the new workshop. The courtyards were to be demolished and incorporated into the street, or used for parking. Susanna Innocenti, a UNESCO consultant resident in Sanaa (an Italian, but not connected with the Quaroni Bonifica project) expressed her concern in a report to the Cultural Heritage Division in Paris. The project was intended to provide a model for future work, she wrote. Permitting traffic through narrow alleyways where none existed would set a bad precedent. These alleyways, along with the courtyards that abutted them, had important social functions, especially for women and children. Ruined areas could be cleaned up and squatters removed, but the courtyards should be left as they were, and the alley reserved for pedestrians:

> The *hawsh* is one of the typical characteristics of the Sanani style and it must be preserved ... It should be stressed that not all the technical staff of EOPOCS agree with the proposed solution ... We must ensure that this example, to be repeated in other areas, is a good one ... the study for the proposal was probably not prepared in an adequate time spent to understand the life in the quarter and with adequate knowledge in local architecture.

Her report was accompanied by a map illustrating the proposed route, and photos of the gated alleyway and courtyards.[109]

Innocenti's views were echoed by conservation professionals I interviewed. The *hawsh* and the alley, said Jamil Shamsan, are part of a spatial system designed to ensure familial privacy. Women use the areas as an extension of their homes for certain household tasks and to socialize with neighbors: they are veiled, but they feel free to spend time in this space, since no strangers pass by. "The Sanaani house doesn't have the bent entrance (*al-madkhal al-munkasir*) that you find in other Arab towns, which prevents passers-by from seeing inside," said Shamsan.

> Instead, we have the court (*hush*); the locked quarter (*al-harat al-mughlaqa*), and the cul-de-sac (*al-shari' al-mughlaq*). If you happen to go into these areas, people will tell you to get out. They'll say, 'this place is locked' (*hina mughlaq*), or 'there is no way through' (*ma fish tariq*).

The Italians had only been in Sanaa for six months, he said. They may have studied the physical layout and the economy, but they didn't understand the city's social relations:

> They wanted to make the neighborhood like the French Quarter in New Orleans ... EOPOCS staff knew that these kind of spaces existed in Europe, and they knew they wouldn't work in Sanaa. They told the Italian team this. But the Italians persisted: they tried to convince the residents of the plan, but they refused. A few told them directly: 'We're not in Europe.'

---

109 Report to M. Bouchenaki, Chief OPS/I, UNESCO, date not visible; written before June 5, 1989, the date of Bouchenki's letter to Lewcock, to which the report is attached.

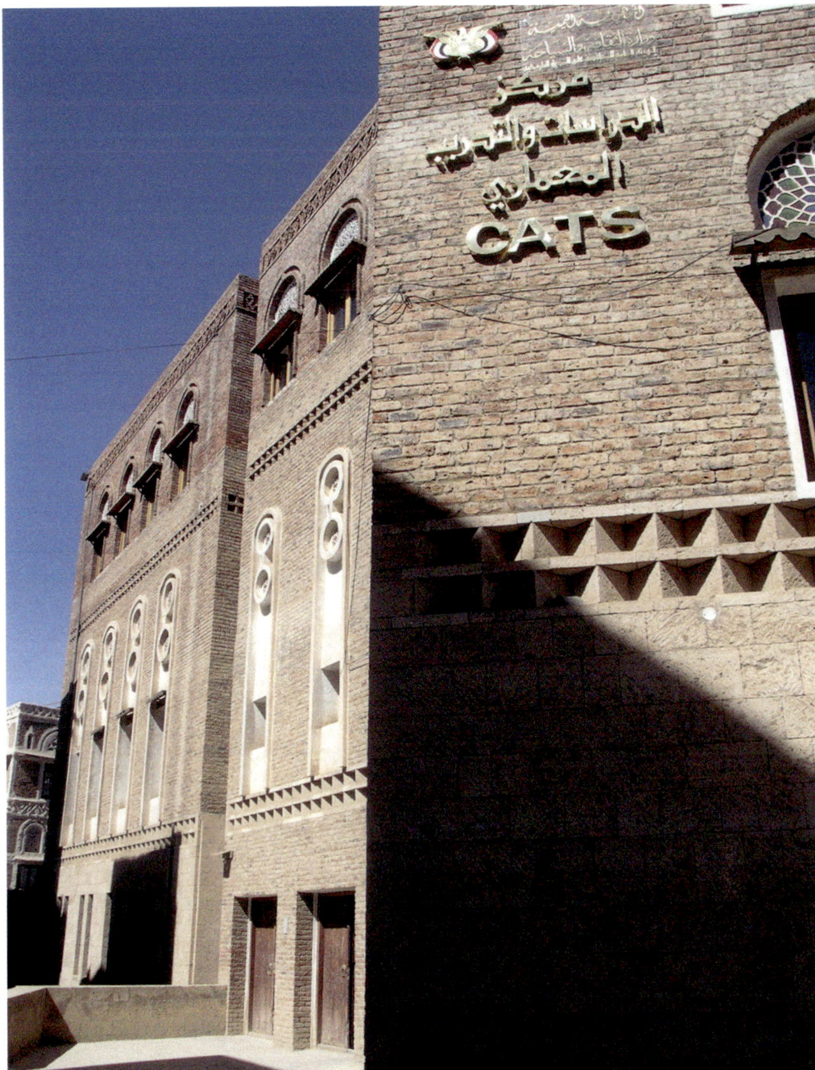

Figure 6.8 GOPHCY's Center for Architectural and Technical Studies (CATS), by Quaroni Bonifica

Shamsan, who is from Lower Yemen, said the plan might have worked in Aden where social relations are more open, but not in Sanaa.

This seemingly closed and hostile environment frustrated the Italian team—as is evident from Boccardi's opinion piece in *Chroniques Yéménites.* Reiterating the goals of the project—to address the problems of the old city, to provide replicable models for restoration and urban conservation, to instill a sense of pride in local heritage—he admitted that now, as the project neared its end, only some of these goals had been realized:

> It is perhaps instructive to emphasize the great difficulties encountered at the local level by the Italian project during the course of its implementation. A sometimes difficult collaboration with the local administration, the mistrust of the population and its acute sense of private property, have largely undermined the successful progress of this project and threatened its realization. All these factors impeded the correct installation of planned urban infrastructure and have compromised the … effective use of public spaces. And if this 'insensitivity' and lack of participation [on the part of the Yemenis] in fact constitute the reason for the Italian presence in Yemen, the main question that poses itself today is … whether a simple maintenance operation really justifies so much expense, the signing of international agreements, and the presence of such a large number of technicians, experts, and supervisors. The misunderstanding is perhaps born of a utopian and ideological vision, a 'pasolinian' vision one could say, concerning an active Yemeni participation in the safeguarding of their own artistic and architectural heritage. In reality, the possibilities for intervening in the whole of the urban fabric are practically nonexistent. The Italian intervention would have doubtless been better perceived if it had been limited to saving a specific monument, concentrating its effort like France and Germany, on particular objectives: the city wall or one of the caravansarays.[110]

Earlier in the same issue, the author described the restoration of Bayt Sari' as "integrating and improving traditional Yemeni construction techniques with the help of modern methods, respecting the original structure of the building and guaranteeing its longest possible conservation." The use of traditional techniques, less and less used nowadays, required research and analysis that allowed the production of a kind of manual, which could be used in the restoration of future buildings.[111] But in the opinion piece that followed this report, he admitted that the high cost of the work and the use of overly complex techniques could not but discourage homeowners from emulating the work. Moreover, they would not be able to visit the project since it was in private hands.

> These economic factors and lack of knowledge about local problems are not the only issues … From a theoretical point of view, this elevated conception of restoration is not suited to the architectural reality of the country. Yemeni architecture makes use of very fragile materials, and has undergone over the course of centuries all kinds of modifications that are not easily dated and are difficult to reference—in conformity with an architectural history that is characterized by little monumentality, and the absence of great technological and typological transformations. It does not require recourse to advanced technologies used in the restoration of monumental buildings.

A decade later, Boccardi reiterated many of these points in an interview. But he also pointed out positive aspects of the project. It anticipated reuse of the tower house as a multi-family structure, transforming each floor into an independent living unit.[112] The interior was carefully designed to incorporate lighting and other modern features in a way that offset traditional finishes and elements.

---

110 G. Boccardi, "Point de Vue" (translation mine).

111 G. Boccardi, "Projet de la Coopération Technique Italienne" (translation mine).

112 See sketch floor plan, "Apartamento autonomo per una famiglia," no date, Fondo Maria Luisa Neri, Archivio Storico Olivetti. Jamil Sari' and Jamil Shamsan both noted, however, that the division of the house by floor does not correspond to ownership, which is usually by room.

Both Neri and Boccardi believe that the heavy-handed structural intervention was designed by Bonifica's engineers, who had little knowledge of Yemeni architecture. The techniques used at Bayt Sari' were projected during the Proposal Phase (figures 6.9, 6.10).[113] Throughout the basement and on the upper floors, the inside faces of walls were stiffened with a layer of plaster or concrete, reinforced with wire mesh; a layer of concrete was applied to floors to stiffen them, embedded with wire mesh or steel elements. The stair core was also consolidated with reinforced concrete, and tied with a steel ring at the basement level. In some areas, exterior brickwork and stonework were "rebuilt section by section" using what was probably the traditional technique of *fals*. This kept the outside surfaces unchanged—even as the interior of the house was essentially rebuilt.

According to al-Hadrami, an old *usta* was employed for the masonry work, one of numerous Yemeni workers on the project. But others, including Muhammad Hamza, the *'aqil* of the quarter, recall that laborers were from Ethiopia. Several informants told me that during the course of the restoration work, no Yemenis—not even employees of EOPOCS staff—were allowed in the building. Jamil Sari' said that even his family was not allowed to enter. But he managed to get in on one occasion, with supervisors (*muraqibin*):

> I saw that they had demolished the ceilings and the walls; I heard that they had put wire mesh in the walls. They put concrete (*ciment kimawi*) and *qaddad* on the roof. They had excavated eight meters under the storage room (*harr*); maybe they were looking for things hidden in 1948 [the siege of Sanaa during the attempted overthrow of the imam]. This house is seven hundred years old. We thought that there must be antiquities (*athar*) in the house, that's why they wouldn't let us in. When I was little, scholars used to enter our house with cameras. Maybe their equipment let them see what was behind the walls.

"Why the secrecy?" asked Hamza. "It made people suspicious." Rumors were still circulating in 2004: while I was photographing the house, a neighbor told me that the Italians had stolen treasure from the basement. The idea that money and valuables are concealed in old houses is a widely held belief; it may derive from the practice of hiding one's wealth behind a stone, especially in the stair pier.[114] It figured prominently in an incident at another house, Bayt al-'Anbasa, which was being restored by a Swiss team. The public scandal that resulted in 1992 was doubtlessly fueled by events at Bayt Sari'.[115] The secrecy that surrounded the project may have come from EOPOCS rather than the Italians. "At this time, we were prohibiting the use of cement in the old city," said a former staff member, "but the Italians built a concrete structure inside the house."

Near the end of the project, in early 1993, the family forcibly reoccupied the house—leading to an official protest by the Italian Foreign Ministry and suspension of project funds. Although several versions of the story circulate, they reveal a common thread. Under the project contract (1986) EOPOCS had agreed to make available two

113 "The Urban Texture, Guide of Standard Measures, Summary of Example D, PIII 28." At this time Quaroni and DeCarlo were directing the Sanaa project. According to Neri, the structural approach contradicts the intentions of Quaroni and DeCarlo and their consultant, Eugenio Galdieri, an expert in earthen architecture with extensive experience in the Middle East (personal communication).

114 Thanks to Amat al-Razzaq for this point.

115 The Swiss team at Bayt al-'Anbasa was accused of stealing trunkloads of treasure ostensibly discovered in the building's foundations. The episode was widely covered in the press and had significant repercussions; it also reveals much about how conservation was perceived. See F. Mermier, "Sana, métaphore de l'Etat yéménite," p. 58, fn. 12.

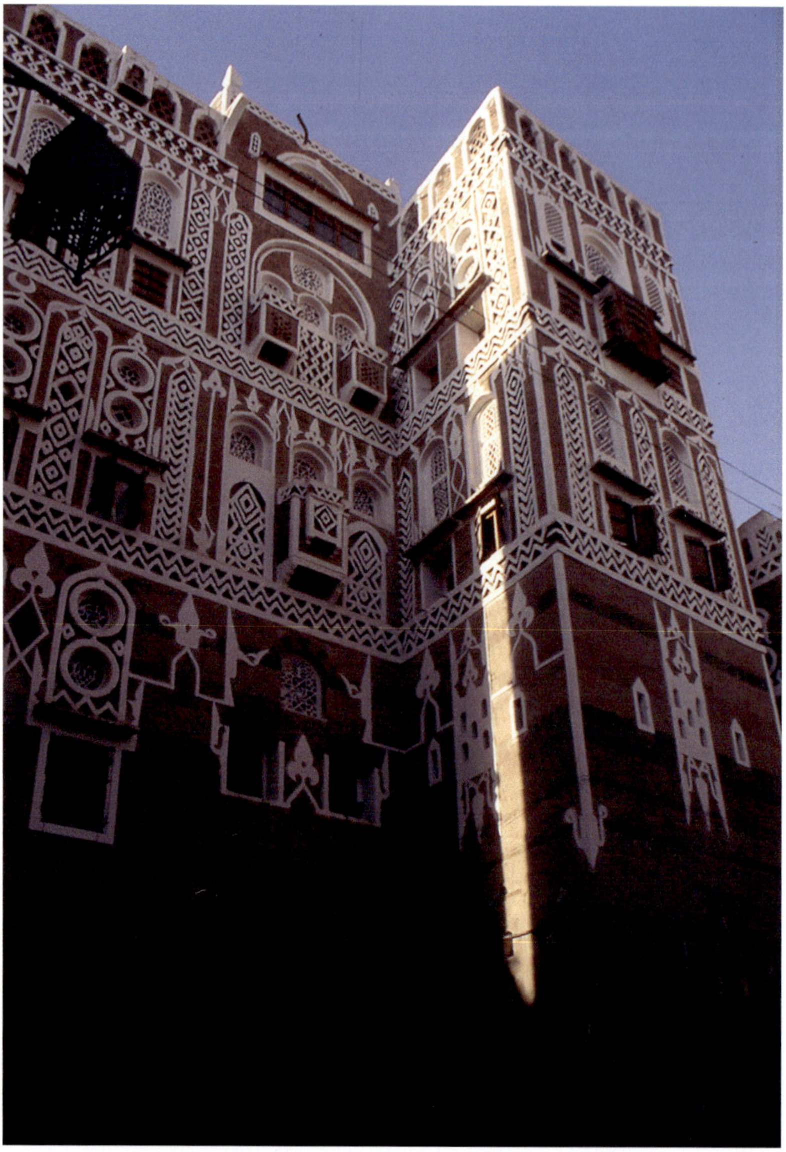

Figure 6.9 Beit Sari', Mu'adh quarter; restored under the Implementation Phase of the Quaroni Bonifica project

to four houses available for restoration. The houses would be acquired by the Yemeni government—since according to the rules of the Italian technical cooperation, work could not be performed on private property. But al-Haddad had a hard time convincing owners to sell. Under pressure, he managed to persuade the owner of Bayt Sari' to allow

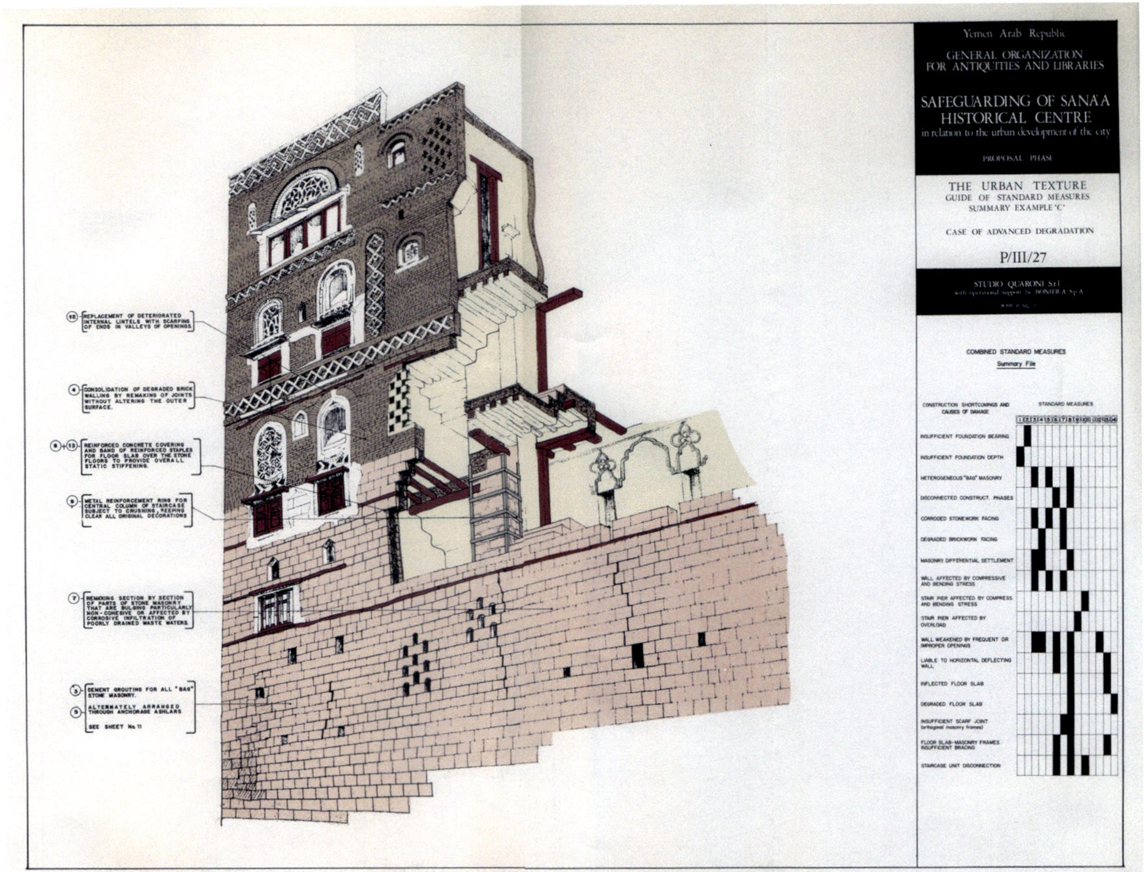

Figure 6.10 Proposed interventions for a house in an "advanced stage of degradation"

Source: Studio Quaroni S.r.l./Bonifica S.p.A, "Safeguarding of Sanaa Historical Center in Relation to the Urban Development of the City," Proposal Phase, 1981; reprinted with permission of Maria Luisa Neri.

his house to be restored.[116] According to one version of the story, the owner agreed to vacate the house and negotiate terms of the sale later; he agreed verbally to a selling price of YR 20,000,000. By the time the project was complete, the owner had died and the heirs did not want to negotiate.

The owner's eldest son, Jamil Sari', told me a different story. The contract signed by his father stipulated that the owner would take back the house once work was complete—with no obligation to sell. Because of this arrangement, EOPOCS paid for temporary housing for the family at a house called Bayt al-Wazir.[117] In 1993, as the project was nearing completion, the family saw an article in the newspaper that said their house was to be used as a center for conservation studies. "As soon as we saw this we rushed to the house with our weapons and stayed there," said Sari'.

> We knew that if the case went to court, we would have no recourse. Al-Haddad sent a soldier [probably a guard posted at the conservation office] to ask why we did this. We said, 'Tell al-Haddad he'd better not come here.' But al-Haddad did come with the Italians, and we showed him the newspaper article. We explained that we were afraid that the house would be taken from us, so we came back. The Italians understood, they were cultured and educated people (*muthaqqifin muta'allimin*).

The Italian team wanted to return to the house to finish outstanding work, but the family refused. They also refused an offer of USD 1,000,000 for the house from al-Haddad, who said the money would come from an Italian company. "We didn't believe it was true," said Sari'. "We thought it was a way for them to get back into the house."

Several informants offered a version of the story that confirmed Sari's. Al-Haddad, they told me, prepared two different contracts. The version presented to the family allowed them to return to the house when restoration was complete. The other version, presented to the Italian counterparts, said that the government of Yemen had secured ownership of the house; when restoration was complete, the house would be used for a public function, as stipulated in the project agreement. My informants believed that the Italian contractor knew there were two different contracts, but that the government of Italy did not.

Lamenting the fact that this project, which was supposed to be a model project open to professionals and the public, would never be seen, Franck Mermier wrote:

> The idea of purchase by the public authorities had always hovered beneath the project, but was never realized … [T]his failure is revealing, above all, of the quasi-improvised character of restoration policy that is designed blow by blow, according to the possibilities of financing. The foreign party, strengthened by its capital and technology, can impose many of its conceptions, but is sometimes confronted by local forces that cause the projects to deviate from their initial objectives.[118]

116 Yemeni informants told me that the Italian team—in particular, Gabriella Esposito—insisted on restoring Bayt Sari'; Maria Luisa Neri recalls that the Yemenis insisted on that house.

117 "Of course al-Haddad promised the family that they could return to the house," says a former staff member. "Why else would we have paid for temporary lodging for the family?"

118 "Sanaa, métaphore de l'état yéménite," in G. Grandguillaume, et al., eds, *Sanaa hors les Murs*, p. 51.

Most professionals disapprove of the techniques used in the project and of the way in which it was conducted. They fault not the Italians but their own leadership, which failed to set conditions on the project. "It's the fault of the administration of the Executive Office," said a former staffmember. "As for the techniques used, I can accept it as an experiment, but it can't be used as a model. That kind of effort can only be justified for an outstanding building." The project sent the wrong message to the community, said another architect, and worked against the goals of the Campaign. It suggested that conservation requires sophisticated techniques and foreign expertise, because local techniques were inadequate. At the same time, the secrecy that surrounded the project contradicted the notion that heritage was a public endeavor, for the benefit of the community. When the project ended, the Italian project director reportedly said with regret: "With that money, we could have restored so many houses."[119]

Later Projects

The UNESCO Campaign ended in 1994, but activities had been waning for several years.[120] There were two main reasons: the withdrawal of donor funding in 1991, after the Gulf War; and the government restructuring of 1990–1991 that transformed the conservation office into GOPHCY, responsible for all historic cities of a newly united Yemen. The office's mandate expanded just as its political influence declined: no longer attached to the Prime Minister's office, it was, in theory, a semi-autonomous agency within the Ministry of Culture. But in practice, like other government offices, it was weighed down by a heavy bureaucratic structure and had little fiscal control. With GOPHCY's decline, other players took up the banner of conservation. One of these was the Capitol Secretariat (*Amanat al-'Asima*, the Mayor's Office), which had been the nemesis of the conservation office during the Campaign. Directly responsible to the Prime Minister, the office became an active and efficient sponsor of conservation. Competition between the Mayor's Office and GOPHCY still existed in the mid-2000s, especially regarding the right to issue building permits in the old city, but there seemed to be an uneasy détente. There was even talk of restricting GOPHCY's role to that of an advisory agency, leaving implementation to the Mayor's Office—a solution that had been proposed by Lewcock and others in the 1980s.

Since the early 2000s, most major conservation projects have been sponsored by Social Fund for Development. Created in 1997 with support from the World Bank and a direct relationship to the Prime Minister, it quickly became the main development agency in the country. In the late 1990s, after a World Bank study for a major cultural heritage initiative was aborted, funds for the project were rerouted through the SFD and other donors began to follow suit. With the renewal of donor interest, there was a modest renaissance in conservation activities. Conservation came to be viewed as a development strategy: it creates jobs, protects cultural assets, and promotes tourism—a potentially important source

119 This would have been the Director of Works (*Direttore dei Lavori*) for the Foreign Ministry, Boccardi's boss.

120 I have not been able to find an official end date for the Campaign, but it appears to have ended 10 years after it was officially announced by UNESCO's General Director in 1984. "We thought the Campaign was supposed to last twenty years, not ten years," a former official told me.

of foreign exchange.[121] Cultural heritage gradually came to account for a greater proportion of the SFD's program activities, and eventually became a separate unit within the agency.

The SFD's cultural heritage projects have been funded through loans and grants from the World Bank, the EU, the Arab Fund (Kuwait), and individual donor nations, often with matching funds from the government. Typically government agencies like GOPHCY, GOAMM, and the governorates submit projects outside their normal operating budgets to the SFD for funding; for these agencies (as for donors), it is a way to bypass the bureaucracy of the ministerial system. These agencies partner with the SFD for implementation; in practice, this means secunding employees as consultants to the SFD. This is the only way to implement special projects given very limited government resources. Most professionals are pleased that the SFD has renewed opportunities in conservation, but some officials complain that it further undermines the efficacy of government.

The SFD has financed and implemented a wide range of cultural heritage projects, beginning with the restorations of urban gardens; these projects also included the restoration of associated structures like mosques, *marna*'s (well enclosures), and *birka*s (retention basins). The SFD has also sponsored the restoration of monuments like the 'Amiriyya *madrasa* in Rada, the Ashrafiyya Mosque in Ta'izz, and most recently, the Great Mosque of Sanaa. At the time of this writing, the Great Mosque restoration was the only ongoing project in Sanaa; others had stopped, largely because of decreased donor funding after the 2011 uprising.[122]

As of the mid-2000s, the SFD had made only limited attempts to address the issue of house maintenance, because donor aid generally disqualifies private property. But professionals pointed out that maintenance of the everyday fabric, including the houses, was critical. This required a change of mindset among donors. 'Abdullah al-Dailami, then Director of the Cultural Heritage Unit, thought that projects could be structured so that donors would be willing to finance them.

> There needs to be a fund to maintain houses, and incentives or subsidies for electricity and water. We should reap the advantage of 'international value.' Why should residents pay for expensive maintenance, especially when they aren't allowed to expand their houses? … Old cities should not be seen as private assets. In the end they become public assets, so they should be eligible for grants.[123]

The idea of the historic city as a public asset figured in an innovative project for Shibam-Hadhramaut, a World Heritage site known for its magnificent earthen tower houses (Figure 6.11). The project, which received an Aga Khan Award in 2007, was launched with funding from the SFD and GTZ, the German state development agency (now GIZ); the local branch of GOPHCY acted as counterpart. This is a complex project with many components; I will discuss only a few here, to provide a basis of comparison with Sanaa.[124]

---

121 "In Italy historic cities are our 'oil'," said Daniele Pini in a conversation with local professionals (Cultural Heritage Inventory Workshop, Sanaa, July 2004).

122 N. al-Makaleh, personal communication.

123 In addition to the project for Shibam-Hadhramaut, to be discussed below, al-Dailami said that homeowner loans were made available in Zabid and Sa'ada through the Dutch-funded UCHP. When that program closed in 2004, the SFD was asked to take it over.

124 This section is indebted to discussions with 'Omar 'Abdulaziz Hallaj, former GTZ project director; Jamal Bamakhrama, director of the local GOPHCY office; and *usta*s Salih Bahadaila, Khamis Baraman, Sa'id Basowaytin, and 'Ali Murbash; and "Rehabilitation of the City of Shibam,"

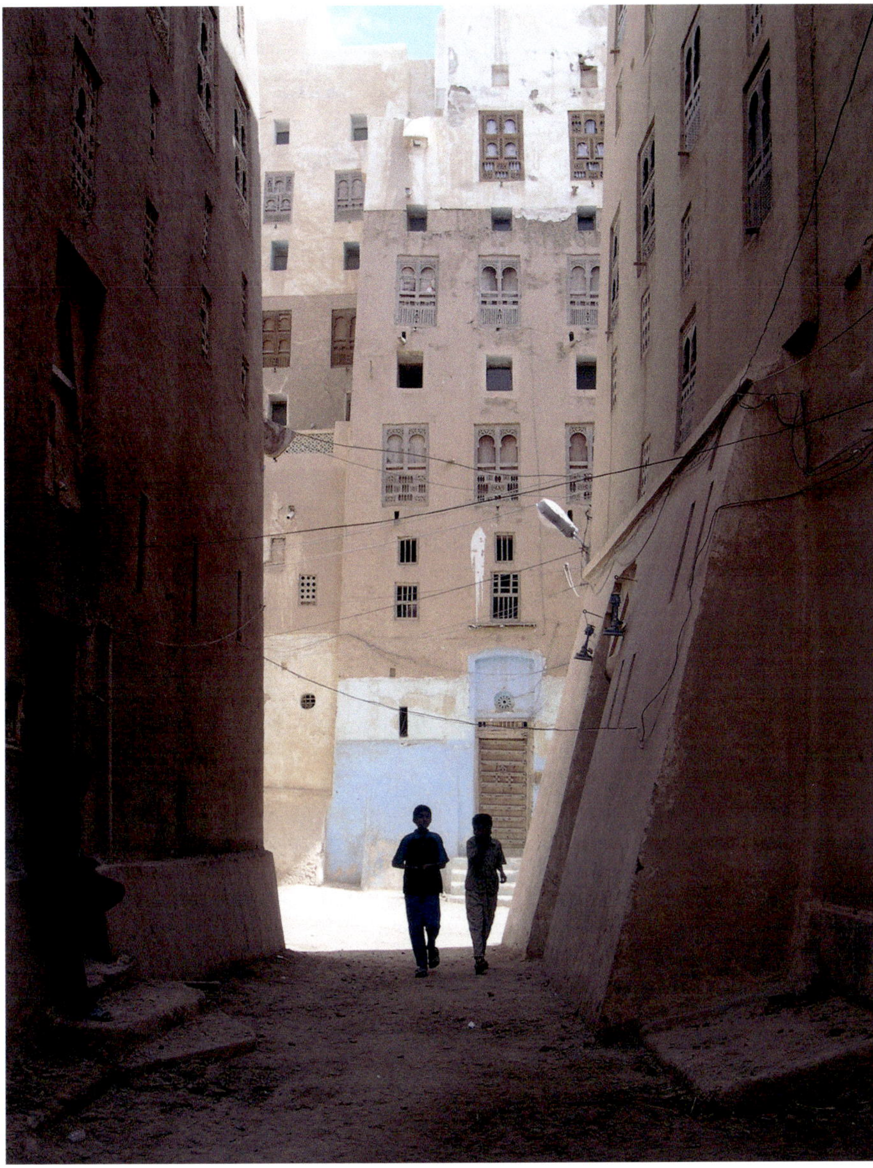

Figure 6.11 A street in Shibam-Hadhramaut

Economist Burkhard von Rabenau developed an argument for assisting homeowners. Listing a site as historic is a kind of indirect tax on residents, he argued, because they are denied development rights and are expected to bear high maintenance costs. It is only right that part of the tax be returned to residents; usually described as a "subsidy," to compensate residents for the costs they bear. The amount of the subsidy is based on a number of factors, including the difference between work inside an historic area and work outside it, and risk involved for the builder. In the case of Shibam, this produced a cost differential of about 50 percent.[125] In the case of the GTZ project, residents were asked to pay 70 percent of restoration costs. Homeowners I interviewed in Shibam felt this cost was too high. Nevertheless, interest in the project seemed strong and *usta*s called it a "benefit to the city (*maksab al-madina*)." As of December 2004 some 60 contracts had been completed, another 20 were underway; 30 homeowners were on the waiting list. In addition, the project had documented around 30 percent of the houses in the city.

According to 'Omar 'Abdulaziz Hallaj, the former GTZ project director, they did not provide "models" for restoration, but rather emphasized conservation as a process. For this reason, they did not begin with historically significant buildings. Work was initiated at the request of the homeowner (or in some cases, the tenant) who determined the scope of the intervention. A team composed of architect, engineer, and *usta* (the latter usually worked as a consultant to the project) assessed the work to be done, and prepared an estimate. The estimate was calculated according to square meters of work, based on a database of prices developed with *usta*s. The idea was to build their confidence in quantity estimating.[126] The owner signed a contract with the project office, and then worked with an *usta* of his choice. The owner began work with his own funds; the project reimbursed him for 35 percent of each draw.

The project employed various strategies to ensure the quality of the work. In one case, the project retained a retired *usta* to supervise the work of younger builders. In other cases, it helped younger men secure contracts by partnering them with experienced *usta*s. The younger builders knew mud construction but had less experience with the buildings of Shibam. The massive tower houses are 100 to 150 years old; as in Sanaa, structural problems are largely due to the introduction of piped water. The project worked with *usta*s to develop standards for maintenance and building reuse, said Hallaji, without imposing norms.[127] Partial demolition and rebuilding, which has always been a normal part of house

---

Intervention Architecture: Building for Change (Aga Khan Award for Architecture, 10th Cycle), London: I.B. Tauris & Co Ltd, 2007, pp. 34–49. See also J. Feiner, "Study Report for the Preservation of Old Shibam," GTZ/GOPHCY, 1995.

125 According to Hallaj, the difference between repair work in historic and new districts was calculated at around 40 percent. The figure includes not only costs of materials and techniques, but transport costs: in historic areas access by vehicles and machinery is often impeded by narrow streets, small windows and doors, and so forth. To this figure of 40 percent, a risk factor of 10 percent was added.

126 The database consists of prices for various types of work, based on time and materials that *usta*s say they use for various tasks. As in Sanaa, *usta*s are accustomed to pricing work by the day, that is, according to times and materials. But the *usta*s I interviewed said that the project team's estimates were relatively accurate. The estimate includes a 20 percent contractor's fee; if the owner managed the work himself, he kept the contractor's 20 percent fee.

127 GOPHCY transcribed these principles in a report of 2001 (in Arabic) with photographs. It covers guidelines for construction and layout; specifications for wall, floor, and ceiling assemblies and finishes; and so forth.

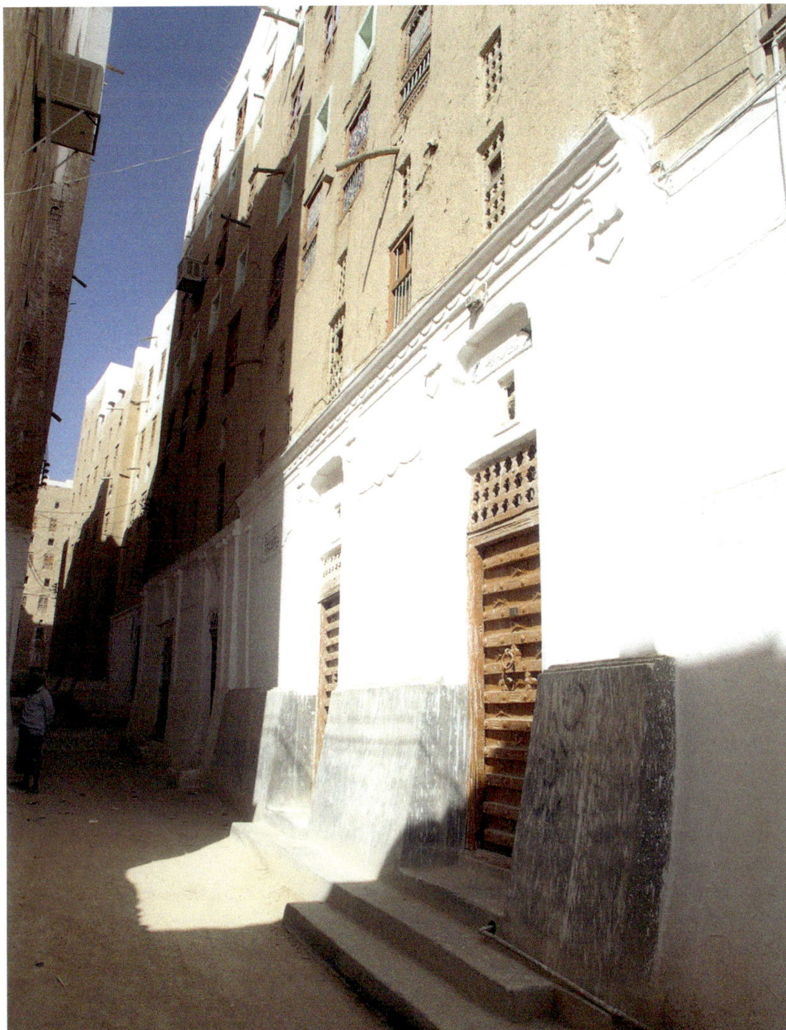

Figure 6.12 A restored house in Shibam with a new neoclassical cornice. Classicizing features have long been part of *ustas*' repertoire in the Hadhramaut

repair, was fairly routine.[128] It was expected that the *usta*'s materials, techniques, and details would change over time and on a case-by-case basis (Figure 6.12). The project encouraged

128 I asked one builder if demolition was prohibited. "No," he replied, "if there's a lot of damage it must be rebuilt." In the old days, said an elder mason, if an owner wanted to demolish his house, the *'aqils* recorded the position of windows, drains, and other features that affected neighboring houses.

retention and restoration of significant historical ornamentation, while allowing flexibility in the majority of cases. The most important thing, said Hallaj, was to guarantee the future supply of master builders, and the viability of mud construction. The project therefore supported the apprenticeship system, with minimal interference.

The project benefited from and supported a strong and self-confident cadre of builders, specialized in earth construction. Largely due to cost and availability, mud has remained the preferred construction method in the Wadi Hadhramaut. New building types in mud dot the valley—large, wide free-standing buildings, with plain, rhythmic openings that are reminiscent of the tower houses. Forms and techniques change according to client preferences and the availability of materials.[129] Because of the fragility of mud, existing structures are often rebuilt—so ongoing maintenance provides a constant source of work. These factors have contributed to the economic strength of mud building, to the evolution of types and techniques, and consequently, to continuing prestige of the *usta*. "It's not like *usta*s are sitting around, waiting for restoration work," said Jamal Bamakhrama, director of the local GOPHCY office. "They will tell you they're too busy." The project helped to establish an association of mud builders, apprentices, and laborers.[130] Yet the future of mud building is uncertain. Although concrete is poorly suited to the extremely hot climate, many people choose it over mud. *Usta*s believe that over time, concrete will displace mud as the preferred building material—in part, because homeowners want more space, and concrete allows much thinner walls (20 cm, as opposed to 45 cm for mud).[131] Some were not anxious for their sons to continue in the trade, which is hard and tiring. "I want them to write," said the older *usta*—that is, to work in an office, as an employee or professional.

Perhaps because of their strong position, *usta*s seemed to be treated as partners in conservation activities to a greater degree than in the north. According to an architect from Lower Yemen who worked on the Sanaa Campaign, *usta*s are treated as equals because there is a more enlightened mentality (*'aqliyya*) in the south, formed under more than two decades of socialism. There may be some truth in this view: Lewcock recalled that when he was laying groundwork for the two UNESCO Campaigns in the early 1980s, community meetings were a normal part of the process in Shibam, but were discouraged by officials in Sanaa.

During the Sanaa Campaign, reports by Lewcock and the International Working Group repeatedly called for a loan bank and other forms of assistance to homeowners. Some officials, however, felt that a loan bank would not work in Sanaa because it would not be respected. Even in Shibam there were problems. "Citizens expected the government to renovate and restore, to do everything for them," said Bamakhrama. "Now they're starting to get the idea that they have to contribute." Professionals at the SFD and GOPHCY doubted

---

The house was then rebuilt exactly as it was. "Now the conservation office (*al-hai'a*) enforces these rules, instead of the *'aqil*s."

129 Concrete beams may be used in mud structures; plumbing pipes (*qasab anabib*) are often used for framing because they resist insects and allow longer spans than the traditional wood *'ilb* (botanical name: *zizyphus spina-christi*). In any case, *usta*s told me, *'ilb* has become scarce.

130 The association was established in 2005. See M.A. al-Kadri, "Final Report: On Performance of Shibam Historic Houses Program 2nd Phase," Housing Unit/GOPHCY-Shibam, April 2006, p. 2; "Rehabilitation of the City of Shibam," p. 38.

131 A resident of Sanaa told me that his father built a new house of concrete in the old city in the 1970s for the same reason: concrete allows thinner walls than stone, and thus more interior space. On *usta*s' views of the future of mud building, see Caterina Borelli's documentary film, "The Architecture of Mud" (1999).

that a participatory project like the one in Shibam would work in Sanaa. In the mid-2000s, homeowners in Sanaa had access only to a limited and informal grant program: funds earmarked for homeowner compensation—in cases of damage from street paving or burst pipes—were diverted to provide small grants for house repair. But the number of grants did not come close to meeting needs; nor were the amounts sufficient to cover repairs. Even if citizens could be persuaded to participate in a project, there would be other problems to overcome. There are many more players and layers of bureaucracy in the capital, and they all expect "cuts" from a funded project. But the biggest problem is scale: Sanaa is 10 times the size of Shibam, with some 5,000 houses to Shibam's 450.

The Restoration of Façades on the Wadi Sa'ila (2003)

Despite these obstacles, a pilot project for house restoration and repair was undertaken in Sanaa in 2002–2003. In contrast to the usual process, whereby projects are proposed to the SFD by GOPHCY or other government agencies, this project was initiated by the SFD's director, 'Abd al-Karim al-Arhabi.[132] It was conceived as a project of "facade improvement" to beautify a stretch of the Wadi Sa'ila—part of preparations for Sanaa's year as Cultural Capital of the Arab world (2004).[133] From the time of the Quaroni Bonifica studies, the Wadi Sa'ila had been seen as an appealing area for tourists. Various proposals for paving and upgrading the Sa'ila had been considered since the mid-1970s, but it was not paved until the mid-1990s. While the upgrading the Sa'ila was primarily to control flooding and improve conditions in the old city, officials frankly admit that it was conceived as a project to attract foreign tourists.[134] The planning of cultural and recreational activities along the wadi were designed, at least in part, with tourists in mind. The houses along the Sa'ila, as we saw in Chapter 4, provided a picturesque backdrop for the controversial open-air theater in the Bustan al-'Amri, created by the Ministry of Culture in preparation for the 2004 events.

A study of existing elevations was carried out in the summer of 2002, with the goal of identifying problems. Under the pilot project a small number of houses would be restored, selected according to the interest of owners and the architectural and aesthetic quality of the structures. Preference would be given to houses overlooking the Sa'ila; if the project was successful, it would be replicated throughout the city. The project was implemented in coordination with the Mayor's Office; the project counterpart was the elected Local

132 Al-Arhabi served as Minister of Labor and Social affairs in the 2001 government; in 2006 he was appointed Minister of Planning and International Cooperation (Burrowes, *Historical Dictionary of Yemen*, pp. 35–6).

133 An article in the English-language newspaper *Yemen Observer* reported: "The cleaning and restoration of these highly visible houses comes amidst preparation for Sanaa as the Arab Cultural Capital, 2004. Historic landmarks around the city will be restored to their original condition" ("Maintaining the historic Al-Saila facades," 5/3/03, p. 1).

134 For example, the director of Planning for GOPHCY (T. Marchand, "Walling Old Sana'a," p. 50). The project was funded through the U.S. government's Title 3 "Food for Development" program. The U.S. donated wheat to the Yemeni government, which sold it on the open market; proceeds were used to finance the project. (Samsarat al-Mansura was financed in part by a similar German program; see Chapter 3.) This first phase of the Sa'ila improvement was for the area within the old city; subsequent phases improved the Sa'ila to the north and south.

Council, which represents most voting districts in the old city. 'Abdullah al-Hadrami acted as supervising architect for the SFD.

The project documents note that Sanaa's status as world heritage (*turath 'alami*) derives from the unique decorations of its house facades and from the planning of its streets and urban gardens, which should therefore be conserved. Articles in the local press emphasized the project as one of beautification: exterior features like *hizam*s (decorative brick courses), *zakhrafa* (ornament, especially plaster), and *qamariyya*s (arched window lites) would be repaired, and unsightly utility lines and pipes would be relocated. In an interview with a local newspaper Fahd al-Alaya, director of the Local Council, said that the goal of the pilot project was

> to return the facades [of four to five houses] to their prior condition and to remove modern innovations (*al-istihdathat al-jadida*) like new openings, windows, and awnings, and defacements resulting from random acts (*'amal ... ashwa'iyya*) … [These changes] have ruined many houses that are considered historic monuments (*al-ma'alim al-athariyya*) … [The project] will restore to old Sanaa the beauty and splendor of its original decoration (*liya'udu li Sanaa al-qadima rawnaqiha wa jamaliha al-zakhrafi al-asil*).[135]

In this article and in local discourse generally, the term *ashwa'i*, "random," is used to indicate contemporary changes; the word, which also means "blind" or "senseless," implicitly contrasts these changes to the inspired, premodern order that produced the old city.

This language is echoed in SFD project documents: in this and other projects, work is justified by the endangered status of cultural heritage, invoking Sanaa's status as a World Heritage site. The project aimed to eliminate defacements that threatened the survival of Sanaa's architectural style. But the intent, al-Hadrami told me, was not just cosmetic: it was to restore the "spirit of the building (*ruh al-mabna*)." The project was not limited to facades, but aimed to correct structural problems that caused damage to the facades. It was a much-needed attempt to address the conservation and maintenance of houses, involving homeowners and the community in the process. In contrast to the GTZ-SFD project in Shibam, owners were expected to pay only 30 percent, with the remainder covered by the project.

In March 2003, applications were solicited from interested citizens; according to al-Alaya, they were evaluated according to their ability to contribute to the costs of restoration.[136] The project began in late May and took several months to complete. Five contiguous houses were selected for the project on the east bank of the Sa'ila, north of Talha bridge (Figure 6.13). The owners of one house were suspicious of the project and did not want to participate; according to the supervising architect, two additional houses were added at the northern end, for a total of six houses.[137] U.S. $30,000 (from the Arab Fund) was available for the project, and of this amount $26,000 was spent, an average of $4300 per house. Stonework and brick were repaired and repointed; plaster and exterior woodwork were repaired. The entrance of one house was stabilized and rebuilt; in another, windows that had been closed with brick were reopened, and a new wooden door in the "old style (*'ala shakl al-qadim*)" was made and installed. Electrical wires were consolidated and, where possible, plumbing pipes were hidden in shafts; in the case of three houses,

135 "Tahsin wajihat al-manazil al-mutilla 'ala al-sa'ila," *26 Septimbir*, 3/13/03, p. 1.
136 "Tahsin wajihat al-manazil al-mutilla 'ala al-sa'ila," p. 1.
137 "Tahlil al-as'ar," SFD document, no date.

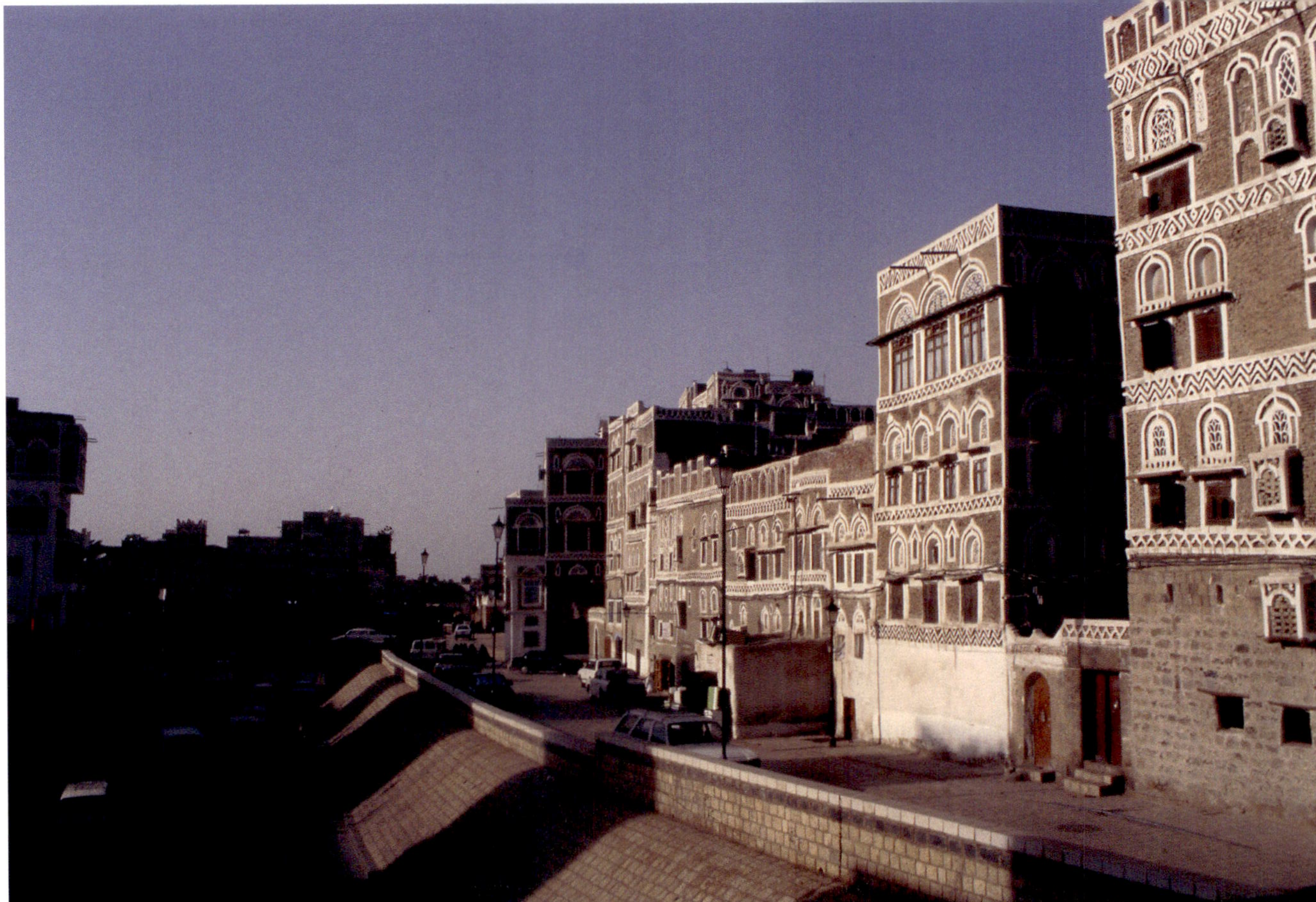

Figure 6.13 Houses on the east bank of the Wadi Sa'ila, restored by the Social Fund for Development under the supervision of 'Abdullah al-Hadrami

toilets were relocated to the rear facade. Water tanks were removed to the rear of the houses, and in one case, an arched parapet wall (*tijwab*) was built to hide it. This was a new use for the *tijwab*, designed to conceal women on the roof: "We adapted an element of the traditional system of the old city," said al-Hadrami. In this house, water damage to the first floor structure had deformed the facade; the affected portion of the floor was rebuilt and a modern bathroom was installed, enlarged by capturing part of the hallway. The owners, who allowed me to visit the house, were happy with the work. But they said that the rear facade, facing the garden (*bustan*), was also in need of repair; unfortunately they did not have the money for this work. "They couldn't even pay their 30 percent on the work we did," said the contractor.[138]

An important part of the project was "capacity building": *usta*s were to be trained as contractors, and the Local Council was to be trained as counterpart (*al-jihha al-kafila*). Both were to be assisted by the SFD's supervising architect.[139] The elected Local Council (*majlis al-mahali*) and various NGO's (*jam'iyyat*) had been involved in conservation-related activities in the old city, but their work had declined in recent years.[140] In the Sa'ila project, the Local Council was supposed to manage construction and secure payments from the homeowners, but according to SFD officials, their performance was disappointing on both counts. Homeowners did not pay their 30 percent contribution, and in the summer of 2004, the Local Council was being held accountable for the funds. Two contractors reported that they had not been paid, more than a year after work had been completed.

As we have seen, government and donors require the use of general contractors, but they are often unfamiliar with traditional building. Most *usta*s cannot qualify as contractors, nor are they familiar with the procedures of competitive tendering. For al-Dailami, this is the greatest dilemma of conservation. "*Usta*s have never submitted tenders, they don't have any bank assets, and so forth," he said. "Yet they are the only ones who can deal with historic buildings." The SFD tried to address this issue by helping *usta*s qualify as contractors. Typically a contractor must put up a security bond (*dhamana*) for 10 percent of the contract price; during the course of work 5 percent of each draw is withheld as security, usually for one year. In the Sa'ila project, the SFD waived the security bond and agreed to withhold 5 percent on only a few items. Al-Hadrami helped the *usta*s prepare bids, interpret drawings and specifications, and assist in scheduling—essentially training them as contractors, and putting them in charge of the process.

Of the three successful bidders, two were *usta*s who had not previously worked as contractors; the third was a general contractor who had no experience in restoration. One of the two *usta*s, a master of *qaddad*, was awarded a contract for three houses, and became the *de facto* supervisor on the other two. In an interview and tour of the site, he spoke of his work proudly. But he said he struggled financially, despite easier terms from the SFD. He did not have capital to start work. He asked for an advance, but they would not provide one without a security bond of $5,000. He finally managed to negotiate a small advance in exchange for a smaller security bond, but it was not enough:

138 The row of house restored under the project have their rear facades on Maqshamat al-Kharraz.

139 "Ittifaq itar al-'aml (agreement for scope of Work)," SFD document, no date.

140 The latter include "Friends of Old Sana'a," created in the early 1990s; but by the mid-2000s it was inactive.

> The money ended, and the work hadn't even begun. I bought supplies—*nura* (plaster), wood, and other materials. Now I need more money, but I can't get any without more security. If I were a real contractor, I would complete the work quickly, with my own money, then I would get paid. There is no easing of the situation (*ma fish tashil*) …
>
> … I got tired from this work. The work was exacting (*daqiq*) … I would bring *ustas* to do the work, and if they didn't do it right, I would bring someone else … The stonemasons did *fals*, and reused the same stone. Even the brick that was damaged, we changed with traditional (*taqlidi*) materials and methods. The doors were iron (*hadid*) so we changed them to *tanab* [the wood traditionally used for exterior woodwork]; some windows were aluminum and we changed them to *tanab*. As for the wooden awnings (*kunan*; sing. *kunna*), we made new ones. We did everything in the traditional style, like it was before (*kullu 'ala al-taqlidi 'ala ma kan al-awwal*). Some windows, we repaired. The [Social] Fund agreed that the windows should cost YR 50,000 but they cost more than that. There were other things we did not expect, like *fals* on the interior walls and arches. [All these things] increased my costs …

The *usta* hired the required trades according to al-Hadrami's recommendations and managed them: masons for stone and brick, carpenters, a tile setter, and a plumber (electrical work was performed under a separate contract). With the exception of the carpenter who was paid by the piece, the *usta* paid his subcontractors by the day; but he had calculated his own bid to the SFD by quantity. The work was very hard and exceeded the scope of his contract:

> I worked seriously—day and night. I even neglected my children. I used to sleep in the building. The other contractors didn't do this … they would take off. I was the responsible one … [The other contractors] brought their own supervisors (*mushrifin*), but they didn't know the work. The architect [al-Hadrami] let me supervise all five houses … Even if there was rain, I would come myself to cover things up and take care of the site. The workers always called me for everything, but nothing came to me [i.e. money].

In August of 2004, a year after the project ended, he had received most of his fee from the Social Fund, but the last five percent was still being held as security for outstanding issues (*dhamanat al-siyana*, "maintenance security"). His subcontractors were asking for their payments, but the money was held up (*muhtajaza*) in the bank. The *usta*'s lot is difficult, he said, because he has few resources (*imkaniyat*) and work is sporadic. "I lost money on the project, I lost my health," he says. "The only thing I got was exhaustion and reputation (*ta'b wa sitt*)"—that is, some people think he made a lot of money. When asked how he felt about the work and whether he would do it again, he replied:

> I am happy [with the work], my conscience is content (*dhamiri murtah*) … Yes, I would do this kind of work again. Now I have more background. I can do traditional work and I can supervise it. But I need encouragement and support. Then I would be ready to compete, I could do anything.

White-Washing the Old City (2003–2004)

In December 2003 the Ministry of Culture (MOC) initiated another and much larger scale project in preparation for the 2004 events: the whitewashing of facades throughout the city. Initially only houses and monuments on the Wadi Sa'ila and along the main thoroughfares were to be whitewashed, but according to the Ministry, the project was so popular it was extended throughout the old city. The budget for the project was YR 65,000,000, or USD 350,000, and the MOC would bear 100 percent of the costs. There were complaints that local experts had not been consulted, but the MOC said it was working with conservation authorities.[141] According to informants, the project may have been designed to compete with the SFD facade project, which the Minister of Culture thought was too expensive.

When I arrived in Sanaa in January 2004 to begin my fieldwork, protests were being heard throughout the conservation community. I toured the old city with architect Saba al-Suleihi, veteran of the UNESCO Campaign, who was dismayed at the city's "Disney-like" quality:

> Plastering reflects a social process: normally houses are plastered on special occasions, like weddings or for the 'Id feast. So the old city is always in a cycle of decay and renewal. To make it white and pristine, all at once, completely destroys its character.

The project was billed as a white-washing project, that is, the renewal of plaster (*guss*) decoration through the application of diluted plaster (*nura*). But in many instances old plaster was removed and new plaster applied. It was applied thickly, with a trowel rather than by hand, and finished against a straight-edge—an effect that contrasted sharply with traditional techniques (Figure 6.14).[142] The work was upsetting to professionals, and to some of the craftsmen who were executing it. When the Minister of Culture came to review the work craftsmen reportedly told him, "This is not how the work should be done." The Minister told the craftsmen to carry on and follow the instructions of their supervisors. Most shocking to professionals was damage to certain monuments, a point that was taken up some months later by a mission from the World Heritage Center. Craftsmen had been instructed to scrape away old plaster—in some cases with metal tools—which destroyed original ornamentation and damaged the surface of the brick. One of the affected monuments was the minaret al-Abhar Mosque (fourteenth century), where bands of geometric decoration and calligraphy were damaged.[143]

Architects said that plastering sent the wrong message to the community: it suggested that conservation is about decoration and image, rather than process. Those who worked on the SFD's Sa'ila project were especially disturbed by the project. "The work they did is

141 "Technical Report," World Heritage Mission to Sanaa, 24–30 July 2004, pp. 1–2. The MOC told the World Heritage mission that it had coordinated the project with two bodies: the High Committee for the Preservation of Old Sanaa (established circa 2000, and responsible to the Prime Minister); and the Coordinating Council of NGO's. It also said that the chairman of GOPHCY was on a panel that directed the project, and that eight GOPHCY technicians were used as supervisors. This was not confirmed by GOPHCY.

142 During construction *guss* is applied by hand at a thickness of .3–1.5 cm; in whitewashing, the diluted solution produces a layer of only .2–.5 cm, which is finished with a piece of leather (*shufra*). According to experts from the World Heritage Center who investigated the MOC project in July 2004 (see below), plaster had been applied up to about 15 cm in thickness ("Technical Report," p. 4).

143 "Technical Report," p. 6.

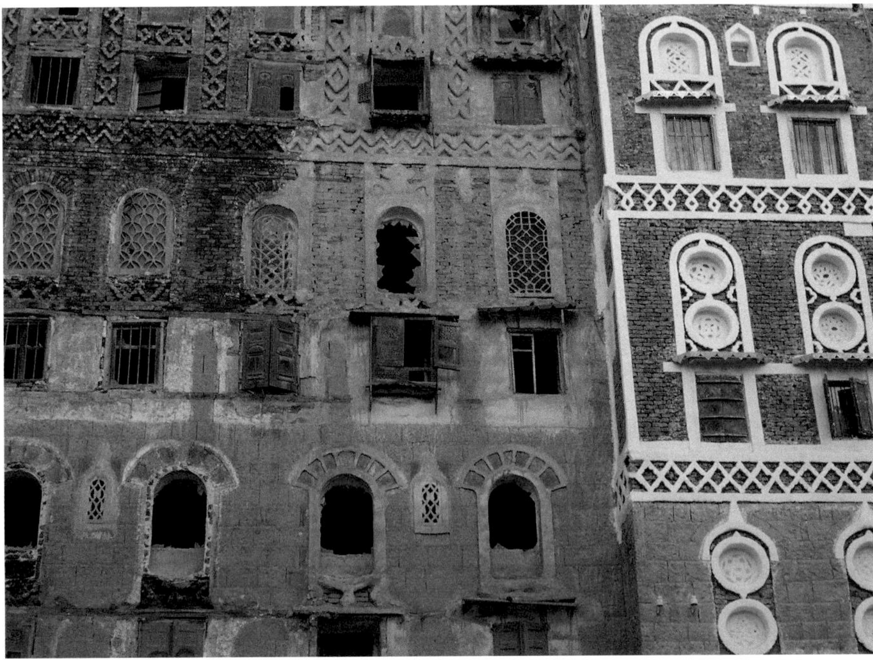

Figure 6.14 Houses at al-Abhar Square during the Ministry of Culture's white-washing project. On the left, plaster around window openings has been scraped; on the right, new plaster has been applied against a straight-edge

all exterior (*khariji*)," said a contractor, the one who had no prior experience in restoration. "We did the work as it should be done." An SFD official lamented that the project would adversely affect future endeavors. "We're trying hard to convince homeowners to contribute to restoration costs," he said. "Now people are spoiled, because the Ministry of Culture paid 100%—but it was all surface work, there was no substance." Al-Hadrami was philosophical. "The good thing is that plaster wears away," he said. "A few rainy seasons and it will be gone."

In February 2004, a noted foreign consultant and later, the World Heritage Center, were notified of the plastering project and were asked to intervene. The World Heritage Center wrote to the Ministry of Culture requesting information about the project, and asking that the work be halted until its methodology and results could be assessed. In June the Ministry agreed to stop work and invited a technical mission from the World Heritage Center to review the project. When a mission of two experts arrived in July 2004, hundreds of buildings along the Sa'ila and major arteries had been plastered.[144] According to several accounts, the visit was tense. Later in the week, the Minister organized and attended a meeting in

144 Some 800 buildings of the projected 3,000 ("Technical Report," p. 4). The two experts were Giovanni Boccardi (then Chief of the Arab States Unit at the World Heritage Center) and Jacques Feiner.

the old city with prominent citizens and the Coordinating Council (*majlis al-tansiq*) of NGO's. The group informed the experts from the World Heritage Center that the plastering project was very popular with residents of the old city, and should be resumed. The experts presumably raised some of the objections that they later outlined in their mission report. According to the World Heritage Convention, the Center should be consulted on any large-scale programs that might affect the image and state of conservation of a World Heritage site, and this had not been done. Moreover, no documentation, research, or technical studies had been undertaken in preparation for the project.[145] The local representatives expressed their desire to cooperate with the World Heritage Center. But plastering, they said, is a traditional practice, and they knew perfectly well how to do it; such work does not require studies or special supervision.[146]

The technical report submitted by the experts was sharply critical of the project, and at the same time conciliatory. The authors commended the Ministry for its work, which had the positive effect of making residents aware of the need to conserve the old city. At the same time, the lack of technical criteria and "an insufficient level of expertise and skills among supervisory staff and master masons" had a negative effect on the project. Monuments, important houses, and minor vernacular buildings were all treated in the same manner. This had a homogenizing effect that "contrasts sharply with the character of Sanaani architecture, which is known for its extreme variety." Significantly, the report challenged the right of the project and its supporters to represent "tradition":

> The simple reference to the 'tradition,' repeatedly evoked (sic) during the mission by the persons involved in the programme, does not seem a sufficient guarantee for the conservation of the outstanding universal value that justified the inscription of the property on the World Heritage List, considering the changes introduced by the rapid modernization of Yemen and the increasing demographic and commercial pressure on the Old City. Indeed, this so-called tradition appears to be subject to rapid evolution, depending on the changes in taste and the introduction of new fashions coming from the outside.

The report recommended that the plastering be resumed, with certain conditions: only accredited conservation architects and *usta*s should be used; buildings should be dealt with on an individual basis; and techniques should conform to traditional practice.[147]

It was clear that the Ministry intended to proceed with the work, and felt that it was authorized to do so. While the experts from the World Heritage Center could not stop the work, they sought various concessions. The Minister of Culture reportedly agreed that plastering methods would be modified: in particular, existing plaster would not be removed and brick would not be scraped. The mission asked for and was assured of cooperation on other matters: the government should act to halt large-scale violations that were appearing throughout the city; and the Ministry of Culture should consult with the World Heritage Center before undertaking future projects.

As signator to the World Heritage Convention, the state party is normally seen as the legitimate representative of heritage. For some critics, such protocols contribute to a reified vision of culture identified with the nation state. But in Yemen the persistence of traditional practice resists this kind of reification; there is thus a more complex dynamic between

145 "Technical Report," p. 2.

146 World Heritage official, personal communication.

147 "Technical Report," pp. 9–12.

local, state, and international actors. In asserting its role as representative and caretaker of Yemeni heritage, the Ministry of Culture undertook a project which, in the view of many, contravened local practice and threatened the "authenticity" of the old city of Sanaa.[148] With little ability to influence policy, local professionals sought the intervention of the body that underwrites the old city's status as "world heritage." In effect, they invoked the authority of the World Heritage Center over that of their own government. In the end, the World Heritage Center managed to gain certain concessions, on the basis of historical authenticity and precedent. In this case, then, an international agency defended local practice and the right of local experts to interpret "tradition." Among these experts were *usta*s and craftsmen, living representatives of the past who knew how the work "should be done."

The Rehabilitation of the *Marna'* of Talha (2002–2004)

As we saw in Chapter 5, local professionals generally see themselves separated from builders by a "cultural gap." Builders, they feel, are not sensitive to *turath*: they have been influenced by new materials and practices, and take liberties with historic monuments in an attempt to improve or beautify them. Architects do not accept international practice uncritically: they quietly dismiss the notion of "material authenticity" since they allow rebuilding, and allow new work to pass for old. But in theory, at least, they uphold the idea that a building is an historical document that should be conserved in its original state. As such, they restrict the kind of creative restorations that builders would undertake if left on their own. In the GTZ project in Shibam-Hadhramaut we saw a move away from this position. Unless a building is considered of exceptional importance, *usta*s are given license to change elements and develop details, as they do in new construction and repair work throughout the valley. The project recognizes that traditional building practice must continue to evolve if it is to remain alive and vital. This suggests a different conception of authenticity, which is closer to that of the builders themselves: faithfulness to the practice of one's predecessors, which includes—even requires—the ability to respond creatively to changing conditions and preferences.

In this final project, architects themselves are the agents of creative "restoration"—although the SFD is careful to refer to the work as rehabilitation (*i'adat ta'hil*) rather than restoration.[149] The project is notable because it concerns a minor monument—the *marna'* or well enclosure of Qubbat Talha—that was conserved through a publically funded project. The mosque of Talha (*qubbat Talha* means the "dome of Talha," that is, a mosque) is thought to be very old, although little of the original structure remains. In the early seventeenth century an Ottoman governor enlarged the mosque and added a minaret; it was renovated again in the first half of the nineteenth century by the ruling imam. The *marna'* is a utilitarian structure, but it was built under official patronage and may predate the seventeenth century expansion.[150] As such one would expect it

148 "… [C]riticisms sparked by the programme within the community of experts and professionals … alleged [a] lack of sensitivity and a loss of authenticity caused by the extensive renovations" ("Technical Report," p. 1).

149 On some project documents the term *ta'hil* is used alone—which means to "make suitable" or "qualify." *I'ada* is the transitive form of "return"; the term *i'adat ta'hil* literally means "requalifying," returning something to a usable state. It is usually translated as "rehabilitation."

150 On Talha mosque see the following articles in Serjeant and Lewcock, eds, *Sanaa*:

to be conserved with strict adherence to international guidelines. Instead, architects reinterpreted the historic structure—which was functionally obsolete—to accommodate a new use.

Most of the urban gardens restored by the SFD are part of endowed complexes that included a mosque and a bathhouse, and were irrigated with water recycled from these structures. Water was drawn from a well by animal power, and channeled to the ablutions pool of the mosque and to the bathhouse. Once used for washing, the gray water was diverted to retention basins (*birak*, sing. *birka*), and then released to the garden. This gravity-based system required a certain topographical relationship of well, buildings, and garden; the garden itself was often several meters below street level. A pulley above the well was attached to the backs of animals (camels, oxen, or donkeys); they walked down a long ramp, the length of which was proportionate to the depth of the well. The ramp was usually under a shelter of some kind, and in many cases the space above was captured for uses related to the endowment—resulting in a unique building type, the *marna‘* (pl. *marani‘*). The type can easily be identified by its narrow plan and stepped roof. The finest examples, such as the one attached to Talha mosque, were built of stone with upper stories in brick.[151]

The *marna‘* of Talha extends along the street that runs along the southern boundary of the mosque complex (Figure 6.15). The well is located at the eastern end of the building, and the ramp slopes down to the west. Until the 1960s or later, the rooms above the ramp were used as a children's Qur'an school (*kuttab*), and one room served as the teacher's residence. The long rear (north) facade of the *marna‘* abuts the ablutions building.[152] From here, gray water was channeled to the *bustan* (orchard) of Talha and the *maqshama* of al-Tawus, both located to the north on the other side of a major east-west artery (Talha Street). Within the *marna‘* at the top of the ramp, there was a basin for the animals and, up a short flight of steps, a small room for the driver. A stable, below the highest part of the ramp, was accessed from the street.[153] In the eastern wall of the complex, near the minaret, there was a second water basin for animals.

The UNESCO Campaign aimed to restore and reuse *marna‘*s and other utilitarian buildings, like water fountains (*subul*, sing. *sabil*), sesame mills (*ma‘asir*, sing. *ma‘sar*), and watchtowers (*maharis*, sing. *mihras*). The Campaign brochure (1989) noted that most of the *marna‘*s had been fitted with electric pumps, and the buildings themselves had fallen into disuse:

> The present state of these fine structures varies considerably, some needing extensive restoration and reconstruction. One of them could be put back into working order and operated by a camel at certain times of the day in a living 'outdoor museum.'

R. Lewcock, "The Urban Development of Sanaa," p. 133; R. Lewcock, et al., "The Smaller Mosques of Sana'a," pp. 381–2; "The Post-Medieval and Modern History of Sanaa and the Yemen," p. 89.

151 R. Lewcock, "The Buildings of the Suq/Market," in Serjeant and Lewcock, eds, *San‘a*, p. 302. Lewcock notes that the length of the ramp was modified over time as the level of groundwater diminished; in some cases ramps were extended underground.

152 Showers (*matahir*, sing. *mathar*) were in the ablutions building, but toilets (*mutakhadhdhat*), which are dry and ventilated, were in a separate building.

153 R. Lewcock, "The Buildings of the Suq/Market," in Serjeant and Lewcock, *San‘a*, p. 302, On the gardens of Talha and al-Tawus and their condition prior restoration see M. Barcelo, et al., eds, *Les jardins de la ville de Sanaa*, pp. 92, 94, 98.

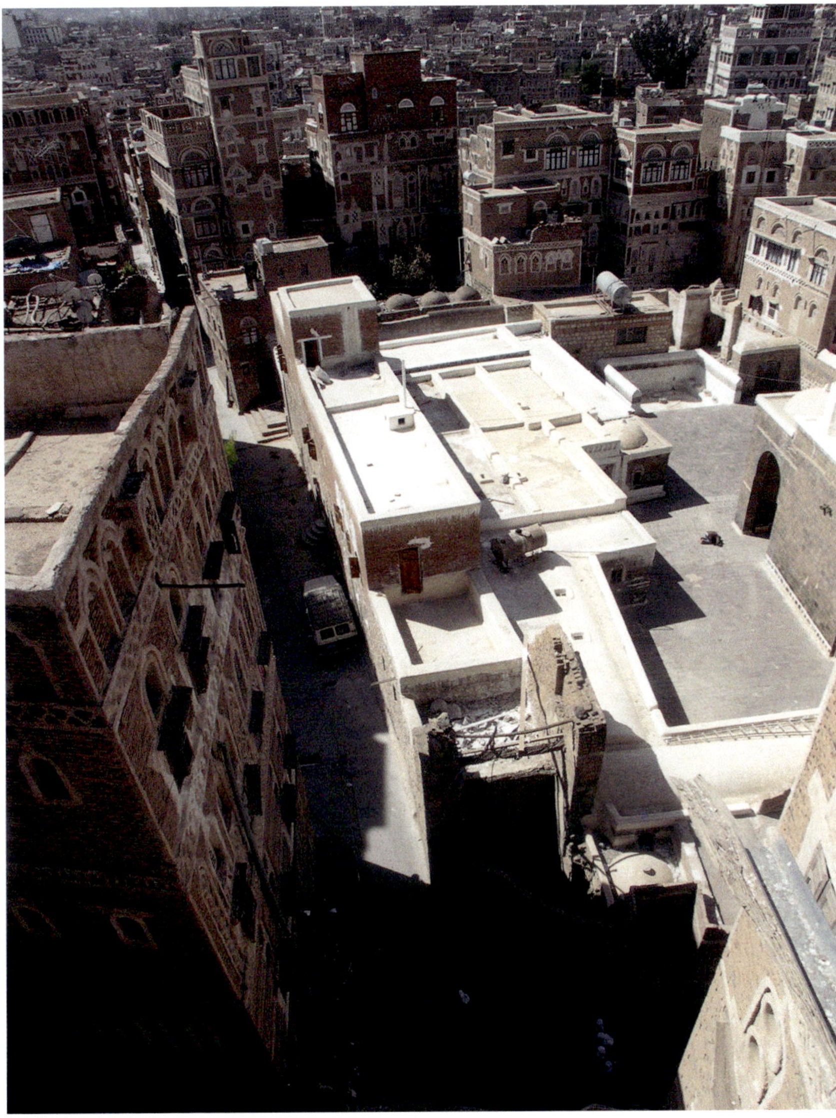

Figure 6.15 Aerial view of the Talha complex

The *marna'* of Talha was to be restored as part of a larger project for the neighboring mosques of Talha and al-Tawus and their respective gardens, but a use for the building was not specified.[154] Only a small number of the many projects envisioned by the Campaign were implemented, but in the late 1990s some were revived by GOPHCY and the SFD. The *bustan* (orchard) of Talha was restored by the SFD circa 2002–2003, along with the

154 B. Lane, "Pilot Projects," pp. 51, 57.

neighboring gardens of al-Tawus and al-Kharraz. Around the same time, a separate project was initiated for the conservation and reuse of the *marna‘* of Talha. Studies for the project were conducted in mid-2002, and final approval came in December of that year.[155]

In the SFD project summary, the *marna‘*s are described as important elements of the urban cultural environment; they are now "exposed to a high degree of danger due to the daily violations" by inhabitants, and by some "traders of the suq who are working hard to change these buildings into private property." The output of the current project would be a building that "serves the needs of the community." The rehabilitated *marna‘* would be used as a sewing center by an NGO called *Jam‘iyyat al-Mar’a al-Muntija* (Society of the Productive Woman). It would include a workshop, an exhibition space for "charity products," and a library for children[156]—perhaps inspired by the former use of the *marna’* as a Qur’an school. The designers faced many challenges—especially the problem of fitting the facility into a long, narrow plan (3.8 by 33 meters). They were also faced by a severely deteriorated building fabric. Since the mid-1970s, when Lewcock documented the *marna‘*, part of the wall along the street had collapsed. A section drawn in 2002 shows that parts of the roof had also collapsed; within the building, accumulated earth and debris obscured the old ramp. According to the *usta* who worked under the contractor, the wall along the street had collapsed due to groundwater and rain infiltration; then the roof followed. The only option was to take the wall down to its foundations and rebuild it, using the salvaged stones.

The design of the project was protracted, involving two different consulting architects and several phases of design. The debate continued throughout construction and involved supervisors working for the SFD, GOPHCY, and the contractor, resulting in long delays.[157] Throughout the process, it was clear that a new building was being created, inspired by the original form of the *marna‘*: in the words of the technical supervisor, "It's a new building in traditional materials." The building had to be faithful to *turath* and, at the same time, it had to suit the new function.

There was no ground floor in the old *marna‘*—just a ramp with an earthen floor. The rooms above stepped down along the street, following the slope of the ramp (Figure 6.16). Given the new program, it hardly made sense to re-excavate the ramp and reconstruct the stepped upper floor and roof. Instead, there would be three full floors (one partially below grade) and a room above the stair. Windows and rooflines would be set at a constant elevation above the floors—in contrast to the old windows, which stepped down with the ramp. In the original building, the Qur’an school on the upper floor was reached via an external stair on the rear facade (accessed from within the complex, not from the street). The new program called for an interior staircase, which was difficult to achieve given the narrow plan. In two different schemes, the stair was placed roughly in the middle of the building, which created a new hierarchy on the facade.

---

155 "Marna‘ Talha" project file, SFD. The project description says that the funds ($110,000) were from the World Bank, but SFD staff said they came from the Arab Fund (Kuwait). According to the architect who drew up site plans for the gardens of Talha, al-Tawus, and al-Kharraz, this was the third group of gardens to be restored. This architect subsequently prepared preliminary plans and elevations for the *marna‘* of Talha, which were later modified by a second architect. I will hereafter refer to these individuals as the first and second consulting architects. The following narrative is based on interviews with the architects with the project manager, and on SFD project files.

156 "Project Summary," SFD, 18 March, 2003.

157 This is clear from a memo of February 2004, written by the contractor to the SFD almost a year after work began. Some of the delay was due to the uncompacted soil, which required a new foundation design. But other delays were due to disputes over historical correctness ("Marna‘ Talha" project file).

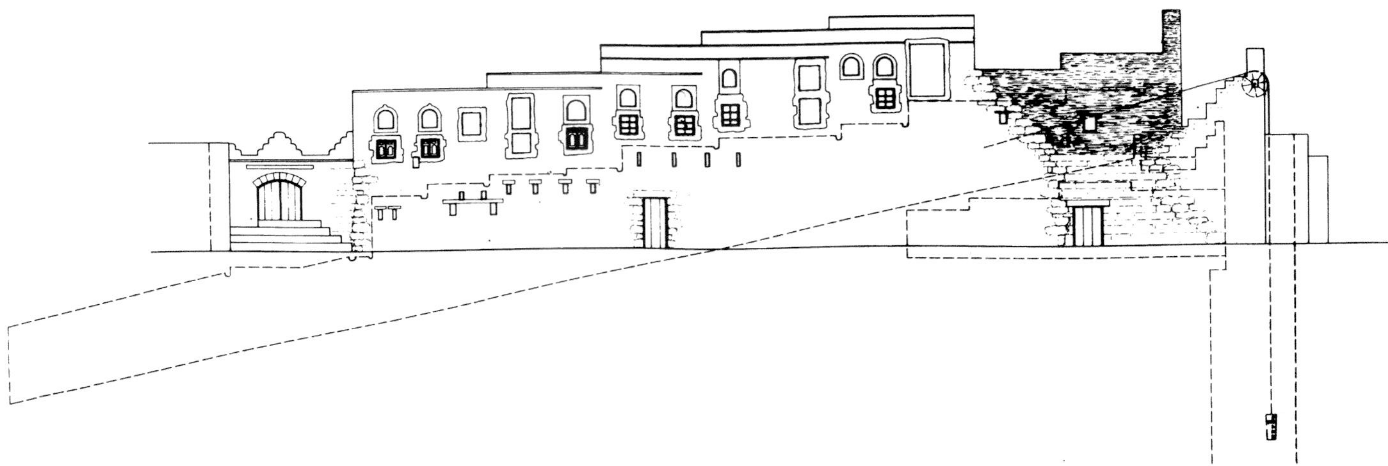

Figure 6.16 **Facade of the Talha *marna‘* as documented by Lewcock in the early 1970s. The ramp and stepped roof are indicated with dashed lines**

Source: Serjeant and Lewcock, *San‘a*, p. 301. Reprinted with permission of Ronald Lewcock.

If one were not familiar with the original form of the *marna'*, however, one might mistake the new building as a restoration of the original. Reinforced concrete was used in some parts of the building: under the rebuilt stone foundations, and in perimeter beams that run along the underside of each floor. These beams are recessed and covered by stone, so that they cannot be seen from the outside.[158] The exterior walls were rebuilt largely with salvaged stone, with new fired brick (*yajur*) for the upper story, as in the original building. Traditional materials were used for floors and ceilings—wooden beams packed with earth—even though these would not be visible from the outside.[159] At the eastern end of the building, the rebuilt wall ties into brick piers that support the pulley; the piers and the curving wall around the well-head appear to have been left in their original state.

A fascinating debate over the design of the building—and thus over the nature of *turath*—emerges from design drawings and other project documents, and from conversations with various actors involved in the project. The first consulting architect drew several iterations of the facade (Figure 6.17). His scheme had a prominent staircase that split in two directions after the intermediate landing. In plan, the stair took up much of the long north wall, and reduced the vestibules on both floors to corridors; it also created a wide stair tower on the facade. Since the roof had to be rebuilt, the architect designed it in concrete—perhaps because concrete perimeter beams were being used. But the design was rejected by the SFD supervisor, who wanted the roof to be built of wood and earth. "We have *turath* of thirty or forty years ago," said the architect, "but the supervisor wanted *turath* from the time of Sayyidna Muhammad!" He wanted to add balconies, but these were also rejected as inconsistent with *turath*.

Then a second architect known for his work on *turath*-related projects, 'Ali Sa'id al-Qubati, was brought in to help with the design.[160] He redesigned the stair as a square well type with short, straight runs. This simplified the plan, reduced the number of level changes, and freed up space. It also reduced the mass of the stair tower: it became narrower and taller, producing a more pronounced stepping effect—similar to the roof of the original *marna'* but in the opposite direction. The first architect had used wider windows in the second floor, like those in a *mafraj* (the sitting room at the top of a house); the arched window lites above were also wide, almost semi-circular. The window sills did not rest on the *hizam* as is usual, but rather were several brick courses above it. In iterations developed with al-Qubati, the windows were brought closer to the *hizam*. The windows became smaller—closer in size to those in the original structure, with the traditional 1:1 proportion of window to lite.

Al-Qubati wanted the new facade to be "similar" to the original one. He indicated this in notes on a sketch that he must have drawn late in the design process.[161] He proposed to simplify the elevation, in conformance with "the simplicity of the old facade." The original building had no *hizam* or *zakhrafiyya* (plaster ornamentation), which is characteristic of residential architecture. So he proposed to eliminate the *hizam* and replace it with wooden

158 A similar technique is used at a smaller scale in new *turathi* facades, especially in the old city: a continuous beam (*corniche*) above the windows is recessed behind half bricks.

159 The contractor complained that he lost a month searching for *'ilb*, the wood traditionally used for beams and required by SFD supervisors. When he finally managed to locate some, the supervisors changed their mind and decided to build the roof in *athal* (tamarisk; memo to the SFD, received 2/22/04).

160 Al-Qubati designed Al-Hawta Hotel in Sayun (Hadhramaut Valley); he also worked on a project for a hotel at the old American Embassy in Bir al-'Azab (Sanaa), which did not go forward.

161 The sketch was signed by the SFD supervisor and submitted to Nabil al-Makaleh, the project manager, for approval ("Marna' Talha" project file, SFD).

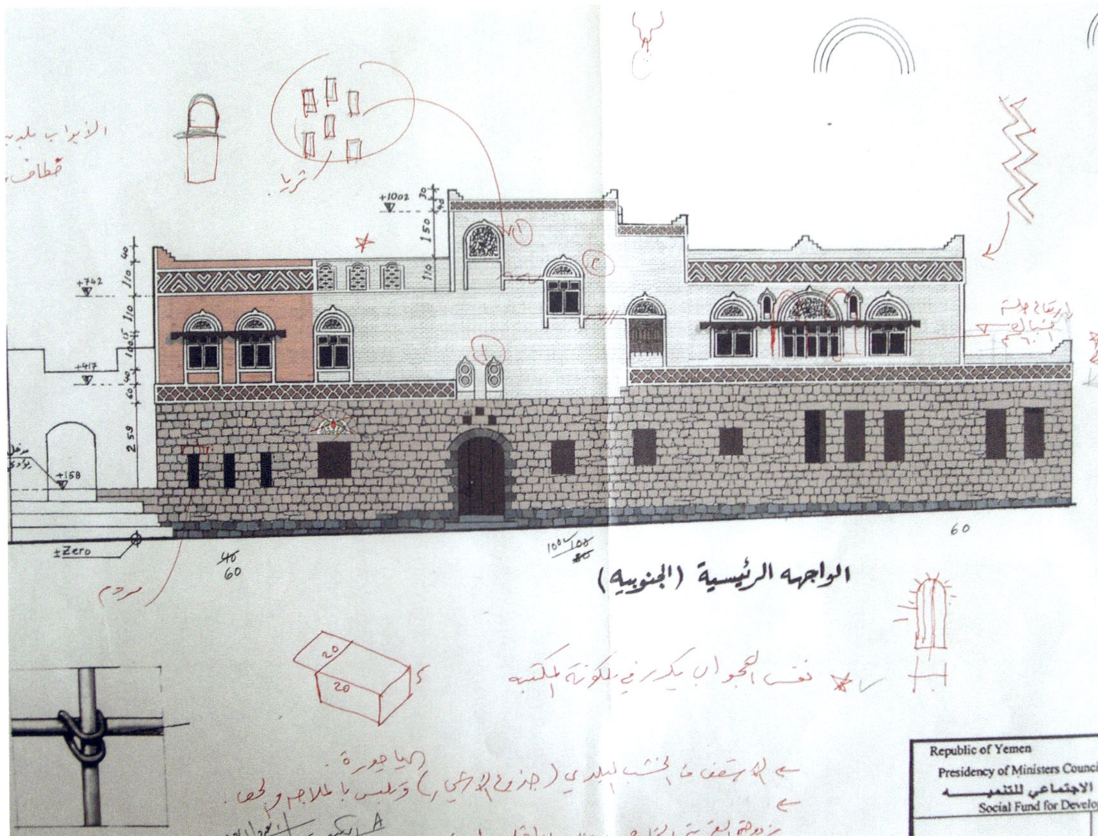

Figure 6.17 **Preliminary design for the adaptive reuse of the marna‘ as a women's sewing center (no date, circa 2003)**

Source: Image courtesy of the Social Fund for Development.

beams (*farashat khashabiyya*), which would separate the brick from the stone, and serve as heads or sills for windows. At the roofline, there would only be a simple coping (*zarwa*) finished with plaster. On the ground floor, original stone should be reused where possible, or new stone of a similar quality; window lintels should also be of stone. For the upper stories, brick should be of the same quality as that in the old building. The new function of the building should be respected, wrote the architect. In particular, natural light should be maximized, to the degree permitted by "the exterior (historic) form of the building (*al-shakl al-khariji (al-turathi) lil mabna*)." The project should be implemented "with sensitivity to the building's site in old Sanaa ... It is recommended that you rely on a supervisor experienced in this kind of work" to ensure that no extraneous openings are made and that natural light is provided to the interior. Nabil al-Makaleh, the SFD project manager, approved the proposal because it "coincides with the spirit of the old building (*tatbiq ruh al-mabna al-qadim*) with regard to stone construction, and [per our] review of the old photos."

The project was executed roughly according to this last sketch, although the placement of openings and detailing continued to evolve during construction (Figure 6.18). Again, the main criterion was similarity to the old building, and to the surrounding context. In a site visit toward the end of construction, the supervising architect for the SFD found certain aspects of the work in conformance with *turath*, and others that were unacceptable. Interior plasterwork was inconsistent, and joints between bricks in the stair tower (*bayt al-daraj*) were not filled with mortar. When he pointed this out to the workers (who were working under the supervision of the *usta*), they said that they had been told to follow the example of surrounding buildings. The technical supervisor (*fanni al-mashru'*) who "has experience in *turath* projects" expressed much the same view. They told the SFD supervisor that "the form of the building required of them is the same as that of surrounding buildings."[162] Perhaps the workers omitted the mortar because it had deteriorated in neighboring buildings!

Architects, supervisors, and workers, in their various ways, aspired not to historical accuracy but to *similarity*—and they seem to have succeeded. As I toured the site with the *usta*, a young man from the neighborhood came up to us and showed the *usta* an old photograph of the site. He congratulated him because the restored *marna'* looked exactly like it did in the photograph—except for the basin in the east wall, which had yet to be restored. At the time I had not yet reviewed the project drawings, so I was not aware of the extent of the changes. In retrospect, I am sure that the view in the photo—the only view possible in the narrow street—distorts the facade, making the changes less obvious. The wellhead in the foreground, which has not changed, furthers the impression that the building beyond is the same. This impression is reinforced by the materials, details, and general massing of the building.

When I asked the young man how he felt about this kind of work, he replied: "All of the old city should be rebuilt exactly as it was." The *usta* told me that all the families in the neighborhood wanted the *marna'* to be rebuilt as it was—except for the family living in the house to the south. They complained that the rebuilt *marna'* would block light and air, and GOPHCY had to issue a stop-work order (although at the time, March 2004, exterior work was substantially complete). The first consulting architect pointed out that light would not be blocked, because the house lies to the south of the *marna'*. "They just want a cut," he told me, "because they know that the money [for the project] comes from abroad." The *usta* was upset by the stop-work order:

162 "... *afadu biann shakl al-mabna' matlub minhum nafs shakl al-mabani al-mujawira*" (report of site visit on June 4, 2004 signed by the supervising architect; "Marna' Talha" project file, SFD).

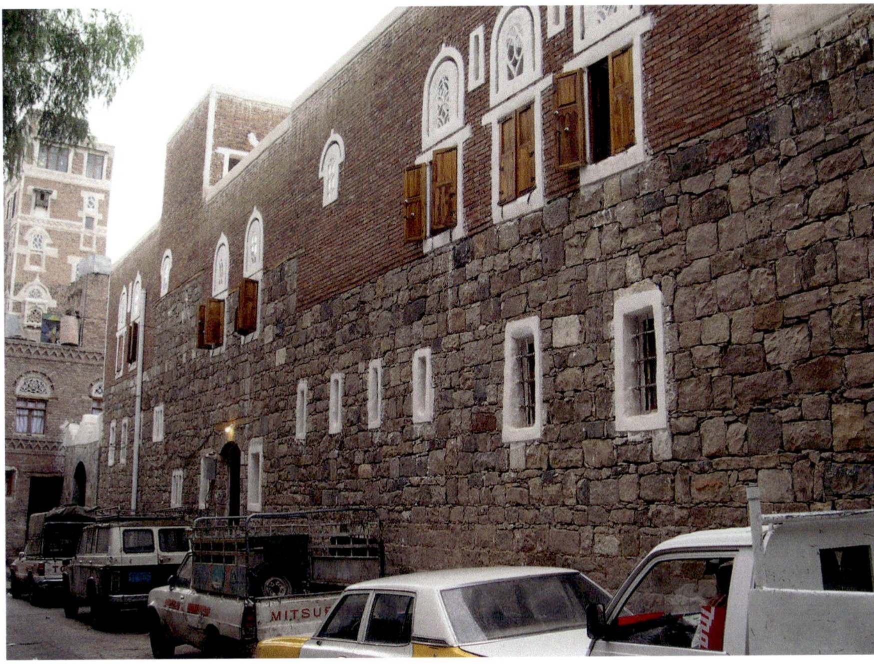

Figure 6.18 View of the *marna'* after renovation

> These people are complaining because they don't remember what the building looked like before it collapsed twenty years ago. The problem is, here people don't appreciate *turath*—not like in your country (*mish mithl 'indkum*, addressing me). There needs to be a law that requires all *turath* to be rebuilt exactly as it was.

On this first visit to the site, the *usta* told me that he had rebuilt the exterior wall in its original form. On a second visit, we went inside—and it became clear the interior had been completely reconfigured. "This was a mistake," said the *usta.* "The idea of a sewing center is good, but the building should have been kept as it was." I asked another *usta* (the *qaddad* master who restored the Sa'ila facades) how he felt about the *marna'* project: should it have been kept as it was? "It doesn't matter," he replied. "The important thing is that the *marna'* still looks the same on the outside." Like al-Qubati who identified the exterior as *turathi*, and like the official at the Ministry of Endowments who saw the exterior of Samsarat al-Bawa'ni as historic (*tarikhi*), for this *usta*, too, *turath* is conveyed primarily through the exterior facade. The *usta* who actually did the work—known among conservation professionals for his sensitivity to *turath*—felt that *turath* included the inside. The whole building should have been kept as it was.

Likeness or resemblance, notes Herzfeld, is culturally constructed: it has rhetorical use in establishing continuity and consolidating the identity of a group. The ability to recognize resemblance is not natural, but rather depends on prior aesthetic criteria and on the politics of a particular situation. "[E]ven when those concerned agree in principle on the criteria of resemblance," writes Herzfeld, "they may disagree about specific cases, and their disagreement will reveal something of the power play among them."[163] In Chapter 5, we saw that both traditional building practice and modern conservation aspire to authenticity: both claim faithfulness to an original, even as they change and reinterpret it. But they differ in their definition of authenticity, and the criteria by which it is established. Modern conservation aspires to historical accuracy in terms of form and materials, based on criteria that are documentary and "scientific." For *usta*s, authenticity means mastery in the tradition of one's predecessors, in a mode (*maqam*) that is unique to Sanaa. The case of the *marna'* is fascinating because it reveals a new, synthetic definition of authenticity that makes use of *both* traditional and documentary criteria. The debate over the appropriate treatment of the *marna'* draws on documentary evidence—drawings and photographs of the building's prior state—and also on personal and popular memory. It also involves judgments of the Sanaani aesthetic by actors with different types of expertise: architects, builders, *usta*s, and craftsmen.

Architects, who in theory represent international practice, used documentary evidence in an unexpected way: not to reconstruct the *marna'* according to scientific standards, but as a guide to recover the "spirit" of the original building. This spirit is defined in aesthetic terms—in massing, materials, and details. But it is also defined in terms of the building's *use*: the new function, like the old, is a charitable one and hence suited to an endowed structure. Perhaps most interesting is the reversal of roles. Architects here take up the position usually identified with *usta*s—approaching conservation as a "creative" rather than a scientific process. Yet they are careful to defend their position in terms of international discourse. "I wouldn't call [the work] restoration," said al-Makaleh, correcting me when I used the term. "As an adaptive reuse project, more liberty can be taken with the building fabric." Yet the building is a minor monument, part of an historic endowed complex. It is the *usta*, known for his sensitivity to *turath*, who takes up the international position: "It was a mistake [to change the building]," he said. "It should have been kept it was."

163 *Cultural Intimacy*, pp. 94–5. The prior point is from p. 28.

Chapter 7
Lessons from Sanaa

What does it mean to conserve in a place where the "historic past" is, in some sense, still alive? What does this mean for international policy and practice—and what are the implications of the case of Sanaa for conservation in general? The stakes of this interaction are raised by the ever-expanding definition of heritage. Originally conservation involved select monuments of special historic or artistic importance. Now, the field claims a mandate for the protection of a "vernacular" heritage that includes entire historic cities and landscapes. As the definition of heritage has expanded, so have the attributes that "carry" authenticity.[1] In the era of values-based conservation these attributes are, in theory, no longer determined exclusively by experts, but by various stakeholders. What happens in a place like Sanaa where so many different stakeholders, with different backgrounds—officials, inspectors, architects, builders, craftsmen, residents—claim to represent the historic past? To be added to the World Heritage list, a site must demonstrate "outstanding universal value," which is assessed based on its significance and authenticity. State actors and various specialists may be the initial arbiters. But who determines authenticity in the day-to-day life of a World Heritage city?

The precepts of conservation, which changed over time and varied from place to place, came to be codified in international charters and administered by agencies that act as new global curators. Despite a wide variety of practices and recent revisionist thinking, international guidelines are still seen to reflect universal values and principles which may be conditioned, but not challenged, by cultural relativism.[2] Like other aspects of Western epistemology and practice, however, these "universal" values are projected from a geographical/cultural center—and thus are themselves culturally relative.

Conservation developed in a specific historical and intellectual climate, as part of Europe's experience of modernization and industrialization—a process that spanned two centuries. Changes in building practice, in land use, in the organization of society were perceived as qualitatively different and incompatible with earlier forms. Without careful conservation, these earlier forms would soon be extinct. In order to protect the forms of the past from extinction or corruption, a clear break had to be imposed between past and present. This sense of rupture was palpably felt by contemporaries, but it was also constructed by new interests and groups. In the nineteenth century and later, many of the craft traditions that had produced old buildings were still alive in Europe and the United States. Not unlike Yemeni builders today, these craftsmen were engaged in a struggle with new professionals, the architect, the contractor, and later, the conservator. As in Yemen, some practitioners restored creatively, inspired by styles of the past and contemporary revivals—as represented by the school of Viollet-le-Duc and his followers. With the emergence of "classic" monuments conservation after the turn of the century, this interplay of past and present was, at least officially, rejected: buildings were now to be conserved

1 See above Chapter 1.

2 See, for example, J.-L. Luxen, "Reflections on the Use of Heritage Charters and Conventions," *Conservation, the GCI Newsletter*, vol. 19, no. 2, 2004.

according to strict historical and scientific criteria, rather than creative interpretation. The new approach in many ways reflected the ideas of architectural modernism: in particular, the notion that the architecture of the present should be as distinct as possible from that of the past.[3]

Yemen's experience of modernity differed greatly from that of Europe and other countries in the region. This experience was not as sudden as often portrayed: it unfolded in a series of "detailed shifts" over an extended time.[4] But especially in the highlands, geographical and political isolation kept much of the material and social fabric relatively intact. When the "shackles" of the imams were thrown off during the revolution and North Yemen opened its doors to the global market, there was the impression of a sudden and dramatic transformation. In the mid-twentieth century, capitalist modernity was at an advanced stage: the sense of radical change was due primarily to the changed conditions of material life. For architects, both foreign and local, the rupture is physically inscribed in the fabric of the city. The contrast between the old city and development outside the walls gives the impression that the city has bypassed certain historical stages: in particular, one misses the nineteenth century suburbs which, in other Middle Eastern cities, mark a gradual transition to contemporary building.

Barely had North Yemen begun to modernize—following mainly Egyptian models—when foreigners and concerned local architects called for the conservation of the old city of Sanaa. The moment was a serendipitous one for Yemen: the World Heritage initiative, reflecting a new interest in regionalism, appreciated precisely the kind of vernacular settlements in which Yemen was so rich. At the time, policymakers and planners elsewhere in the region showed little interest in old cities: unlike remote antiquities, they were seen to have little value. They were unwanted reminders of a backward past, which modernizing elites wanted to forget. But in Yemen, when the idea of a safeguarding campaign for Sanaa was proposed, policymakers had already started to address the problems of the old city. Although they may have felt ambivalent toward it, they were closer to it: many of them had grown up in the old city, and had relatives living there; the houses still bore their family names. The past it represented was their own.

The UNESCO Campaign in Retrospect

In 1995, a decade after it was launched by UNESCO's General Director, the Campaign to Save Old Sanaa was awarded the prestigious Aga Khan Award for Architecture. The award was made not for building conservation, but for urban revitalization—and indeed, this is widely seen as the greatest success of the Campaign. "The main point is that the old city is much better than it was," said Rémy Audoin, former director of the French Institute, who was an early promoter of the Campaign. Street paving, utilities, and services like garbage collection have made a tremendous difference in the old city, making it one of the best serviced sectors of Sanaa. The Campaign was both a national and an international effort. But it is important to note that street paving and infrastructure were largely financed by the Yemeni government, not by foreign donors.

---

3 M. Petzet, "The Venice Charter—forty years later," pp. 21–2; and A. Colquhoun, "Newness and Age Value in Alois Riegl," p. 220.

4 B. Messick, describing the relationship between education and social transformation in Yemen (*The Calligraphic State*, p. 107).

Professionals credit the Campaign with fostering a sense of pride in local architecture, and "raising awareness" of the need to conserve. Conservation built on local pride, and on the marked cultural confidence of a country that had never been controlled by a European power.[5] A decade after the UNESCO Campaign ended, conservation seemed to have taken root to a remarkable degree. "If the old city is conserved as well as it is, it's certainly not because of GOPHCY," said one architect. "It's because the people conserved it." This observation contrasts sharply with public and private comments by some officials and intellectuals, who blame the "decay" of the old city on "the people's ignorance." Conservation professionals are disturbed by such comments: it is the officials themselves, they say, who allow the large-scale violations. While some members of the political elite have provided critical support for conservation, they do so from the comfortable distance of their offices and villas in the new districts. "The officials are strangers to Yemen," said a local professional. "Even their money is elsewhere."

The relatively good state of conservation of the old city is due, in no small part, to poverty. Tower houses are not easily adaptable to modern lifestyles or new uses; this, along with difficult access, led most people to build outside the old city rather than inside it. Newcomers lacked the resources to make major changes to the buildings. "Poverty is both our blessing and our curse," said one architect. But the idea of *turath* has also been assimilated to a remarkable degree: it resonates with local concepts and practices that have not died out. Support for *turath* may also be seen as part of a more general desire to maintain a local sensibility in architecture; this was already evident in the 1970s, when Egyptian modernism was rejected by architects and residents alike. This local sensibility can be seen in virtually all new building, in the old city and new. While foreigners and some local architects see much of the new work as "kitsch," residents and builders see it as new examples of heritage (*turath*)—in the words of one builder, "*turath mutahaddath* (modernized heritage)."

In other areas, the UNESCO Campaign is seen as less successful. It failed to create a viable institutional structure and legal framework for conservation—which became especially evident after the conservation office's relationship with the Prime Minister ended. More than two decades later, the basic tools of heritage management were still missing. As one architect put it:

> The Campaign was only 35% successful. If I do a project on my own, no one knows about it; under the Campaign it's different. That's where the Campaign succeeded. But what were the tools, what were the means? What is the old city of Sanaa, what do I want from it? Until now, we don't know. The Campaign gave me the bread, but not the tools to make the bread.

Yemeni professionals faulted their own leadership for this: especially since the end of the Campaign, they said, the conservation effort suffered from a lack of political will. Some also faulted UNESCO for not insisting on a conservation plan and regulatory measures as first steps in the process.[6] But according to Lewcock, UNESCO *did* see the need for such mechanisms—and indeed, the Campaign strategy and his reports repeatedly refer to

5 G. Grandguillaume, "Sanaa, ville d'Arabie," pp. 4–6.

6 This was acknowledged by Anna Paolini, then UNESCO Desk Office for Arab States, at a 2005 symposium marking the opening of the 'Amiriyya *madrasa* in Rada. "In the early years of the World Heritage Convention, the inscription process did not foresee the need to have management

an upcoming plan and legislation. The conservation plan was postponed to ensure that emphasis would be on infrastructure. Building regulations were proposed, he said, but they were blocked. (Indeed, multiple iterations of draft laws were stalled or rejected by the government—probably because of restrictions on the use and disposition of private property.) In the opinion of a World Heritage official, the Quaroni Bonifica studies of the early 1980s should have formed the basis of a conservation plan:

> There were two major problems [with the Campaign]: a conceptual one, on the part of UNESCO and donors; and a lack of local capacity. UNESCO's strategy was to compile a list of projects to sell to donors. But the donors left, disappointed, because there were no institutions to bring the work forward.

The money spent on these projects, he said, should have been used for a conservation plan, and for the development of a technical office that could offer meaningful assistance to homeowners. While these criticisms are valid, we should see them in a wider perspective. Ineffective institutions, lack of legislation, and a dearth of local capacity are problems that plague the whole Yemeni government. Dynamic, semi-autonomous entities like EOPOCS—that bypass the ministerial system—are periodically created; they get things done, and are eventually absorbed within the system.[7] We should remember too that planning for the city generally—in terms of tools, enforcement, and local capacity—is still lacking. This makes the accomplishments of the dedicated professionals working in the old city all the more striking.

The restoration of key historic buildings was seen as an essential part of the Campaign effort; many are important from an historical point of view, and represent unique typologies. UNESCO asked foreign donors to sponsor these projects, because it felt that they would have little interest in infrastructure. But in the 1970s the first infrastructure projects for the old city had been sponsored in part by foreign funds; indeed, this happens all the time, under the rubric of "development assistance." What seems to have changed is not the nature of the work, but the way it was framed—as part of a project to conserve a World Heritage City. It was expected that local embassies, in particular, would want to sponsor glamorous projects that carried the cachet of conservation. The restored buildings were given new functions, but in most case these were designed with tourists in mind; even the project to revive handicrafts, for which professionals had great hopes, did not build on the resources of the community, nor was it driven by the community's needs.

In designing uses for the restored buildings, there were competing notions of culture at work. Lewcock had a deep respect for both the people of Sanaa and its architecture, which he knew so well. Throughout his consultancy for UNESCO, he insisted that the old city of Sanaa was a *living* historic city: as such, he argued for projects that were relevant for the residents of the old city. Many local professionals and foreign consultants shared this view. But another view of culture came into play—one that had both foreign and local roots. In the late 1970s and '80s, urban renewal had largely been defeated in western Europe. Historic centers were conserved in part as cultural symbols: although they would lose many of their original inhabitants, they would take on "new uses"—for example, as venues for cultural events. Some consultants thought that the same would be true for Sanaa. For Ludovico

---

mechanisms in place," she said. "Now UNESCO and the World Heritage Center are encouraging this, also as a means to manage change."

7 Robert Burrowes, personal communication; see Chapter 3.

Quaroni's team, the old city was the historical and moral center of the Yemeni capital; it would also "nourish Yemen's relationship with countries in the world, in the East, and in the West."[8] It was therefore important to conserve the old city not only for its own sake,

> but also, and perhaps mainly, in [anticipation of] the future role that the Historical Centre will be able to fill at [the] international level. We are thinking particularly of a cultural role which the historical city could undoubtedly take on in the future.[9]

For some Yemeni professionals and policymakers, these sentiments echoed the "cultural revival" envisioned by the revolution. The revival included both an elite "Arab culture" long suppressed by the imams, and a popular culture that expressed the national spirit. While the former derived from Yemen's links to the wider Arab (and modern) world, the latter was local and traditional. Old Sanaa would be a place to "stage" an authentic Yemeni culture—for foreign audiences and for modernizing elites, who wanted to return to their roots.[10] In the early years of the Campaign, recalled one architect,

> many at the conservation office were politically on the left; they had a rigid, old school approach to architecture. Their view of the old city was romantic, impractical—and undemocratic. It didn't account for people's needs. An engineer at the office thought we should consider the Italian idea of demolishing the shops of the *suq* and putting parking underground. 'We can't be romantic about the shops,' he said. 'We need to think about economic needs, safety, and so forth. We need a vision for the future.' Maybe he was right.

The "romantics" were former revolutionaries, and also deeply attached to the old city. Its medieval form, and the traditional people who inhabited it, seem to confirm its authenticity—as described by the World Heritage Convention. The view persists among both foreigners and Yemenis: as an official put it, "even the people in the old city are *athar* [antiquities]." The old city embodied all that was virtuous and moral in Yemeni culture; but also the "backward" ways that the revolution was supposed to end.[11] In Sanaa as elsewhere, heritage provided a way to come to terms with the past, to domesticate it and idealize it. A certain ambivalence—pride in a local genius that is recognized by the international community, and discomfort at persistent backward ways that are resistant to modernization—underlies the "cultural gap" between these professionals and the builders and residents with whom they share the responsibility of conserving the old city.

It is interesting that the success of the Campaign is attributed less to monuments conservation, but rather to modernizing operations—which were, in fact, needed to remediate earlier, failed modernizing efforts (the introduction of piped water). For Giovanni Boccardi, the experience in Mu'adh quarter showed that urban recovery (*recupero urbano*) was virtually impossible in Sanaa, given the closed and at times hostile nature of local society. Indeed, working in Sanaa was not easy: the first Yemeni surveyors and their UNDP supervisor feared for their lives when laying out roads in the early 1970s; as recently as 2002, conservation work in a *maqshama* in the old city was challenged with gunfire. But

8 C. LaRocca, "Studio per la salvaguardia del centro storico di Sanaa," p. 58.

9 "Action Plan," January 1986, p. 20.

10 On the idea of "staging" the traditional see T. Mitchell, "The Stage of Modernity," in T. Mitchell, ed., *Questions of Modernity*, Minneapolis: University of Minnesota Press, 2000.

11 T. Marchand, "Walling Old Sanaa," p. 49.

after many difficulties, street paving and infrastructure in the old city were accepted and ultimately appreciated by residents. Urban upgrading was driven not by a romantic notion of the past, but rather by the desire to modernize and improve. In simply bettering quality of life, these operations validated the everyday environment—and perhaps more effectively communicated the notion of "heritage."

The Campaign brought together two different concepts of conservation, which have their own histories and logics: the careful preservation of monuments, according to historical and "scientific" criteria; and an urban strategy that tries to reconcile old city fabrics with the requirements of modernization. In the post-war era, these two understandings of conservation converged in international protocols and national and local laws, but the differences between them were not clearly articulated. Almost by default, the conservation of historic cities came to be guided by principles and standards that were originally developed for artworks and monuments: they prohibit change, which is necessary for cities to remain alive.[12] A number of international charters have dealt with historic cities, but most fail to question the applicability of concepts like authenticity and significance to these complex "artifacts," which are also living environments.[13] Officials at UNESCO and the World Heritage Center know that standards for cities have to be different: "Integrity is more important than authenticity," one official told me, off the record. The conservation of World Heritage Sites is entrusted to "scientific" bodies, since they are considered collective human masterworks. This creates an aura of untouchability—condemning the site to museification or to decay. "ICOMOS and ICCROM managed to avert many disasters—this must be stated up front," said an Italian consultant with long experience in the region.

> But conservation implies rehabilitation, revitalization—not just restoration. Many people at UNESCO feel this way. But UNESCO has no money, so most of the strategies are designed by ICOMOS and ICCROM. They are very conservative: they're only concerned with the restoration of buildings. The approach works for monuments, but not for cities. There is a kind of conceptual fight.

It's easier for agencies and municipalities to simply prohibit change, he said, since bureaucracy has a difficult time with complexity: "For public bodies, it's easiest to say, 'let's restore.' It's more difficult to judge buildings and interventions on a case-by-case basis."

The Documentary Project

The modern approach to conservation—what I will call the "historicist approach"—relies on various tools to document and classify cultural products. Like other bureaucratic tools, they translate local context-based knowledge into theoretical knowledge; once ordered and rationalized, complex social and material phenomena become legible and easier to manage.[14] As such, heritage is a modernizing phenomenon, part of the apparatus of the nation state. Like other modernizing schemes it is often seen as hegemonic, reshaping the

12 See M. Lamprakos, "The Idea of the Historic City."

13 Some have hopes that this will change with the adoption of UNESCO's Recommendation on Historic Urban Landscapes (2012); see G. Araoz "World-Heritage Historic Urban Landscapes: Defining and Protecting Authenticity," *APT Bulletin*, 39: 2–3, 2008, pp. 33–7.

14 J. Scott, *Seeing Like a State*, pp. 1–13.

past according to an official notions of history. But complete hegemony is rarely realized: such schemes involve subjects, conditions, and materials that ultimately force bureaucracy to modify its methods and even its aims.[15] The field of cultural heritage—charged with managing and conserving all that is "pre-modern"—seems particularly unsuited to a top-down approach. "When the World Bank got involved in cultural heritage, it was a whole new world for them," said another Italian consultant. "They were used to big infrastructure projects. Conservation is small-scale, complex, local." In Yemen, where government institutions are often ineffective, the interpretation of conservation guidelines is even more multi-vocal—like the democracy experiment itself.[16] Authenticity is, in effect, continually *negotiated* by the various actors in the heritage enterprise: all of whom see themselves as legitimate representatives of the past, and exercise agency by choosing among a range of practices and discourses.

Traditional tools for documenting and classifying heritage are based on rigorous and formalized procedures designed to minimize the subjectivity of the surveyor. In case of the inventory—which the eminent conservator J. Marston Fitch calls the *sine qua non* of conservation planning—a trained surveyor documents a building or site, and evaluates its integrity and significance based on extant material fabric. It is acknowledged that judgments involve some degree of subjectivity, but procedures are designed to minimize it. Based on the surveyor's evaluation, the building or site may be ranked according to local, regional, or national significance (in the case of the World Heritage sites, international significance or "outstanding universal value"). The inventory is thus a highly charged process that determines the degree of protection accorded to a building, complex, or district.[17] Inventory forms are usually based on certain assumptions about the nature and historicity of the artifact. It is assumed, for example, that the artifact was created by a known author at a given moment in time, and that its original form can be determined; and that the artifact is a discrete object (or in the case of a site or "cultural landscape," that it has clear boundaries). It also assumed that the artifact can be located within a wider chronology (on the "historical time line"), and that it is representative of a type, a style, a culture, or a time period (and thus reflects a stage in artistic or human development).

The expanding definition of heritage has challenged these tools and assumptions in various ways—and Yemen, in particular, challenges them on virtually all counts. In the first place, the country's historic settlements—and even its buildings—are difficult to assign to the historical time line. The practices that built Yemen's cities and buildings are, to a great extent, still alive, although the social and institutional forms that shaped them have changed or disappeared. Second, because modernization is recent, much of the built heritage of Yemen is relatively well conserved. Unlike the old cities of Cairo and Damascus where much of the residential fabric was made of earth and has decayed, the building materials used in the Yemeni highlands were durable ones, stone and brick. In other areas, like the Wadi Hadhramaut, the earthen architecture has until recently been subject to continual maintenance; in many places it is relatively well conserved. The project of documenting, let alone conserving, this heritage is daunting even to foreign consultants: there are

15 W. Denslagen, an advocate of the Venice Charter, notes that there is an enormous gap between theory and practice even in Europe—and even in the restoration of important monuments. See "Restoration theories, East and West," *Transactions*, vol. 18, 1993, p. 6.

16 On this point see S. Philips, *Yemen's Democracy Experiment*, p. 49.

17 On the role of the inventory in the conservation process, see J.M. Fitch, *Historic Preservation*, pp. 307–8. Typically, the inventory precedes, and serves as a basis for, a conservation plan.

thousands of exceptional settlements, many of which are partially or fully abandoned due to migration and the depressed agricultural sector (Figure 2.2). The immense scope of the undertaking was obvious at a 2005 workshop to design a comprehensive cultural heritage inventory—which also envisioned the documentation of "intangible heritage" like poetry, music, dance, and even cultural practices. "The complexity of an inventory derives from the quality and quantity of cultural heritage," said architect Vincent Sommella. "All of Yemen could be considered as cultural landscape."[18]

The sheer quantity of heritage in Yemen challenges the capacity of a classificatory system to document, contain, and explain it. This problem was foreseen by Alois Riegl, who prophesied the coming age of values-based conservation: he realized that the multiplication of values, and in particular the criterion of age value, would lead to the infinite expansion of the canon. An age limit (usually 50–60 years) is included in most national conservation laws, and in most of the regional laws reviewed by Yemeni professionals in the mid-1980s. But they realized that an age limit would not work in Yemen: "It works in the Gulf, because their *turath* is new," said lawyer Tarek al-Hammadi. "In Yemen, most things are over fifty years old." This led them to distinguish between the "archeological monument" (*ma'lam athari*) and the "historic monument" (*ma'lam tarikhi*) which would in theory be accorded different levels of protection. While neither term excluded the possibility of habitation, the "historic monument" was defined as one that continues to support life.

The massive heritage of Yemen cannot be ordered according to conventional periodization; nor can its prodigious forms easily be classified according to stylistic or formal criteria. In Europe, the documentary practices of conservation developed alongside historical studies; even in Sanaa, the best documented of Yemeni settlements, such studies are still in their infancy, so the basic data required in inventories is often lacking. The continuous use of materials and recycling of forms over long spans of time make dating difficult; moreover, several historical layers are often contained in the same building, with the lower floors many centuries older than the upper floors. "In Sanaa, we can't specify historical features, as in the Gothic, Renaissance, and so forth," said Yassin Ghalib. "[Historical] layering (*tarakum)* for us is not like the European. In Yemen it unfolds over the long term (*'ala al-mada al-tawila*)." In conservation, as in the art historical methodology on which it draws, a building's value is determined by its position in an evolutionary sequence, defined by style or period; this in turn is closely related to notions of monumentality. Although recognition of the vernacular has, in theory, widened the scope of values in architecture, it is largely valued as the *inverse* of the monumental tradition—that is, for the absence of change and for anonymity. In Yemen, said a World Heritage official, the kind of great technological changes that gave rise to new monumental types in Europe are generally lacking. Over the years foreign experts have been fascinated not by change in Yemeni architecture, but rather by its continuity.

Indeed, the old city of Sanaa was listed as a World Heritage site on the basis of continuity—even "timelessness"—which is best represented in its residential architecture. This is perhaps why the Yemeni house is seen to symbolize the nation's heritage. The Quaroni Bonifica team saw the Yemeni house as the perfect type of a society that remained largely undifferentiated:

18 Presentation at the UNESCO/GOPHCY Cultural Heritage Inventory Workshop, Sanaa, May 2005.

> … [T]he Yemenite house, the most important and most typical expression of southern Arabian architecture, in the architectural perfection of which it is an expression, seems to bear out the hypothesis that its origins go far back in time, 'worked' by an extremely long experience inherited—for how much longer?—by the highly skilled nameless masters, the usta, who work freely without the constraints and conditioning factors of an architecture that is the expression of an official power, which in the long and troubled history of Yemen, has never in practice taken the form of 'designed' civil architecture. The anthropological basis of this is to be sought in the substantial lack of contrast between the central power and the power of the mainly mercantile citizenry, in the special status of the cities whose main impetus towards aggregation was their commercial function. When power launched its architectural messages, it was a foreign power, and it did so by way of religious buildings such as mosques and funereal architecture, practically without any consequences on the architecture of civil dwellings.[19]

In architecture, as in other spheres, great changes are attributed to foreigners—and these changes have little impact on building practice and society. *Usta*s are "nameless masters" of a craft that is somehow natural, exercised "without constraints." This contrasts markedly with builders' own account of their training and practice—and of buildings that bear the distinctive stamp (*basma*) of their makers, who are remembered as masters and teachers. Studies of residential architecture in Sanaa have begun to document, or at least suggest, patterns of change and development that occurred over time.[20] As we have seen, however, the Orientalist trope of an unchanging, "timeless" society and architecture is used by local actors—including *usta*s themselves, for whom it affirms the historical nature of their practice.

Two Approaches to the Past

The groundwork for modern conservation—the creation of a documentary and national history—has to a great extent been laid in Yemen. It dates to the imams' sponsorship of historical studies; in recent decades, the process has been nurtured by foreigners, who undertook archeological missions and pressed for the conservation of Yemen's historic cities. As in Europe, the construction of modern history and heritage has gone hand in hand with modernization and nationalism. It was initially driven by an elite, and gradually came to involve other segments of society—particularly as representations of the past came to be circulated via print, and more recently virtual, media. But in Yemen, the process has been compressed into one or two generations; indeed, it is still unfolding. In the Middle East the derivation of heritage (*turath*) from inheritance (*wirth* or *wiratha*), both conceptually and etymologically, mirrors and essentially replicates the development of the term *patrimoine*: the idea of family, or patrilineal, inheritance is transferred to the nation state.[21] Adopted

19 "Safeguarding of Sanaa Historical Centre," Proposal Phase, P0: Foreward and Summary, p. 24.

20 This kind of slow, incremental development, observed Lewcock, is also characteristic of the mass of "minor" architecture in Europe, which tends to be overlooked by evolutionary models of art history (personal communication).

21 See M. Herzfeld, "Mere Symbols," *Anthropologica*, vol. 50, 2008, pp. 141–55; and "The European Self: Rethinking an Attitude," in A. Pagden, ed., *The Idea of Europe: from Antiquity to the European Union*, Cambridge: Cambridge University Press, 2002. On *patrimoine* and the link to

from other Arab countries, *turath* has come to encompass a wide variety of practices. "The concept of heritage is different in Yemen," said 'Omar 'Abd al-'Aziz Hallaj, the Syrian architect who directed the Shibam-Hadhramaut project. "In Aleppo, heritage is about *athar* (antiquities); for Shibamis, heritage is *turath, taqalidna*—our traditions. It's the way we live our life, they say. It's not the house."

In Yemen, the official project of heritage draws on local understandings of the past in a variety of learned, artistic, and popular traditions. It has begun to transform these traditions, and has also been transformed by them. These traditions share a genealogical model, in which origin (*asl*) is seen as the source of family standing, expert knowledge, authority, and inspiration. This has historically been true not only of oral and artistic practices, but of text-based practices—until the latter were gradually transformed in the early twentieth century with the advent of printing. In these various forms the authenticity of one's claims is substantiated by a human link. "Between me and the author there are two men," said al-Nahi of Ibb, a renowned religious scholar.[22] Similar statements are made by builders: "I learned the work from my father, who learned it from his father, and from his father before him, like that (*hakadha*)," said a builder, gesturing with a hand over his shoulder to indicate the remote past. In various areas of cultural life and production, then, historicity is associated not with material artifacts or texts, but with the persons and practices that create and sustain them.

"I understand my work through *wiratha*," said 'Abdullah al-Rawdhi, using a form of *w-r-th* that can be translated as "legacy," rather than material inheritance. "They say they're preserving *turath*, but they're treating it like a frozen thing." These words capture, in a way that is both incisive and poignant, the conflict between two notions of the past: the one performative in nature, the other historicist. The builder's practice is historical because it is inherited, and because it *requires* artistry and improvisation, in the tradition of his predecessors. Supervisors needed al-Rawdhi's expert skills for the restoration of the bridge at Damt, but they insisted that he adhere to documentary criteria: "He wanted to beautify the bridge; we wanted to make it a monument." When supervision is less stringent, builders sometimes seize the opportunity to beautify or improve certain elements—and are pleased when their work is seen as "original." During a visit to a house that he had restored, an *usta* pointed out a change he had made: he had rebuilt two arches supported on a single column, adding a decorative base that was not part of the original structure. "Why did you do this?" I asked. "Art (*fann*)," he replied with a smile—clearly pleased that he could get away with it. Later, as we walked past a near-by Ottoman fountain (*sabil*), he told me that a pilaster on its neo-classical facade had inspired the improvisation.

In building as in other practices, faithfulness to one's predecessors is affirmed while the composition itself—whether building, poem, or text—is interpreted in the act of performance. This improvisation occurs within culturally accepted frameworks of practice and aesthetics; it is through improvisation that the aesthetic is understood and appreciated as traditional.[23] Although a builder may affirm the superiority of his predecessors, he

---

property, see R. Handler, "Who Owns the Past? History, Cultural Property, and the Logic of Possessive Individualism," in R. Williams, ed., *The Politics of Culture*, Washington, DC: Smithsonian Institution Press, 1991, pp. 63–74.

22 Cited in Messick, *The Calligraphic State*, p. 15. The authority of texts, writes Messick, derives from the spoken word. This in turn produces a kind of "genealogy of texts" in which the copy (*nuskha*) has a particular relationship to the original (*asl*; p. 29).

23 What Herzfeld calls the "creative deformation of conventions" (personal conversation). The former point was made by Caton.

improves on the past according to new needs and an evolving aesthetic that he shares with his peers. Performance of the past is, in effect, synonymous with "tradition." Tradition, writes Richard Handler, is not the mechanical reproduction of past forms; rather, it is an "interpretive process that embodies both continuity and discontinuity. [It is] a process of thought—an ongoing interpretation of the past."[24]

This performative view of the past differs from the modern historicist approach in fundamental ways. In the former, artistic practice actively constitutes culture; in the latter, culture is represented by, or embodied in, the object or text.[25] In the former, the past is valued for its continuing relevance in the present; in the latter, as a document of a distant past—that is, for its essential otherness.

A New Synthesis?

In a place like Yemen, where traditional building practice is still vital, it is tempting to see the historicist approach to conservation as superfluous, and ultimately destructive. But because of the rapid pace of change, conditions are different today—and *usta*s as well as architects are aware of this. Invoking the superiority of one's predecessors—doubtlessly a very long tradition—has taken on a new urgency, as it finds disturbing confirmation in dramatic material and social changes. On my visits to building sites, *usta*s immediately pointed out the inferior materials they had to work with: brick that was of a different size and quality; new types of stone, because traditional quarries have closed; lime that was fired with a different type of fuel, and therefore substandard. In the capital and other cities, *usta*s' skills have been influenced by their work in concrete construction; before conservation lent new value to the old skills, some *usta*s no longer appreciated them. While asserting their role as worthy heirs of their trade, *usta*s fear that "the experience of fifteen hundred years has been lost ... We builders are goats without a shepherd: each works the way he wants."

In the context of capitalist modernity, change is no longer a matter of assimilating new concepts and practices within an existing framework: rather, new systems are embraced that undermine the framework itself. Anthropologist Steve Caton noted that tribal poetry may die out in Yemen because of competing values transmitted through the modern educational system, such as the privileging of text over spoken language, and standard Arabic over dialect. Earlier in the century, religious scholarship was transformed under the impact of printing. Manuscripts, as both material products and discursive objects, had been open-ended and fluid; the new printed text "contained a new authority, a new truth value, enhanced by the definitiveness of the technology."[26] In a similar way, local building practice is being transformed by a new professionalism that privileges the drawing over embodied practice. This transformation is still incomplete: in new construction, owners encourage *usta*s to improvise freely on architects' plans. But as in societies that modernized earlier, a variety of economic and social factors in Yemen now favor the progressive "rationalization" of building. Ironically conservation, which aims to preserve cultural forms from extinction,

24 R. Handler and J. Linnekin, "Tradition, Genuine or Spurious?" *Journal of American Folklore*, vol. 97, 1984, pp. 273–4.

25 Following Bakhtin and his circle, Caton argues that art is "*constitutive* of social phenomena, particularly ideology. Artworks as practices are active agents, not just passive reagents, in history" (*Peaks of Yemen I Summon*, p. 250).

26 B. Messick, *The Calligraphic State*, p. 127.

contributes to the demise of traditional building practice by adopting similar, rationalizing methods: for example, the competitive tendering system, and the privileging of the text—drawings, historical documents, material analyses—over embodied practice. It would appear, then, that local building practice is threatened on two fronts: by new constructive systems that undermine its conceptual, social, and aesthetic framework; and on the other by conservation, which tends to reify it.

What, we may ask, is the future of traditional building practice in this context? In many areas of the world that experienced modernization, historicism has tended to replace traditional building practices as the latter were subsumed within modern constructive systems. Because of the rapid pace of change in Yemen, the situation is somewhat different: traditional builders continue to practice, and have found validation for their work in historicism. They often use the language of historicism, affirming their work as faithful to the original—and thus affirming their role as historical agents. Masons who designate themselves as traditional (*taqlidi*), writes Marchand, "in marked distinction to those working in the city with modern materials and equipment … advantageously manipulate this identity through the discourse."[27] My fieldwork suggests, however, that there is not a clear-cut distinction between "traditional" and "non-traditional" builders, at least among those working in conservation. These builders have assimilated the notion of the old city as a special place, requiring careful adherence to the old ways. But they also consider their work outside the old city, using modern materials and techniques, as *turath*. In some cases, they have different crews for work inside and outside the old city. Like residents who move easily between the old city and the new districts, these builders move easily between the old and new, framing both in terms of the "traditional."

Builders use not only the language of historicism: in some cases, they also use its tools. Sultan Sallam's exceptional work on traditional building—the result of many months in the field with *usta*s—is used by at least one builder in his own practice. He admires Sallam for his documentation of the essential elements of the Sanaani style, for example, the various types of *hizam*s (decorative brick courses); knowledge of *hizam*s is considered essential for mastery, but has been forgotten by most *usta*s. He has used Sallam's drawings in restorations and in new construction. In one case, he rebuilt the facade of a house not far from the Great Mosque, which had been damaged by settlement. He used a photo of the old building to do the work, and reused the same stone. But in place of the old doorway—a simple one, with a wooden lintel—he built a larger and more elegant opening, framed by a stone arch. His intent, he told me, was to "fill the facade." The new doorway was based on one of Sallam's drawings (figures 7.1, 7.2). For this builder, Sallam's work plays a role similar to that of the nineteenth-century pattern book—blurring the line that exists, at least in theory, between conservation and "revivalism."

If historicist tools and concepts have to some extent been absorbed in the performance of tradition, the inverse is also true: the existence of living builders conditions and limits historicism. Local architects and even foreign consultants generally ignore certain aspects of international practice—especially the injunction against rebuilding, and the directive to distinguish new work from old. According to a former GOPHCY official, rebuilding was standard practice until it came to be abused by property owners who built entirely new, and incompatible, structures. Now, he says, the process is more carefully controlled. In project files at the Inspections Department, buildings are carefully documented with photographs and, in some cases drawings; the intent, say inspectors, is to ensure that buildings are

27 T. Marchand, "Process over product," p. 143.

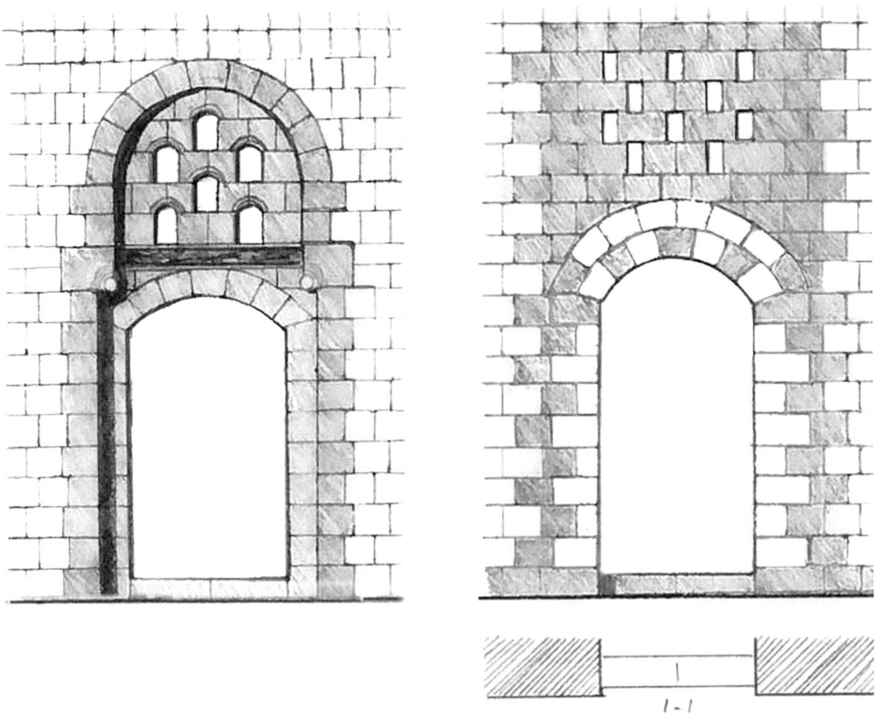

Figure 7.1 Door elevations from Sultan Sallam's work on Yemeni building crafts
Source: After S. Sallam, "Al-hiraf al-taqlidiyya al-islamiyya fi al-'amara al-yamaniyya."

conserved or rebuilt "exactly as they were." But many sets of before and after photographs show that substantial changes are allowed; since the work is often done using reused or traditional materials, and by *usta*s using the same techniques, it is often difficult to tell that the work is new. If there is little prejudice against rebuilding, there is virtually no attempt to distinguish new work from old, even when the process is managed by architects. Depending on the importance of the building, historical accuracy is not always a priority. The *marna'* at Talha is an interesting case, since it is part of a mosque complex that has the status of a monument. Although residents and builders say that it was rebuilt "as it was," the building was substantially reconfigured. The architect and project manager were concerned not with historical accuracy: rather, they tried to be faithful to the "spirit (*ruh*)" of the original and respectful of its historic context. In practice, then, both architects and builders continue to fabricate the "traditional," preserving the "Sanaani style" while addressing new needs.

These views are echoed by foreign consultants, who might be expected to insist on international standards. "Rebuilding in the old city is fine," said Lewcock, "as long as traditional materials are used." A similar approach was applied in much of the work conducted under foreign supervision in Sanaa. It is shared by the Italian team that ran a workshop at CATS in inventory methods: reconstruction is allowed, as long as it is done with the proper materials and methods (Figure 7.3). The survey form itself represents

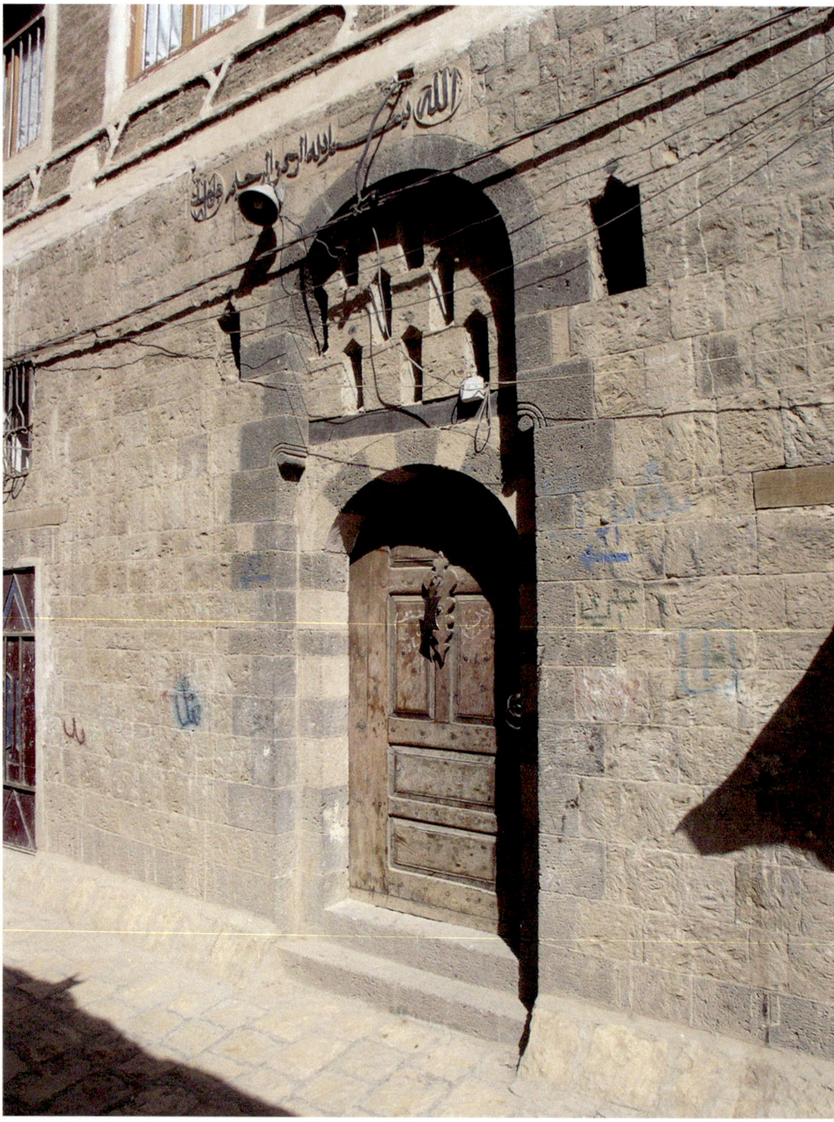

Figure 7.2 **A new door opening in a house restored for a private client, mid-2000s. The *usta* based the design of the door on Sallam's drawing in Figure 7.1**

an interesting adaptation of international methods to the local context. Developed by consultants and GOPHCY staff, the survey form avoids notions of period and style. Rather, surveyors are asked to identify the number of "historical layouts" on the facade of each building. The concept was not an easy one for trainees to grasp, in part because of translations between Italian, English, and Arabic. Daniele Pini, who led the team,

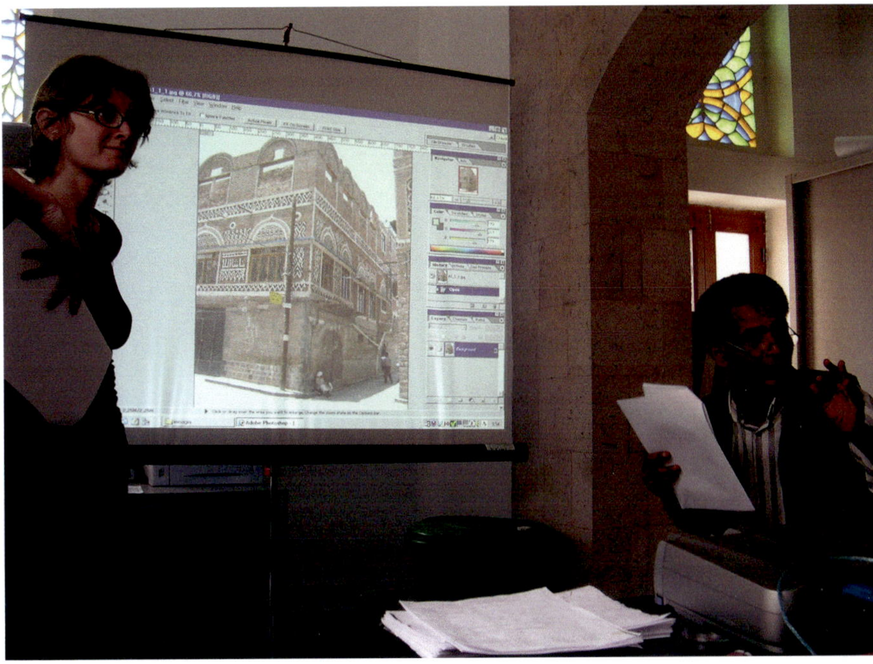

Figure 7.3 Training seminar in inventory methods at the Center for Architectural Studies (CATS), July 2004. On the left, Saveria Teston, with the Italian team; on the right, Yassin Ghalib

tried to clarify it in a meeting with trainees: “layout” means, in effect, “composition.” A typical building of old Sanaa consists of three “historical layouts”: the base of the building, executed in stone, with small window openings; the upper floors, in brick with varied window openings; and a *mafraj*, a narrow sitting room, on the top. If a building has only one layout—a single material, like stone, with uniform window openings on all floors—it is an indication that it is modern. “Richness,” rather than historicity *per se*, thus becomes the basis for conserving Sanaani architecture; it is further elaborated according to surveyors’ judgments of architectural quality.

This system was developed in part because, as we have seen, notions of period and style are not easily applied in Sanaa. But more importantly, if such categories were invented, they could be used to justify “stylistic restoration”—that is, restoration according to idealized styles and periods. This would ultimately destroy the richness of local architecture, as it has in other parts of the world. In a presentation to GOPHCY officials, architect Saveria Teston explained the team’s approach:

> Each building has its own personality. We want to preserve the complexity and richness of Sanaani architecture; we don’t want to simplify it on a survey sheet. This system leaves doors open … We are learning together how to read the buildings of Sanaa.

> We don't want to say there's a Sanaani type: it will end up homogenizing the architecture. The idea is, you give people the freedom to manage the elements, according to their social and economic needs. You only have to control the way the elements are executed. It's dangerous to say this window is good, this window is bad ... People will show you how they want to change their houses. GOPHCY needs the tools to guide this process.

The approach was questioned by some officials, but it seemed to be appreciated by most of the surveyors: they probably heard similar ideas in their training sessions with Yassin Ghalib, in preparation for the inventory. "Every place, every building has its own character," Ghalib once told me. "It speaks to you like a human face speaks." At the UNESCO workshop to launch a nation-wide cultural heritage inventory, Ghalib said: "We don't want our conceptions of beauty to be translated through the eyes of others ... There is not one *turath*, there are many, varied *turath*s." The survey form for old Sanaa aims at the most basic level of classification, leaving open the possibility of readings internal to the culture.

The direction is promising, but it does not go far enough. Despite world recognition of Sanaani architecture, "we still have no idea what it is," observed one architect. "We would do better in Florence: that's where our understanding of architecture comes from." In contrast to architecture of the distant past, built by builders who died long ago, there are living builders in Sanaa: their vital and subtle discourse can reveal much about the nature and value of architecture in *local* terms. This kind of discourse also exists among builders in the Hadhramaut and in other parts of the country. With all the international funds and expertise that have been devoted over the years to the study and conservation of Yemen's heritage, why have builders themselves not been studied and included? Some years ago Najwa Adra, an anthropologist known in particular for her work on Yemeni dance, was asked to help design an inventory for dances in Qatar. It soon became clear that the foreign experts' categories did not work: instead, they asked people who performed the dances to classify them as they understood them. They tended to classify a dance not according to formal criteria, but criteria of practice—for example, according to the social context in which a dance was performed.[28] A similar strategy should be urgently pursued among *usta*s: they should be asked to help develop classification systems, based on their own experience and understanding of building.

Such endeavors are essential if sites of "outstanding universal value" are to be conserved: for if "universal value" exists, it is produced and conserved through local values. Moreover, an investigation of building practice as "local knowledge" would greatly enrich historical studies, perhaps shedding light on pre-industrial architecture elsewhere in the world. Rather than forcing living experience into historical schemas, there is a unique opportunity to enrich our understanding of architecture through an inter-disciplinary approach that engages both living builders and the artifacts of the past (Figure 7.4). This kind of strategy follows logically from the unique synthesis that is emerging on the ground in Sanaa and elsewhere in Yemen. It is the result of a dynamic interplay between the global and the local, between an academic conception of history and one that is performative, rooted in local practice. Historicist tools and concepts have to some extent been absorbed in the performance of "tradition"—not only by builders, but by residents who would like to see the old city rebuilt "exactly as it was." This is consistent with the definition of tradition, which has always absorbed concepts and tools, and has in turn been transformed by them. At the same time, living practices resist the reifying effects of historicism. It is significant

28 Najwa Adra, personal communication.

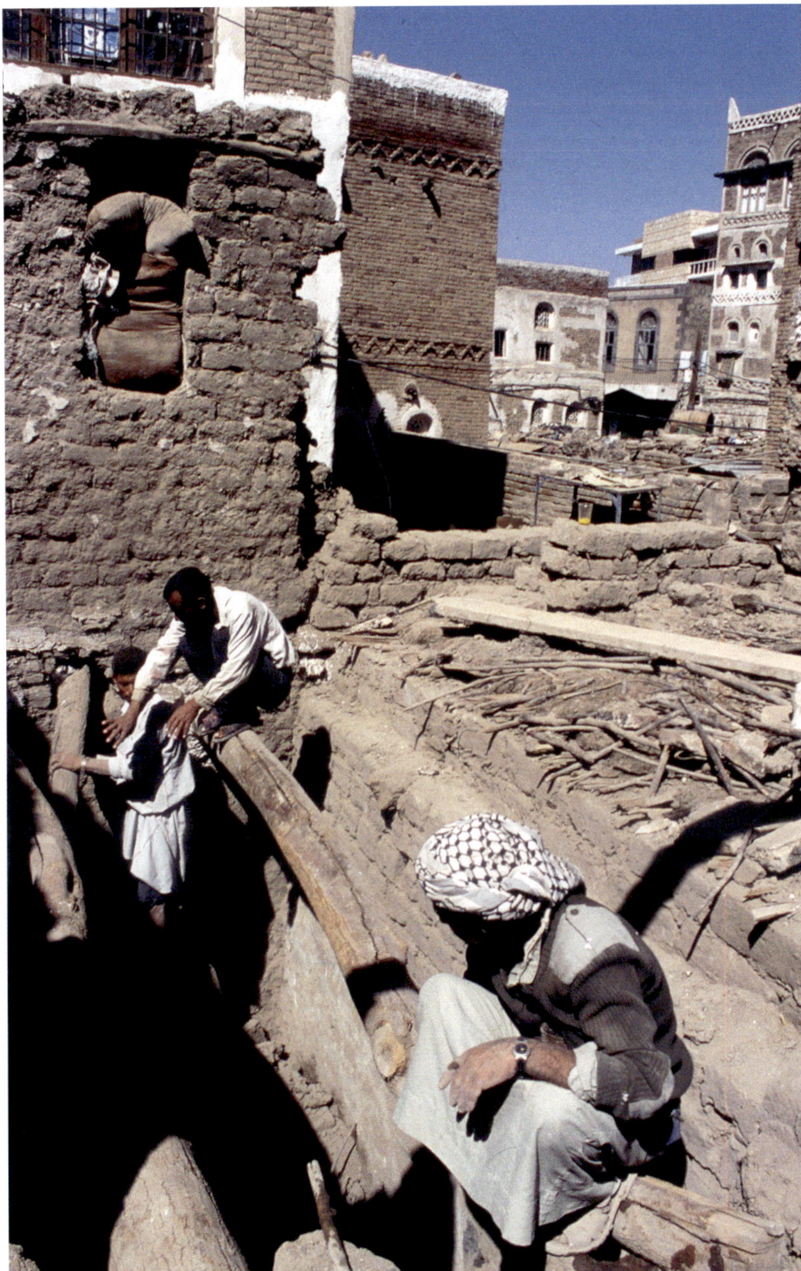

Figure 7.4 The restoration of a *samsara* near Bab al-Sabah that houses migrant workers, sponsored by Oxfam. 'Abdullah al-Hadrami was the supervising architect (October 2003)

that, despite certain differences, architects and builders seem to agree on a culturally constructed framework for tradition: for both groups, historicist methods help to establish the parameters of that framework.

This synthesis of international and local practice produces, in effect, a new definition of authenticity: that which is authentic is "true to the original" in terms of practice, materials, and aesthetics. Authenticity means faithfulness to one's predecessors—in terms of inherited knowledge and practice, and a shared aesthetic that is embodied in the buildings they left behind. This definition of authenticity can, and should, be cultivated as the basis of practice and policy. Policies would be prescriptive as well as proscriptive in nature, based on the idea of conservation as a creative process rather than simply as a means to preserve artifacts. They would recognize and promote builders as independent agents and full partners in conservation. At the same time, historicism would help to establish the framework for change—not by reifying forms, but by demonstrating their continuing relevance for the present.

Appendix
Major Projects Implemented Under the UNESCO Campaign to Save Old Sanaa, 1984–1994

Appendix
Major Projects Implemented Under the UNESCO Campaign to Save Old Sanaa, 1984–1994

Notes

The UNESCO Campaign for Old Sanaa has a complicated history, involving multiple agencies, donor nations, and consultants. Some of the agencies and offices are defunct; written records of the Campaign are incomplete, both at GOPHCY and at UNESCO/World Heritage offices in Paris. The account of projects and costs provided here is based primarily on lists provided by GOPHCY. These lists contain numerous discrepancies, especially with regard to project costs; I have synthesized the lists based on the advice of informants, and according to my own judgment. The information provided in this appendix must thus be seen as an approximation.

Project cost includes contract amount and supplemental work, where these figures were available. Some projects were jointly sponsored by a foreign government or agency and the government of Yemen, but a breakdown of contributions was not always available.

Unless otherwise noted, project costs are in Yemeni riyals as valued at the time of the project. The value of the riyal dropped dramatically during the period in question; for approximate figures see Dresch, *Modern Yemen*, p. 186.

For a complete list of UNESCO expert missions and technical studies, see "Preservation of the Old City of Sana'a: Project Findings and Recommendations", UNDP/UNESCO Restricted Technical Report, UNDP/YEM/88/006, Paris 1996, Annex A.

PROJECT	LOCATION	SPONSOR	COST	START	END	SCOPE
Paving and infrastructure						
Phase 1	Al-Abhar to Great Mosque	Republic of Yemen	10,000,000	Dec–86	Jul–88	13,000 SM, paving and upgrading
Bab al-Yaman Square	Bab al-Yaman	Republic of Yemen	1,965,000	Aug–88?	N/A	1,900 SM, paving and upgrading
Phase 2	Talha/al-Zumar to Bab al-Shu‘ub	Republic of Yemen	12,000,000	Mar–88	Jan–90	18,000 SM, paving and upgrading
Phase 3	Great Mosque to Maydan al-Luqayh	Republic of Yemen, Netherlands	9,000,000	Oct–88	Aug–91	11,700 SM, paving and upgrading
Phase 4	Al-Nusayr quarter	Republic of Yemen, Germany	5,000,000	Jun–89	Jun–09	6,000 SM, paving and upgrading
Phase 5	Bustan Sharib to Bahr Rajraj	Republic of Yemen	13,609,109	Jul–91	Apr–93	13,144 SM, paving and upgrading
Phase 6	Al-‘Alami, al-Quzayli, al-Fulayhi	Republic of Yemen	11,871,631	Jul–91	May–93	10,368 SM, paving and upgrading
Phase 7	Bustan al-Sultan quarter	Republic of Yemen	56,242,402	Oct–90	Nov–97	21,000 SM, paving and upgrading
Suq al-Milh	Suq al-Milh	Republic of Yemen	35,695,490	Jul–92	Aug–97	7,000 SM, paving and upgrading
Northern al-Qasimi Street	Al-Qasimi quarter	Republic of Yemen	1,500,000	Mar–92	N/A	500 SM, paving and upgrading.
Wadi Sa’ila	Old city	Republic of Yemen	72,060,950	1995	Jun–98	12,000 SM, paving and upgrading

PROJECT	LOCATION	SPONSOR	COST	START	END	SCOPE
Restoration, renovation, neighborhood upgrading						
Dar al-Jadid	Mu‘adh quarter	Republic of Yemen	2,192,000	Mar–87	1989	Renovation of former palace of the imam for use as EOPOCS headquarters
Bayt Sunaydar	Al-Qasimi quarter	Netherlands	300,000	1987	1988	House restoration, for use as commercial center and guest house for the Dutch Embassy
Samsarat al-Nahhas	Suq al-Nadhara	Norway, Republic of Yemen	5,767,320	Aug–88	Jul–89	Restoration of a caravansaray to serve as National Center for Revival of Yemeni Handicrafts
Al-Madhhab Hammams	Suq al-Milh	Republic of Yemen	914,809	Sep–89	Dec–90	Renovation and upgrading of ten bathhouses
Southern Wall, Phase 1	Bahr Rajraj, al-Zubayri	France, Republic of Yemen	13,638,270	Sep–89	Aug–90	Restoration/reconstruction of city wall, upgrading of adjacent areas
Bayt al-‘Amri	Wadi Sa’ila	Switzerland	N/A	Late 1989	1990	Restoration of house, to be used as residence for foreign team working on Bayt al-‘Anbasa
Model Neighborhood/ Beit Sari‘	Mu‘adh quarter	Italy	L 778,588,030*	Dec–90	Apr–93	Rehabilitation of a model neighborhood, house restoration
Bayt al-‘Anbasa	Suq al-Milh Street	Switzerland	152,460,000	1991	1995	House restoration, archeological investigations
Samsarat al-Mansura	Suq al-‘Aynab	Germany	4,623,000	Oct–91	Apr–93	Restoration of a caravansaray to serve as gallery for artists’ cooperative
Bayt ‘Atiya	Al-Zumar	Republic of Yemen	3,498,030	Aug–92	Jan–95	Restoration of house for use as GOPHCY guest house.

PROJECT	LOCATION	SPONSOR	COST	START	END	SCOPE
Bab al-Yaman, Phase 1	Bab al-Yaman	Republic of Yemen	2,500,000	Aug–92	1993	Restoration of the southern city gate
Southern Wall, Phase 2	Bahr Rajraj, al-Zubayri	Republic of Yemen	24,650,720	1994	Jan–96	Restoration/reconstruction of city wall, upgrading of adjacent area
Bab al-Yaman, Phase 2	Bab al-Yaman	Republic of Yemen	17,189,814	1995	1997	Restoration of the southern city gate, upgrading of adjacent area
New construction						
Madrasa of Ibn al-ʻAmir	Al-Zumar	Germany	9,500,000	Nov–89	Feb–91	Construction of a 4-story school on site of an old school
Center for Architectural and Technical Studies	Muʻadh Quarter	Italy	N/A*	N/A	1994	Construction of a center for restoration studies, training, and technical assistance to homeowners
Suq al-Baqar Clinic	Suq al-Baqar	Germany, Republic of Yemen	169,961,195	1991	1997	Construction of a 3-story facility specializing in women's and children's health, on site of a former clinic
Major technical studies						
Technical studies Studio Quaroni S.r.l. and Bonifica S.p.A	Old City	Italy	USD 1,200,200	1981	1990	Studies for the conservation of the old city (for GOAMM)
Technical studies and support	Old City	Democratic People's Republic of Korea	N/A	1987	1994	Survey of the old city (map, 1:1000, 1989); technical studies and construction documents for numerous projects

PROJECT	LOCATION	SPONSOR	COST	START	END	SCOPE
UNESCO/UNDP projects						
Development of Handicraft Industries (UNDP YEM/87/008)	Old City/Samsarat al-Nahhas	UNDP, ILO	USD 600,000	Apr–88	Jul–92	Add-on training project to the restoration of Samsarat al-Nahhas, programmed as National Center for Revival of Yemeni Handicrafts
Assistance for the Preservation of the Old City of Sanaa (UNDP YEM/88/006)	EOPOCS	UNDP	USD 374,800	Jun–88	Jul–91	Institution building assistance to EOPOCS; included promotion of Campaign activities, coordination of external assistance, technical assistance and training to EOPOCS staff

* Cost of construction of CATS may be included in figure for Mu‘adh quarter upgrading

Bibliography

1. Reports and Other Primary Sources

The primary sources used in this work are mostly reports, drawings, and maps produced on the old city of Sanaa. There are three main locations for this material:

General Organization for the Historic Cities of Yemen (GOPHCY), Sanaa, Yemen.
The Center for Architectural Studies (CATS), as well as the Department of Projects and Planning, hold primary material on the UNESCO Campaign and subsequent conservation work in Sanaa.
UNESCO: Office of Arab States, Place Fontenoy, Paris.
This office has a comprehensive collection of reports produced for the Sanaa and Shibam campaigns and other projects in Yemen.
Olivetti Foundation Historical Archive, Ivrea, Italy.
This archive holds documentation on Ludovico Quaroni's work in Sanaa, which he conducted with Bonifica S.p.A. in the 1980s. Only the Maria Luisa Neri Collection was available for consultation.

In addition, I consulted personal files of numerous individuals in Sanaa and elsewhere, including the files and field notebooks of Ronald Lewcock.

Reports for UNESCO, UNDP, and ICOMOS (partial list)

'Abdelkafi, J. "Town Planning, Development, and the Creation of Infrastructure," ICOMOS mission report, 12 September–1 October, 1982.

Barbato, G-C, "Restoration Project of the Samsarah al Bowani: Centre for Cultural Studies," UNDP/UNESCO Project YEM/88/006, Final Report, March 1989.

Bertaud, A., "Housing in Sana'a," UNDP?, no date.

Bertaud, A., "Master Plan for Sana'a Valley," UNDP?, no date.

Bertaud, A., "Report on planning, building, and related fields in the municipality of Sana'a," Information Paper no. 1, UNDP?, July 1970.

Binous, J., "Development Regulations, Promotion of the Crafts and Tourism, Elements of the Plan of Protection," ICOMOS mission report, 12 September–1 October, 1982.

Clark, I.C. et al., "International Campaigns for the Preservation and Safeguarding of the Cultural Heritage of Mankind," Report of the Special Committee for the Executive Board, Paris, September 1985.

Daolatli, A-A, "Summary Report on the Restoration, Rehabilitation, and Enhancement of the Old City of Sana'a," ICOMOS mission report, 12 September–1 October, 1982.

Eckert, H., "The Chief Urban Functions of the Old City of San'a Today," Technical Report, Paris: UNDP, 1992.

"Essential Description of the Historical City of Sana'a for the Purpose of Inclusion in the World Heritage List," Technical Office for the Preservation of the Old City of Sanaa (TOPOCS), no date.

Feiner, J., "Study Report for the Preservation of Old Shibam," GTZ/GOPHCY, 1995.

Helmore, p., "Water Supply, Storm Water, and Drainage," ICOMOS mission report, 12 September–1 October, 1982.

"The International Campaign to Safeguard the Old City of Sana'a: Campaign Strategy and Action Plan, 1988–1997," EOPOCS and UNESCO, August 1988.

"International Campaign for Safeguarding the City of Shibam and the Valley of Wadi Hadramaut: Project Findings and Recommendations," UNDP/YEM/92/019 (UNDP/PDY/88/008, Terminal report, Paris 1995)

Lane, Michael Barry, *Pilot Restoration Projects for the International Campaign to Safeguard the Old City of San'a,* General Organization for the Protection of the Historic Cities of Yemen (GOPHCY) and UNDP-UNESCO, no date, ca. 1990.

Lanzoni, L. and Teston, S., "UNESCO-World Heritage Center Mission Report," January 2004.

Lewcock, R., "The Campaign to Preserve the Old City of San'a," report on UNESCO mission of 9–16 April, 1982.

Lewcock, R., "The Campaign to Preserve the Old City of San'a," Report no. 2, UNESCO, Paris, 1984.

Lewcock, R., "The Campaign to Preserve the Old City of San'a," Report no. 3, UNESCO, Paris, 1986.

Lewcock, R., "The Campaign to Preserve the Old City of San'a," Report no. 4, UNESCO, Paris 1987.

Lewcock, R., "Consultant Report on the Conservation of Monuments and the International Campaign for the Conservation of the Old City of Sana'a," UNESCO, December 1983.

Lewcock, R., "Report of Mission to the Republic of Yemen, 8th–16th July, 1990," 1990.

Lewcock, R., "Restoration and Preservation of the Historic Architecture of the Old City of Sana'a," report on ICOMOS mission, 12 September–1 October, 1982.

Lewcock, R. "Shibam and the Wadi Hadramawt," report on UNESCO mission of 27 March–9 April, 1982.

Nardella, B-M, "Mission Report," prepared for UNESCO—Culture, Tangible Heritage Department, January 2004.

"Patrimoine et développement durable dans les villes historiques du Maghreb contemporain: enjeux, diagnostiques, et recommandations" UNESCO-Rabat, 2003 http://unesdoc.unesco.org/images/0018/001865/186523fo.pdf (accessed 3/4/15).

Pini, D., *Inventory of the Historic City of Sana'a: a tool for urban conservation*, Paris: UNESCO Press, 2008.

Pini, D. and Miglioli, F., "UNESCO-World Heritage Centre Mission Report," July 2003.

"Preservation of the Old City of Sana'a: Project Findings and Recommendations," UNDP/UNESCO, restricted Technical Report, UNDP/YEM/88/006, Paris 1996.

"Symposium on the Integrated Urban Policy for the Conservation of the Old City of Sana'a, Yemen," GOPHCY and UNESCO, December 1991.

"Umbrella Project—Preservation of the Old City of Sana'a," UNDP, revised 27/4/87.

Welbank, M. and Figueiras, J./Shankland Cox Partnership, "Sanaa: a Report to UNESCO," London, 1978.

Other Reports:

Association of Netherlands Municipalities, “Fact-Finding Mission on Preservation of Historic Cities of Yemen,” mission report, April 1997 (for the Ministry of Construction, Housing, and Urban Planning).

Berger Louis, International and Kampsax, International, A/s, “Urban Planning Study of San‘a, al-Hudaydah, Ta’izz, Ibb, Dhamar,” 1979 (for the Ministry of Municipalities and Housing).

Bigio, A. and Licciardi, G., “The Urban Rehabilitation of Medinas: the World Bank Experience in the Middle East and North Africa,” Urban Development Series Knowledge Papers, May 2010, no. 9.

Brizzi, G., “Economic Growth in the Yemen Republic,” Washington, DC: World Bank, 2002.

Eckert, H. and al-Maqrami, A., “Cultural Heritage Protection Project: the Social Assessment Report for the Historic Cities of Old Sana‘a, Shibam-Hadramaut, and Zabid,” May 1999.

Feiner, J., “Study Report for the Preservation of Old Shibam,” GTZ/GOPHCY, 1995.

Gilmore, Hanky, and Kirke, Ltd/PanYemen Consulting Services, “Cultural Heritage Protection Project,” volumes 1–4, August 1999 (for the World Bank).

Intervention Architecture: Building for Change (Aga Khan Award for Architecture, 10th Cycle), London: I.B. Tauris & Co. Ltd, 2007.

LeBlanc, F., “Heritage Conservation Terminology and Definitions of Terms Gathered from Various Sources,” 2012. http://www.icomos.org/~fleblanc/documents/terminology/doc_terminology_e.html (accessed 3/4/15).

Meijer, S., “Rehabilitation of Residential Buildings in the Old City of Sana‘a: Creating Opportunities for Progress.” Development Planning Unit, University College, London, September 2000.

Monasser, N. and Heepen, V., “The Boundaries of the Quarters in the Old City of Sana‘a,” Sana‘a: General Organization for the Preservation of Historical Cities of Yemen, 1995 (English and Arabic).

‘Oshaish, A., “Mashru‘ khitta wa barnamaj al-‘amal wa mashru‘ al-hifath ‘ala Sana‘a al-qadima,” 1987.

‘Oshaish, A., “Toward a strategy for the preservation of the urban cultural heritage of Yemen and the experience of the preservation of the old city of Sana‘a,” Urban Cultural Heritage Project, 2002.

Rethinking Development: World Decade for Culture and Development, 1988–1997, Paris: UNESCO, 1994.

Sanaa, a City Development Strategy, World Bank/Cities Alliance, 2009, pp. 19–20, 35–6, http://citiesalliance.org/node/787 (accessed 3/4/15).

Studio Quaroni S.r.1./Bonifica S.p.A, “Action Proposal: Implementation of a Conservation Reinstatement Measure for a Sample Nucleus of the Historical Centre of Sanaa and the Establishment of a Restoration Workshop/School,” January 1986.

Studio Quaroni S.r.1./Bonifica S.p.A, “Safeguarding of Sanaa Historical Center in Relation to the Urban Development of the City,” Rome, 1981–1990, for the Yemen Arab Republic/General Organization for Antiquities and Libraries (Reconnaissance Phase, 1981; Proposal Phase, 1983; Implementation Phase, 1989–1990).

Yoshida, M., "A Concept on 'Conservation-Based Development' for Historic Cities—Problems Related to the Preservation of the Old City of Sana'a in Yemen," report for the United Nations Centre for Regional Development, 1994.

Yoshida, M., "Study Report on the Wadi al-Saila Improvement Project," report for GOPHCY, September 1990.

2. Studies

'Abdelkafi, J., "La Médina de Tunis: l'éspace historique face au processus d'urbanisation de la capitale," *Présent et avenir des médinas (de Marrakech à Alep)*, Tours: C.N.R.S n° 706, "Urbanisation au Maghreb," fascicule de recherches n° 10–11, 1982.

Abu-Lughod, J., "Culture, modes of production, and the changing nature of cities in the Arab world," in Agnew, J., Mercer, J., and Sopher, D., ed., *The City in Cultural Context*, London: Allen and Unwin, 1984.

Abu-Lughod, J., "The Islamic City—Historic Myth, Islamic Essence, and Contemporary Relevance," *International Journal of Middle Eastern Studies*, 19, 1987, pp. 155–76.

Abu Luhum, S., *Haqa'iq wa watha'iq 'ishti, Sanaa: mu'assasat 'afif al-thiqafiyya*, Sanaa: 2005.

Alsayyad, N., ed., *Consuming Tradition, Manufacturing Heritage: Global Norms and Urban Forms in the Age of Tourism*, London: Routledge, 2001.

Alsayyad, N., ed., *Forms of Dominance: On the Architecture and Urbanism of the Colonial Experience*, Aldershot: Avebury Press, 1992.

Alsayyad, N., Bierman, I., and Rabbat, N., ed., *Making Cairo Medieval*, Lanham, MD: Lexington Books, 2005.

Anderson, B., *Imagined Communities: Reflections of the Origin and Spread of Nationalism*, London and New York: Verso, 1991.

Andrews, M., "The Metropolitan Picturesque," in S. Copley and P. Garside, ed., *The Politics of the Picturesque: Literature, Landscape, and Aesthetics since 1770*, Cambridge: Cambridge University Press, 1994.

Appleyard, D., ed., *The Conservation of European Cities*, Cambridge, MA: M.I.T. Press, 1979.

Araoz, G., "World Heritage Historic Urban Landscapes: Defining and Protecting Authenticity," *APT Bulletin*, vol. 39, nos 2/3, 2008, pp. 33–7.

Arce, A. and Long, N., *Anthropology, Development, and Modernities*, London: Routledge, 2000.

Arrif, A., "Le paradoxe de la construction du fait patrimonial en situation coloniale. Le Cas du Maroc," *Revue du monde musulman et de la Méditerrannée*, vol. 73–76, 1994/3–4, pp. 153–66.

Askew, M., "The magic list of global status: UNESCO, World Heritage, and the Agenda of States," in S. Labadi and C. Long, ed., *Heritage and Globalization*, New York: Routledge: 2010, pp. 19–44.

Balbo, M. and Pini, D., ed., "La città antica nei paesi in via di sviluppo," *Urbanistica*, vol. 92, 1988, pp. 30–74.

Barcelo, M., ed., *Les jardins de la vielle ville de Sanaa*, Yemen, Bellaterra: Centre Français d'Archéologie et de Sciences Sociales, Sanaa et Universitat Autonoma de Barcelona Servei de Publications, 2004.

Bastea, E., *Architecture and Memory*, University of New Mexico Press, 2004.

Batisse, M. and Bolla, G., *L'Invention du 'patrimoine mondial,'* Les Cahiers du Club Histoire, no. 2, Paris: UNESCO, 2003.

Bel, J.-M., *Architecture et Peuple du Yemen*, Paris: Conseil international de la langue française, 1988.

Bel, J.-M., et al., "Patrimoine et Restauration," *Chroniques Yéménites*, 1993, pp. 52–5.

Bell, D., *The Historic Scotland Guide to International Conservation Charters,* Edinburgh: Historic Scotland, 1997.

Bendix, R., *In Search of Authenticity: the Formation of Folklore Studies*, Madison: University of Wisconsin Press, 1997.

Bertaud, A., "Economics of Traditional Buildings in Yemen," in R. Holod, ed., *Toward an Architecture in the Spirit of Islam*, Philadelphia: The Aga Khan Award for Architecture, 1978, pp. 91–3.

Bianca, S., *Urban Form in the Arab World: Past and Present*, London: Thames and Hudson, 2000.

Bierman, I., "Urban Memory and the Preservation of Monuments," in J.L. Bacharach, ed., *The Restoration and Conservation of Historic Monuments in Egypt*, The American University in Cairo Press, 1995, pp. 1–12.

Bissell, W.C., "City of Stone, Space of Contestation: Urban Conservation and the Colonial Past in Zanzibar," unpublished dissertation, December 1999.

Bisson, J. and Troin, F., eds, *Présent et avenir des médinas (de Marrakech à Alep),* E.R.A. 706, CNRS/Institut de géographie, 1982.

Bluestone, D., "Academics in Tennis Shoes: Historic Preservation and the Academy," *Journal of the Society of Architectural Historians*, special issue 1999/2000, vol. 38, no. 3, September 1999.

Boccardi, B., "Projet de la Coopération Technique Italienne pour la conservation du Centre Historique de Sanaa," *Chroniques Yéménites*, 1993, p. 52.

Boccardi, G., "Point de Vue," *Chroniques Yéménites*, 1993, pp. 54–5.

Bonnenfant, G. and P. Bonnenfant, *L'Art du Bois* à *Sanaa: architecture domestique,* Paris: Edisud, 1987.

Bonnenfant, P., "La maison dans la peninsule Arabique," in Garcin, J.C., ed., *L'habitat traditionnel dans les pays musulmans autour de la Méditerranée*, vol. 3, Cairo: IFAO, 1991, pp. 715–79, 932–9.

Bonnenfant, P., *Les Maisons Tours de Sanaa*, Paris: Presses du CNRS, 1989.

Bonnenfant, P., ed., *Sanaa: Architecture Domestique et Société*, Paris: CNRS Editions, 1995.

Bonnenfant, P., "Unité, yéménité, et modernité dans l'architecture domestique," *Revue du Monde Musulman et de la Méditerranée*, Aix-en-Provence, 1994, no. 67 (special issue, "Le Yémen, présent et passé de l'unité"), pp. 141–60.

Bonnenfant, P., "Zabid, anti-développement et potentialités," *Peuples Méditerranéens*, nos 72–3, juillet-décembre 1995, pp. 219–42.

Bonnenfant, P., *Zabid au Yémen: Archéologie du vivant*, Aix-en-Provence: Edisud, 2004.

Bonnenfant, P. and Bonnenfant, G., ed., *Les Vitraux de Sanaa*, Paris: Editions du CNRS, 1981.

Bourdieu, P., *Distinction: A Social Critique of the Judgement of Taste*, trans. Richard Nice, Cambridge, MA: Harvard University Press, 1984.

Bourdieu, P., *Outline of a Theory of Practice*, Cambridge: Cambridge University Press, 1977.

Bozdogan, S., "The Lincoln Cathedral versus the Bicycle Shed," *Journal of Architectural Education*, vol. 36, no. 4, 1983, pp. 10–15.

Boyer, C., *The City of Collective Memory: its Historical Imagery and Architectural Entertainments*, Cambridge, MA and London: M.I.T. Press, 1994.

Burckhardt, T., *Fez, City of Islam*, trans. W. Stoddart. Cambridge, England: Islamic Texts Society, 1992.

Burrowes, R., "The Famous Forty and their Companions: North Yemen's First-Generation Modernists and Educational Emigrants," *Middle East Journal*, vol. 59, no. 1, Winter 2005, pp. 81–97.

Burrowes, R., *Historical Dictionary of Yemen*, 2nd edition, Lanham and Toronto: Scarecrow Press, 2010.

Burrowes, R., *The Yemen Arab Republic: the Politics of Development, 1962–1986*, Boulder, CO: Westview Press, 1987.

Cameron, C., "From Warsaw to Mostar: the World Heritage Committee and Authenticity," *APT Bulletin*, vol. 39, nos 2/3, 2008, pp. 19–24.

Casey, E., *The Fate of Place: A Philosophical History*, Berkeley: University of California Press, 1997.

Caton, S., *"Peaks of Yemen I Summon": Poetry as Cultural Practice in a North Yemeni Tribe*, Berkeley: University of California Press, 1990.

Caton, S., *Yemen Chronicle: An Anthropology of War and Mediation*, New York: Hill and Wang, 2005.

Cernea, M., "At the Cutting Edge: Cultural Patrimony Protection through Development Projects," in Serageldin, I., et al., ed., *Historic Cities and Sacred Sites: Cultural Roots and Urban Futures,* Washington, DC: World Bank, 2001, pp. 67–88.

Choay, F., *Invention of the Historic Monument*, trans. O'Connell, L.M., Cambridge: Cambridge University Press, 2001.

Choay, F., *The Modern City: Planning in the 19th Century*, New York, George Braziller, 1969.

Choay, F., "Riegl, Freud et les monuments historiques: pour une approche sociétale de la preservation," in *World Art: Themes of Unity in Diversity*, University Park, PA and London: Pennsylvania State University Press, 1989, pp. 799–807.

Colquhoun, A., "The Concept of Regionalism," in G.B. Nalbantoglu and C.T. Wong, eds, *Postcolonial Spaces*, New York: Princeton Architectural Press, 1997, pp.13–23.

Colquhoun, A., *Modernity and the Classical Tradition: Architectural Essays, 1980–1957,* Cambridge, MA and London: M.I.T. Press, 1989.

Connerton, P., *How Societies Remember,* Cambridge: Cambridge University Press, 1989.

Costa, P., "Sana," in Serjeant, R.B. ed., *The Islamic City*, Paris: UNESCO, 1980.

Costa, P., *Studies in Arabian Architecture*, Norfolk, UK: Variorium, 1994.

Coy, M.W., ed., *Apprenticeship: from Theory to Method and Back Again*, Albany: State University of New York Press, 1989.

Creighton, O., "Contested townscapes: the walled city as world heritage," *World Archaeology*, Vol. 39, No. 3: The Archaeology of World Heritage, September 2007, pp. 339–54.

Daftary, F., et,. al., *Living in Historic Cairo: Past and Present in an Islamic City*, London: Azimuth Editions, 2010.

Damluji, S., *The Valley of Mud Brick Architecture: Shibam, Tarim, and Wadi Hadramut*, Reading, UK: Garnet Publishing Limited, 1992.

Daum, W., ed., *Yemen*, Innsbruck: Pinguin-Verlag, 1988.

Davidson, C. and Serageldin, I., "Conservation of Old Sanaa, Yemen," in Davidson, C. and Serageldin, I., eds, *Architecture Beyond Architecture: Creativity and Social*

Transformations in Islamic Cultures: the 1995 Aga Khan Award for Architecture, London: Academy Editions, 1995.

Davis, E. and Gavrielides, N., "Statecraft, Historical Memory, and Popular Culture in Iraq and Kuwait," in Davis and Gavrielides, eds, *Statecraft in the Middle East: Oil, Historical Memory, and Popular Culture*, Miami: Florida International University Press, 1991, pp. 116–48.

Day, S., "Power-Sharing and Hegemony: a Case Study of the United Arab Republic of Yemen," unpublished dissertation, Georgetown University, 2001.

Delbos, G. and Jorian, P., *La transmission des savoirs,* Paris: Editions de la maison des sciences de l'homme de Paris, 1984.

Denslagen, W., *Architectural Restoration in Western Europe: Controversy and Continuity*, Amersterdam: Architectura & Natura Press, 1994.

Denslagen, W., "Restoration Theories, East and West," *Transactions*, vol. 18, 1993, pp. 3–7.

Development and Urban Metamorphosis, Proceedings of Seminar Eight in the Series "Architectural Transformations in the Islamic World," 2 volumes, Aga Khan Award for Architecture, Singapore: Concept Media, 1983.

Dirlik, A., "Globalism and the Politics of Place," in Dirlik, A., ed., *Places and Politics in the Age of Globalization*, New York: Rowman and Littlefield, 2001.

Dresch, P., "A Daily Plebiscite: Nation and State in Yemen," *Revue du monde musulman et de la Méditerranée*, vol. 67, 1993, pp. 67–77.

Dresch, P., *A History of Modern Yemen*, Cambridge: Cambridge University Press, 2000.

Dresch, P., "The Significance of the Course Events Take in Segmentary Systems", *American Ethnology*, vol. 13, 1986, pp. 309–24.

Dresch, P., *Tribes, Government, and History in Yemen*, New York: Oxford Press, 1989.

El-Habashi, A., "*Athar* to monuments: the intervention of the *Comité de Conservation des Monuments de l'Art Arabe*," unpublished dissertation, University of Pennsylvania, 2001.

El-Habashi, A. and Elzeyadi, I., "Consuming built patrimony, social perspectives, and preservation strategies in modern Cairo: A chronological and cross-disciplinary analysis," in *Manufacturing heritage through policies of preservation.* Traditional Dwellings and Settlements Working Paper Series, vol. 101, Berkeley, 1998, pp. 65–88.

Escobar, A., "Culture sits in places: reflections on globalism and subaltern strategies of localization," *Political Geography*, vol. 20, issue 2, pp. 139–74.

Escobar, A., *The Culture of Development: the Making and Unmaking of the Third World*, Princeton: Princeton University Press, 1995.

Fabian, J., *Time and the Other: How Anthropology Makes its Object*, New York: Columbia University Press, 1983.

Fawcett, J., ed., *The Future of the Past: Attitudes to Conservation, 1174–1974*, New York: Watson-Guptill Publications, 1976.

Fayya', M., "Conserving the Built Heritage of the Walled City of Sanaa, Republic of Yemen," unpublished doctoral thesis, Department of Architecture, Mehran University of Engineering and Technology, Jamshoro, Pakistan, 2003.

Feilden, B. and Jokilehto, J., *Management Guidelines for World Heritage Cultural Sites*, Rome: ICCROM, 1993.

Feiner, J., "La Vielle Ville de Sanaa: analyse morphologique comme fondement de la sauvegarde patrimoniale," unpublished doctoral thesis no. 1652, Ecole Polytechnique Fedérale de Lausanne (EPFL), 1997.

Finster, B., "An Outline of the History of Islamic Religious Architecture in Yemen," *Muqarnas*, vol. 9, 1992, pp. 124–47.

Fitch, J.M., *Historic Preservation: the Curatorial Management of the Built World*, Charlottesville: University Press of Virginia, 1990.

Gause, F.G., "The Idea of Yemeni Unity," *Journal of Arab Affairs*, vol. 6, no. 1, 1987, pp. 55–81.

Gieser, T., "Enskillment inhibited: 'industrial gardening' in Britain," *Journal of the Royal Anthropological Institute*, 20, 2014, pp. 131–49.

Govin, L. and Fromont, M-C, *Thula. Architecture et urbanisme d'une cité de haute montagne en République Arabe du Yemen*, Paris: Editions recherches sur les civilisations, 1984.

Grabar, O., *The Mediation of Ornament*, Princeton: Princeton University Press, 1992.

Grandguillaume, Gilbert, et al., ed., *Sanaa hors les murs: une ville arabe contemporaine*, Collection Villes du Monde Arabe, vol. 1, Tours: URBAMA, CNRS-University of Tours, 1995.

Grandguillaume, G., "Valorisation et dévalorisation liées aux contacts de cultures en Arabie Saoudite," in Bonnenfant, P., ed., *La Peninsule Arabique d'Aujourd'hui*, tome II, Aix-en-Provence, CNRS, 1982, pp. 621–53.

Grandison, K.I., "Challenging Formalism: the Implications of Contemporary Cultural Theory for Historic Preservation," *Landscape Journal*, spring 1999, vol. 18, no. 1, pp. 30–40.

Grigor, T., *Building Iran: Modernism, Architecture, and National Heritage Under the Pahlavis*, New York: Periscope Publishing, 2009.

Gupta, A., and Ferguson, J., "Discipline and Practice: 'The Field' as Site, Method, and Location in Anthropology," in Gupta, A. and Ferguson, J., ed., *Anthropological Locations: Boundaries and Grounds of a Field Science*, Berkeley: University of California Press, 1997, pp. 1–46.

Al-Haddad, A., *Cultural Policy in the Yemen Arab Republic*, Paris: UNESCO, 1982.

Al-Hadrami, Abdullah, "The Revival of the Old City of Sana," *Yemen Observer*, March 2005.

Al-Hadrami, Abdullah, "Yemeni Architecture and the Process of Architectural Change in Urban Forms," *Yemen Observer*, November 2001.

Haidar, L. and Talib, A., "Adaptive Reuse in the Traditional Neighbourhood of the Old City of Sana'a - Yemen," *Procedia—Social and Behavioral Sciences* 105, 2013, pp. 811–22 http://www.sciencedirect.com/science/article/pii/S1877042813044595 (accessed 3/5/15).

Hall, M., ed., *Towards World Heritage: International Origins of the Preservation Movement 1870–1930*, Farnham, Surrey and Burlington, VT: Ashgate, 2011.

Hall, S., "The Local and the Global: Globalization and Ethnicity," and "Old and New Identities, Old and New Ethnicities," in A. King, ed., *Culture, Globalization, and the World System*, Binghamton: State University of New York, 1991.

Hall, S., "Whose Heritage? Unsettling the heritage, re-imagining the postnation," in Littler, J., and Naidoo, R., ed., *The Politics of Heritage: the Legacies of Race*, London: Routledge, 2005, pp. 23–35.

Handler, R., "Authenticity," *Anthropology Today*, vol. 2, no. 1, 1986, pp. 2–4.

Handler, R., "Who Owns the Past? History, Cultural Property, and the Logic of Possessive Individualism," in R. Williams, ed., *The Politics of Culture*, Washington, DC: Smithsonian Institution Press, 1991, pp. 63–74.

Handler, R., "Heritage and Hegemony: Recent Works on Historic Preservation and Interpretation," *Anthropological Quarterly*, July 1987, vol. 60 no. 3, pp. 137–41.

Handler, R. and Linnekin, J., "Tradition, Genuine or Spurious?" *Journal of American Folklore*, volume 97, 1984, pp. 273–90.

Hansen, E., "Sanaa Rising," *Aramco World*, January/February 2006, vol. 57, no. 1, pp. 24–33.

Hardy, M., *The Venice Charter Revisited: Modernism, Conservation, and Tradition in the 21st Century*, Newcastle upon Tyne: Cambridge Scholars Publishing, 2008.

Haughey, P., "Oppression and Resistance: Historic Preservation in the Politics of Modernity," unpublished Master's thesis, Department of Architecture, University of Washington, 2001.

Hehmeyer, I., "Mosque, Bath, and Garden: Symbiosis in the Urban Landscape of San'a," *Seminar for Arabian Studies*, Brepols, vol. 28, 1998.

Herzfeld, M., *The Body Impolitic: Artisans and Artifice in the Global Hierarchy of Value*, Chicago: University of Chicago Press, 2004.

Herzfeld, M., *Cultural Intimacy: Social Poetics of the Nation State*, New York: Routledge, 2005.

Herzfeld, M., *Evicted from Eternity: the Restructuring of Modern Rome*, Chicago: University of Chicago Press, 2009.

Herzfeld, M. "Intangible Delicacies: Production and Embarrassment in International Settings," *Ethnologies*, vol. 36, 2014, pp. 41–57.

Herzfeld, M. "Mere Symbols," *Anthropologica*, vol. 50, 2008, pp. 141–55.

Herzfeld, M., "On Some Rhetorical Uses of Iconicity in Cultural Ideologies," in Bouissac, P., et al., eds, *Iconicity: Essays on the Nature of Culture*, Tübingen: Stauffenburg Verlag, 1986.

Herzfeld, M., *Ours Once More: Folklore, Ideology, and the Making of Modern Greece*, Austin: University of Texas, 1982.

Herzfeld, M., "Patrimoine, histoire et personne dans les lieux de l'Antiquité classique aujourd'hui," in D. Laborde, ed., *Désirs d'Histoire: Politique, mémoire, identité*, Paris: L'Harmattan, pp. 159–82.

Herzfeld, M., *A Place in History: Social and Monumental Time in a Cretan Town*, Princeton: Princeton University Press, 1991.

Hewitt, M.A., "Architecture for a Contingent Environment," *Journal of Architectural Education*, vol. 47, no. 4, May 1994.

Hobsbawm, E. and Ranger, T., eds, *The Invention of Tradition*, Cambridge: Cambridge University Press, 1992.

Iamandi, C., "La Théorie de la Restauration de Cesare Brandi: Validité et Limites pour la Restauration des Monuments Historiques," unpublished Master's thesis, Faculté des Etudes Supérieures, Université Laval, 1994.

Ibn al-Qasim, *Wasf San'a: mustall min kitab al-manshurat al-jaliya*, Ṣan'a: al-Markaz al-Faransi lil-Dirasat al-Yamaniya, 1993.

Jeffrey, R.B., *Yemen: A Culture of Builders*, American Architectural Foundation, 1989.

Jokilehto, J., *A History of Architectural Conservation*, Oxford: Butterworth-Heinemann, 1999.

Jokilehto, J., "Authenticity in Restoration: Principles in Practice," *APT Bulletin*, nos 3–4, 1985, pp. 5–11.

Jokilehto, J., "Conservation Principles and their Theoretical Background," in *Durability of Building Materials*, 5, 1988, pp. 267–77.

Jokilehto, J., et al., *What is OUV? Defining the Outstanding Universal Value of Cultural World Heritage Properties*, Paris: ICOMOS 2008. http://www-docs.tu-cottbus.de/interkulturalitaet/public/BoK/ICOMOS_What_is_OUV.pdf (accessed 3/5/15).

Joy, C., *The Politics of Heritage Management in Mali, from UNESCO to Djenné*, Walnut Creek, CA: Left Coast Press, 2012.

Kavuri-Bauer, S., *Monumental Matters: The Power, Subjectivity, and Space of India's Mughal Architecture*, Durham, NC: Duke University Press, 2011.

Al-Kawkabani, N., "Al-ittijahat al-mi'mariyya al-haditha fi madinat Sanaa ma bayn 1962–90," unpublished Master's thesis, University of Sanaa, 2000.

Khalaf, S., "Gulf Societies and the Image of the Unlimited Good," *Dialectical Anthropology*, vol. 17, 1992, pp. 53–84.

King, G., "Some observations on the Architecture of Southwest Saudi Arabia," *Architectural Association Quarterly*, vol. 8:1, 1976, pp. 20–29.

Kirkman, J., ed., *City of Sanaa: Nomad and City Exhibition*, Museum of Mankind, Department of Ethnography of the British Museum, London: World of Islam Festival Publishing Co. Ltd., 1976.

Kopp, H. and Wirth, E., *Sanaa: Développement et organisation de l'espace d'une ville arabe*, Cahiers de l'Institut de Recherches et d'études sur le Monde Arabe et Musulman (IREMAM), CNRS-Universités d'Aix-Marseilles, 1994.

Kühn, T., "Ordering Urban Space in Ottoman Yemen, 1872–1914," in J. Hanssen, et al., eds, *The Empire in the City: Arab Provincial Capitals in the Late Ottoman Empire*, Beirut: Orient Institut der Deutschen Morgenländische Gesellschaft, 2002, pp. 329–47.

Labadi, S., "World Heritage, authenticity, and post-authenticity," in S. Labadi and C. Long, ed., *Heritage and Globalization*, Routledge: 2010, pp. 66–84.

Lambert, J., "Saba et le barrage de Mareb: récit fondateur et temps imaginé," *Peuples Méditerranéeans*, no. 56–7, July–Dec. 1991, pp. 141–76.

Lamprakos, M., "The Idea of the Historic City" in *Conserving the City*. Special Issue of *Change Over Time*, vol. 4.1, April 2014, pp. 8–39.

Lamprakos, M., "Old Heritage, New Heritage: Building in Sanaa, Yemen," in *Architecture and Urbanism in the Middle East*. Special issue of *Viewpoints*, Middle East Institute, November 2008, pp. 34–7 http://www.mei.edu/content/architecture-and-urbanism-middle-east (accessed 3/5/15)

Lamprakos, M., "Rethinking Cultural Heritage: Lessons from Sanaa," *Traditional Dwellings and Settlements Review*, vol. XVI, no. 11, spring 2005, pp. 17–37.

Lamprakos, M., "Riegl's 'Modern Cult of Monuments' and the Problem of Value," in *The Venice Charter at 50*. Special Issue of *Change Over Time*, vol. 4.2, September 2014, pp. 418–35.

LaRocca, C., "Studio per la salvaguardia del centro storico di Sanaa," *Urbanistica*, vol. 92, 1988, pp. 49–60.

Larkham, P., *Conservation and the City*, London and New York: Routledge, 1996.

Larson, M.S., *The Rise of Professionalism*, Berkeley: University of California Press, 1977.

"La place de la vielle ville dans l'expansion urbaine de Sanaa." Proceedings of a colloquium organized by the Cultural Office of the French Embassy and GOPHCY, December 1990; published in *Dirasat al-Yamaniyya*, no. 45, January–March 1992.

Lave, J. and Wenger, E., *Situated Learning: Legitimate Peripheral Participation*, Cambridge: Cambridge University Press, 1991.

Leveau, R., et al., ed., *Le Yémen Contemporain*, Paris: Editions Karthala, 1999.

Lewcock, R., "Conservation in Islamic Cairo," in *The Expanding Metropolis—Coping with the Urban Growth of Cairo*, Geneva: Aga Khan Award Seminar, 1985.

Lewcock, R., "Conservation in Practice," in Imamuddin, A.-H., and Longeteig, K., ed., *Architectural and Urban Conservation in the Islamic World*, Geneva: Aga Khan Trust for Culture, 1990.

Lewcock, R., *The Old Walled City of San'a*, Paris: UNESCO, 1986.

Lewcock, R., "Preservation versus modernization: realities in Islamic architecture," in *The challenge to our cultural heritage: why preserve the past? Proceedings of a Conference on Cultural Preservation*, Washington, DC April 8–10, 1984, Washington, DC: Smithsonian Institution Press and UNESCO, 1986, pp. 31–40.

Lewcock, R., "The problems of subterranean water in the old urban areas of Arab cities," in Serageldin, I. and el-Sadek, S., *The Arab City: its Character and Islamic Heritage*, Proceedings of a Symposium held in Medina, Saudi Arabia, 28 February–5 March, 1981; Arlington, Va., 1982.

Lewcock, R., "Three Problems in Conservation: Egypt, Oman and Yemen," R. Holod ed., *Conservation as Cultural Survival*, Philadelphia: The Aga Khan Award for Architecture, 1980, pp. 67–77.

Lewcock, R., "The Technical Problems of Urban Rehabilitation and Upgrading," in *Designing in Islamic Cultures 3: adaptive reuse, integrating traditional areas into the modern urban fabric*, Cambridge, MA: Aga Khan Program for Islamic Architecture at Harvard University and the Massachusetts Institute of Technology, 1983.

Lewcock, R., "Working with the Past," in *Theories and Principles of Design in the Architecture of Islamic Societies*, Cambridge, MA: Aga Khan Program in Islamic Architecture, 1988, pp. 87–96.

Lindholm, C., *Culture and Authenticity*, Malden and Oxford: Blackwell Publishing, 2008.

Low, S., "The Anthropology of Cities: Imagining and Theorizing the City", *Annual Review of Anthropology*, vol. 25, 1996, pp. 383–409.

Lowenthal, D., "Art and Authenticity," in *World Art: Themes of Unity in Diversity*, University Park, PA and London: Pennsylvania State University Press, 1989, pp. 843–7.

Lowenthal, D., *The Past is a Foreign Country*, Cambridge: Cambridge University Press, 1985.

Luxen, Jean Louis, "Reflections on the Use of Heritage Charters and Conventions," in *Conservation, the GCI Newsletter*, vol. 19, no. 2, 2004.

MacCannell, D., *The Tourist: A New Theory of the Leisure Class*, Berkeley: University of California Press, 1999.

Mackintosh-Smith, T., *City of Divine and Earthly Joys: the Description of Sanaa*, American Institute of Yemeni Studies, 2001.

Mackintosh-Smith, T., "The Secret Gardens of Sanaa," *Aramco World*, January/February 2006, vol. 57, no. 1, pp. 34–7.

Mackintosh-Smith, T., *Yemen, the Unknown Arabia*, Woodstock and New York: The Overlook Press, 2000.

Macro, E., *Yemen and the Western World*, London: C. Hurst and Co., 1968.

Maguire, R., "Conservation and Diverging Philosophies," *Journal of Architectural Conservation*, no. 1, March 1997, pp. 7–18.

Marchand, T., "The Lore of the Master Builder: Working with Local Knowledge in Sanaa, Yemen," *Traditional Dwellings and Settlements Working Paper Series*, vol. 137, 2000, pp. 1–17.

Marchand, T., *The Masons of Djenné*, Bloomington: Indiana University Press, 2009.

Marchand, T., *Minaret Building and Apprenticeship in Yemen*, Richmond, Surrey: Curzon Press, 2001.

Marchand, T., "A possible explanation for the lack of explanation: or, 'Why the master builder can't explain what he knows'," *SOAS Working Papers in Linguistics*, vol. 10, 2000, pp. 301–14.

Marchand, T., "Process over Product: Case Studies of Traditional Building Practices in Djenne, Mali and Sanaa, Yemen," in Teutonico, J.M. and Matero, F., eds, *Managing Change: Sustainable Approaches to the Built Environment*, proceedings of the 4th Annual US/ICOMOS Symposium, Philadelphia, April 2001, Los Angeles: the Getty Conservation Institute, pp. 137–59.

Marchand, T., "Walling Old Sanaa: Reevaluating the resurrection of the city walls," *Terra 2000: 8th International Conference on the study and conservation of earthen architecture*, London: James and James, 2000, pp. 46–51.

Maréchaux, P., et al., *Sanaa, parcours d'une cité d'Arabie* (catalogue d'exposition), Paris: Institut du monde arabe, 1987.

Maréchaux, P. and Maréchaux, M., *Yemen*, London: Thames and Hudson, 1998.

Matero, F., "Ethics and Policy in Conservation," *Conservation, the Getty Conservation Newsletter*, Millennium Edition II, vol. 15, no. 1, 2000, pp. 5–7.

Menon, A.G. Krishna, "Conservation in India: a Search for Direction," *Architecture + Design*, 1989, pp. 22–7.

Menon, A.G. Krishna, "Rethinking the Venice Charter: The Indian Experience," International Association for the Study of Traditional Environments Working Paper Series, volume 40, 1992.

Mermier, F., "Des Artisans face aux Importateurs ou l'Ange Maudit du Souk," *Peuples Méditerranéens*, no. 46, janvier-mars, 1989, pp. 155–64.

Mermier, F., "Les Fondations Mythiques de Sanaa et d'Aden" in *Le Yemen, présent et passé de l'unité*. Special issue of *Revue du monde musulman et de la Méditerrannée*, vol. 67, 1994, pp. 131–40.

Mermier, F., *Le Sheikh de la Nuit, Sanaa: organization des souks et societe citadine*, Arles: Actes Suds, 1997.

Mermier, F., "The *Suq*s of Sanaa: Changing Functions and Symbolic Centrality," in M. Gharipour, ed., *The Bazaar in the Islamic City: Design, Culture, and History*, Cairo: American University in Cairo Press, 2012.

Mermier, F., "De l'Usage d'un Concept: la Citadinité a Sanaa," *Peuples Méditerranéens*, no. 46, January–March, 1989, pp. 31–48.

Messick, B., *The Calligraphic State: Textual Domination and History in a Muslim Society*, Berkeley: University of California Press, 1993.

Mignolo, *Local Histories, Global Designs: Coloniality, Subaltern Knowledges, and Border Thinking*, Princeton: Princeton University Press, 2000.

Mitchell, T., "The Stage of Modernity," in Mitchell, T., ed., *Questions of Modernity*, Minneapolis: University of Minnesota Press, 2000, pp. 1–35.

Mundy, M., *Domestic Government: Kinship, Community, and Polity in North Yemen*, New York: I.B. Tauris, 1995.

Mundy, M., "Sanaa in the Nineteenth Century," unpublished M.A. thesis, Faculty of Political Science, Columbia University, 1972.

Naim-Sanbar, Samia, "L'Habitat Traditionnel à San'a: Sémantique de la Maison," *Journal Asiatique*, nos 1–2, 1987, pp. 79–113.

Neri, M.L., "Occidente/Oriente. Un corto circuito per la salvaguardia di Sana," *Innumerevoli paesaggi: Scritti di storia economia per Roberta Morelli*, in D. Felisini, et al., Bari: Cacucci Editore, 2013, pp. 163–74.

Nielson, H-C.K., "The Appearance and Disappearance of Public Space: Sanaa during the First Part of the Century," in Nielson, H.-C.K., and Peterson, J.S., ed., *Middle Eastern Cities: 1900–1950*, Proceedings of the Danish Institute in Damascus I, Aarhus: Aarhus University Press, 2001.

Page, M. and Mason, R., *Giving Preservation a History: Histories of Historic Preservation in the United States*, Routledge: New York and London, 2004.

Peterson, J.E., *Yemen: the Search for a Modern State*, Baltimore and London: John Hopkins University Press, 1982.

Petzet, M., "The Venice Charter—forty years later," *World Heritage Review*, vol. 38, December 2004, pp. 20–23.

Petzet, M. and Koenigs, W., ed., *Sanaa: the Restoration of Samsarat al-Mansurah*, Munich: Bayerisches Landesamt für Denkmalplege, 1995.

Peutz, N. "Reorienting Heritage: Poetic Exchanges between Suqatra and the Gulf," *Revue des mondes musulmans et de la Méditerranée* 121–12, 2008, pp. 163–82.

Peutz, N. "'Shall I Tell You What Soqotra Once Was?' World Heritage and Sovereign Nostalgia in Yemen's Soqotra Archipelago," *Transcontinentales* 11 octobre 2011 http://transcontinentales.revues.org/1135 (accessed 3/5/15).

Philips, S., *Yemen's Democracy Experiment in Regional Perspective: Patronage and Political Authoritarianism*, Palgrave Macmillan, 2008.

Porter, G.D., "Unwitting Actors: the Preservation of Fez's Cultural Heritage, *Radical History Review*, issue 86, spring 2003, pp. 123–48.

Postel, T., "L'UNESCO et la sauvegarde du patrimoine culturel et naturel mondial: Convention du Patrimoine Mondial et campagnes internationales de sauvegarde," unpublished doctoral thesis, Université de Droit, d'Economie, et des Sciences d'Aix-Marseilles/Institut d'Etudes Politiques d'Aix en-Provence, 1986.

Price, N.S., et al., ed., *Historical and Philosophical Issues in the Conservation of Cultural Heritage*, Los Angeles: Getty Conservation Institute, 2000.

Pridham, B.R., ed., *Economy, Society, and Culture in Contemporary Yemen*, London: Croom Helm, 1985.

Rab, S., "The 'Monument' in Architecture and Conservation—Theories of Architectural Significance and their Influence on Restoration, Preservation, and Conservation," unpublished dissertation, Department of Architecture, Georgia Institute of Technology, June 1997.

Rabbat, N., "Islamic Architecture as a Field of Historical Inquiry," *AD Architectural Design*, Special Issue *Islam+Architecture* 74, 6 (Nov.–Dec. 2004): 18–23.

Rabbat, N., "The Formation of the neo-Mamluk Style in Modern Egypt," in M. Pollak, ed., *The Education of the Architect*, Cambridge, MA and London: M.I.T. Press, 1997, pp. 363–86.

Rabinow, P., *French Modern: Norms and Forms of the Social Environment*, Cambridge, MA and London: M.I.T. Press, 1989.

Al-Radi, S., et al., *The 'Amiriya in Rada': The History and Restoration of a Sixteenth-century Madrasa in Yemen*, Oxford Studies in Islamic Art, 13, Oxford: Oxford University Press, 1997.

Al-Radi, S., "Qudad, the Traditional Yemeni Plaster," in *Yemen Update* (American Institute of Yemeni Studies), no. 34, Winter/Spring 1994.

Radoine, H., "Conservation-Based Cultural, Environmental, and Economic Development: the Case of the Walled City of Fez," in L.F. Girard, ed., *The Human Sustainable City:*

Challenges and Perspectives from the Habitat Agenda, Burlington, VT: Ashgate, 2003, pp. 457–77.

Raymond, A., "Islamic City, Arab City: Orientalist Myths and Recent Views," British Journal of Middle Eastern Studies, Vol. 21, No. 1, 1994, pp. 3–18

Reid, D.M., "Cultural Imperialism and Nationalism: the Struggle to Define and Control the Heritage of Arab Art in Egypt," *International Journal of Middle Eastern Studies* 24 (1992), pp. 57–76.

Riegl, A., "The Modern Cult of Monuments: its Character and its Origin," trans. K. Foster and D. Ghirardo, *Oppositions*, Fall 1982, vol. 25, pp. 21–51.

Rist, G., *The History of Development: from Western Origins to Global Faith*, London and New York: Zed Books, 2002.

Rosaldo, R., "Imperialist Nostalgia," *Representations*, Spring 1989, no. 26, pp. 107–22.

al-Sabahi, Hatim, *Sanaa, Yemen: Tradition and Modernity in Sanani Architecture*, Eschborn: Deutsche Gesellschaft fur Technische Zusammenarbeit (GTZ), 1996.

Sallam, S., "Al-hiraf at-taqlidiyya al-islamiyya fi al-'amara al-yamaniyya" unpublished Master's thesis, Department of Fine Arts, University of Cairo, 1988.

Sallam, S., "Ta'thir technologia al-bina al-hadith 'ala al-turaz at-taqlidiyya," unpublished doctoral dissertation, Department of Fine Arts, University of Cairo, 1996.

Al-Sallal, K., "Sana'a: Transformation of the Old City and the Impacts of the Modern Era," in Elsheshtawy, Y., *Planning Middle Eastern Cities: An Urban Kaleidoscope in a Globalizing World*, London and New York: Routledge, 2004, pp. 85–113.

Saqqaf, A., ed., *The Middle Eastern City: Ancient Traditions Confront a Modern World*, New York: Paragon House Publishers, 1987.

Scarrocchia, S., *Alois Riegl: teoria e prassi della conservazione dei monumenti: antologia di scritti, discorsi, rapporti 1898–1905, con una scelta di saggi critici*, Bologna: CLUEB, 1995.

Scarrocchia, S., "L'autonomia della conservazione in forma di colloquio con Alois Riegl," *Casabella* 1991, v. 55, no. 584, novembre-dicembre, pp. 29–33.

Schwarzer, M., "Myths of Permanence and Transience in the Discourse on Historic Preservation in the United States," *Journal of Architectural Education*, vol. 48, no. 1, September 1994.

Scott, J. *Seeing Like a State*, New Haven: Yale University Press, 1998.

Sennett, R., *The Craftsman*, New Haven and London: Yale University Press, 2008.

Serjeant, R.B. and Lewcock, R., ed., *San'a: An Arabian-Islamic City*, London: World of Islam Festival Trust, 1983.

Al-Shawish, A., "Conservation Policies in a Historic City: A Case Study of Sanaa," unpublished Master's thesis, Department of Town Planning, University of Wales Institute of Science and Technology, CARDIFF, October 1987.

Shryock, A., *Nationalism and the Genealogical Imagination: Oral History and Textual Authority in Tribal Jordan*, Berkeley: University of California Press, 1997.

Silberman, N.A., *Between Past and Present: Archeology, Ideology, and Nationalism in the Modern Middle East*, New York: Henry Holt and Company, 1989.

Slyomovics, S., ed., *The Walled Arab City in Literature, Architecture, and History*, London: Frank Cass Publihshers, 2001.

Smith, L., *Uses of Heritage*, Abingdon: Routledge, 2006.

Smith, N., *The New Urban Frontier: Gentrification and the Revanchist City*, New York: Routledge, 1996.

Smith, V., ed., *Hosts and Guests: the Anthropology of Tourism*, Philadelphia: University of Pennsylvania Press, 1989.

Stadnicki, R., "Des portes de Sanaa aux nouvelles entrées de la ville: re-formulation d'espaces de sociabilités," *Espaces et Sociétés*, no. 126, 2006, pp. 119–38.

Stadnicki, R. "Vers un aménagement participative à Sanaa? L'implication des citadins dans les projets de sauvegarde et de réaménagement de la vielle ville," in Legros, O., *Participations et Action Publique: Dakar, Cotonou, Rabat, Tunis, Jérusalem, Sanaa*, Paris: Editions Y. Michel, 2008, pp. 49–68.

Stadnicki, R. and Touber, J., "Le Grand Sanaa. Multipolarité et nouvelles formes d'urbanité dans la capitale du Yemen," *Annales de géographie*, no. 659, 2008, pp. 32–53.

Stovel, H., "Origins and Influence of the Nar Document on Authenticity," *APT Bulletin*, vol. 39, nos 2/3, 2008, pp. 9–17.

Al-Suleihi, S., "The Lyrical Facades of Sanaa," unpublished Master's thesis, Department of Architecture, M.I.T., 1992.

Taher, 'A. and 'Abdullah, Y., *Dirasat Tahliliya 'ala al-'Asima Sanaa, al-Jumhuriya al-Yamaniyya,* Sanaa: Markaz al-Tahir lil-Istisharat al-Handasiyya, 2005.

Tainter, Joseph A. and Lucas, G. John, "Epistemology of the Significance Concept," *American Antiquity*, vol. 48, no. 4, October 1983, pp. 707–19.

Thornhill, C., "Historicism," in Craig, E., ed., *Routledge Encyclopedia of Art,* vol. 4, London: Routeldge, 1998, pp. 443–6.

Tilley, C., et al., eds, *Handbook of Material Culture*, London: SAGE Publications, 2006.

Tipps, D., "Modernization Theory and the Comparative Study of Societies: A Critical Perspective" in *Comparative Studies in Society and History,* vol.15, 1973, pp.199–226.

Titchen, S.M., "On the construction of 'outstanding universal value'; some comments of the implementation of the 1972 World Heritage Convention," in *Conservation and Management of Archeological Sites*, 1996, vol. 1, pp. 235–42.

Titchen, S.M., "On the construction of outstanding universal value: UNESCO's World Heritage Convention (Convention Concerning the Protection of the World Cultural and Natural Heritage, 1972) and the identification and assessment of cultural places for the inclusion in the World Heritage List," unpublished doctoral thesis, Department of Anthropology, the Australian National University, Canberra, 1995.

Toxey, A., *Materan Contradictions*, Surrey, UK: Ashgate, 2011.

Turtinen, J., "Globalising Heritage—On UNESCO and the Transnational Construction of a World Heritage," Stockholm Center for Organizational Research (SCORE), http://www.score.su.se/polopoly_fs/1.26651.1320939806!/200012.pdf (accessed 3/4/15)

Um, N., "Eighteenth Century Patronage in Sanaa: building for the new capital during the second century of the Qasimi imamate," *Proceedings of the Seminar for Arabian Studies*, vol. 34, 2004, pp. 1–15.

Um, N., *The Merchant Houses of Mocha: Trade and Architecture in an Indian Ocean Port*, Seattle: University of Washington Press, 2009.

Varanda, F., *The Art of Building in Yemen*, Lisbon: Argumentum, 2009.

Varanda, F., "Horizontality and Verticality: Architecture and the Environment in Yemen," *Traditional Dwellings: Case Studies in Different Cultures*, Center for Environmental Design Research, University of California, vol. 13, Berkeley, 1989, pp. 33–43.

Varanda, F., "Tradition and Change in the Built Space of Yemen," unpublished dissertation, Department of Geography, University of Durham, UK, 1994.

Vattimo, G., "The End of Modernity," in N. Leach, *Rethinking Architecture: A Reader in Cultural Theory*, New York: Routledge, 1997.

Vom Bruck, G. "A House Turned Inside Out: Inhabiting Space in a Yemeni City," *Journal of Material Culture*, vol. 2, no. 2, July 1997, pp. 139–72.

Vom Bruck, G., "Evacuating Memory in Post-Revolutionary Yemen," in Al-Rasheed, M. and Vitalis, R. eds, *Counter-Narratives: History, Contemporary Society, and Politics in Saudi Arabia and Yemen,* London: Palgrave MacMillan, 2004, pp. 229–45.

Vom Bruck, G., *Islam, Memory, and Morality in Yemen*, New York: Palgrave Macmillan, 2005.

Welbank, M., "Conservation and Development," in *Development and Urban Metamorphosis*, Singapore: Concept Media, 1984.

Whitrow, G.J., *Time in History: Views of Time from Prehistory to the Present*, Oxford: Oxford University Press, 1988.

Wright, Gwendolyn, *The Politics of Design in French Colonial Urbanism*, Chicago: University of Chicago Press, 1991.

Yarwood, J., "Traditional Building Construction in an Historic Arabian Town," *Construction History*, vol. 15, 1999, pp. 55–77.

Yoshida, M., "Conservation-Based Development of Historic Cities: the Case of the Old City of Sanaa in Yemen," *Regional Development Studies*, vol. 1, winter 1994/5, pp. 143–70.

Index

Note: References to figures are listed in *italics*.